AF361911

Process Control and Smart Manufacturing for Industry 4.0

Process Control and Smart Manufacturing for Industry 4.0

Editor

Sergey Y. Yurish

MDPI • Basel • Beijing • Wuhan • Barcelona • Belgrade • Manchester • Tokyo • Cluj • Tianjin

Editor
Sergey Y. Yurish
International Frequency
Sensor Association
Spain

Editorial Office
MDPI
St. Alban-Anlage 66
4052 Basel, Switzerland

This is a reprint of articles from the Special Issue published online in the open access journal *Processes* (ISSN 2227-9717) (available at: https://www.mdpi.com/journal/processes/special_issues/process_control_smart_manufacturing_ARCI_2021).

For citation purposes, cite each article independently as indicated on the article page online and as indicated below:

LastName, A.A.; LastName, B.B.; LastName, C.C. Article Title. *Journal Name* **Year**, *Volume Number*, Page Range.

ISBN 978-3-0365-6596-5 (Hbk)
ISBN 978-3-0365-6597-2 (PDF)

Cover image courtesy of IFSA Publishing S.L.
(design by Tetyana Zakharchenko)

Contents

About the Editor

Sergey Y. Yurish

Dr. Sergey Y. Yurish has served as the president of the International Frequency Sensor Association (IFSA)—one of the major professional associations serving the sensor industry and academia—for more than 23 years. Dr. Yurish is a founder of three companies. He is editor-in-chief of the international peer-reviewed journal Sensors & Transducers and editor of several open access multivolume book series.

Dr. Yurish obtained his PhD degree in 1996 from the National University Lviv Polytechnic (UA). He has published more than 170 articles and papers in international peer-reviewed journals and conference proceedings. Dr. Yurish holds nine patents and is an author and co-author of 12 books. He has delivered more than 90 speeches, tutorials and keynote presentations at industries, peer institutions, and professional conferences in over 30 countries.

Dr. Yurish was a Marie Curie Chairs Excellence Investigator at the Technical University of Catalonia (UPC, Barcelona, Spain) from 2006–2009, where he led and developed one of the most successful projects in the UPC on Smart Sensors Systems Design (SMARTSES), totaling EUR 425,000. His professional accomplishments also include a Senior Research Fellowship at the Open University of Catalonia (UOC, Barcelona, Spain) where he spent a year in 2009–2010. Dr. Yurish has over 25 years of research and academic experience, during which he has developed numerous projects on an international level in frames of various programmers, including NATO, FP6 and FP7.

Preface to "Process Control and Smart Manufacturing for Industry 4.0"

Industry 4.0 holds a great deal of potential and is expected to register a substantial growth in the near future. According to modern market study, the global Industry 4.0 market size is projected to reach USD 377.30 billion by 2029, at a CAGR of 16.3% during the forecast period 2022–2029.

Industry 4.0 is an integrated system which consists of an automation tool, robotic control, communications and big data analytics. The increased adoption of industrial robots is one of the main driving factors of this market, while the data risks associated with the integration of advanced technologies are the restraining factors.

This book, entitled "Process Control and Smart Manufacturing for Industry 4.0", contains the extended papers from the Series of annual IFSA conferences on Automation, Robotics and Communications for Industry 4.0/5.0 (ARCI) on the following topics: Process Automation, Process Control and Monitoring, Design Principles in Industry 4.0, Smart Manufacturing and Technologies, Smart Factories, Machine Learning and Artificial Intelligence in Manufacturing.

The book contains 13 chapters written by 54 ARCI conference participants from seven countries: China, Croatia, Denmark, Germany, Italy, Poland and Romania. This book will inform readers of cutting-edge developments in the field and provide effective starting points and a road map for further research and development. All chapters follow the same structure: firstly, an introduction to the specific topic under study; secondly, a description of the field, including sensing or/and measuring applications. Each chapter ends with a curated list of references, including books, journals, conference proceedings and websites.

The book is intended for researchers and scientists from academia and industry, as well as for postgraduate students.

Sergey Y. Yurish
Editor

Article

Modeling of Erosion in a Cyclone and a Novel Separator with Arc-Shaped Elements

Elmira I. Salakhova [1,*], Vadim E. Zinurov [2], Andrey V. Dmitriev [2] and Ilshat I. Salakhov [3]

1   Department of Processes and Devices of Chemical Technologies, Nizhnekamsk Institute of Chemistry and Technology, Subdivision of the Kazan National Research Technological University, 423570 Nizhnekamsk, Russia
2   Department of Theoretical Foundations of Heat Engineering, Kazan State Power Engineering University, 420066 Kazan, Russia
3   JSC Taneco, 423570 Nizhnekamsk, Russia
*   Correspondence: salahova.elmira@gmail.com

Abstract: Modeling of the separation of catalyst particles from gas using two devices, a cyclone and a novel separator with arc-shaped elements, was performed for fluidized-bed dehydrogenation of C_4–C_5 paraffins to isoolefins as an example. The proposed dust collector allows one to reduce erosive wear by several times (~6.5-fold) in identical regimes and at identical parameters of the process. The effect of particle size on erosive wear was analyzed under near-industrial conditions; the regions most susceptible to wear in the analyzed devices were identified, as well as the functions describing the dependence between the erosive wear rate and particle diameter for the cyclone and separator with arc-shaped elements, making it possible to predict wear in the devices were obtained.

Keywords: gas–solid; cyclone; separator; gas dynamics; erosion

Citation: Salakhova, E.I.; Zinurov, V.E.; Dmitriev, A.V.; Salakhov, I.I. Modeling of Erosion in a Cyclone and a Novel Separator with Arc-Shaped Elements. *Processes* **2023**, *11*, 156. https://doi.org/10.3390/pr11010156

Academic Editor: Sergey Y. Yurish

Received: 6 November 2022
Revised: 27 December 2022
Accepted: 31 December 2022
Published: 4 January 2023

1. Introduction

Petrochemical companies are paying much attention to the improvement of equipment reliability, reduction of operating expenses, and equipment maintenance. It is particularly relevant for existing dust collection systems [1–9], where the fundamental components are cyclones characterized by relatively high performance, as well as simple operating principle and design. Their drawbacks include the relatively high hydraulic drag and susceptibility to abrasive wear of the walls [5–8]. Thus, cyclones used for coarse and medium-degree gas purification are commonly used in petrochemistry for production lines performing fluidized-bed dehydrogenation of C_4–C_5 paraffins to isoolefins. A typical feature of these cyclones is that the contact between particles and cyclone walls alters the fractional size distribution of the catalyst due to particle crushing and abrasion [9–12].

Erosion affecting the upper inlet section and the lower cone of the cyclones is one of the most common problems related to this type of cyclone. F. Fulchini [13] identified the highest-stress areas in the fluidized bed where mechanical strain causes particle abrasion: gas dispensing nozzles, the bubbling layer, cyclones, and bending sections. They also suggested that each of these abrasion sources should be analyzed individually, as there are different abrasion mechanisms (surface abrasion, chipping, and fragmentation). A combination of particle properties and operating conditions, as well as device geometry, are responsible for the predominating mechanism [14–17]. J. Werther and J. Reppenhagen [18] mentioned that the cyclone is one of the most significant factors contributing to particle abrasion, especially at high flow rates of the surface gas. J. Werther et al. [19] proposed a model of particle abrasion caused by cyclones under conditions of surface particle abrasion for catalyst particles in fluid catalytic cracking (FCC). An analysis of the results revealed that the abrasion rate depended on the material properties, kinetic energy of the gas, and particle size. J. Reppenhagen et al. [20] tested the model reliability by analyzing nine different cyclone geometries. Their model was based on pure abrasion; however, J. Werther et al. [21] mentioned that if the inlet flow rate was increased and/or particle loading was reduced, particles would undergo intensive chipping and/or fragmentation, especially for fresh catalysts [21]. The hydrodynamics of a cyclone (being responsible for the particle speed and residence time) and the physical properties of particles (affecting the dependence between abrasive wear and operating conditions) are factors contributing to abrasive wear in the cyclone at a specified loading of solid particles [22,23]. Numerical modeling using the CFD-DEM approach, with allowance for the dynamics of particles, was performed in these studies.

The gas flow in cyclone separators is usually very unstable and highly swirling. Numerous researchers have studied various geometric and operating parameters using computational methods to enhance cyclone performance [24–28].

A.J. Hoekstra et al. [29] used three turbulence models to study air flows inside a cyclone with three different swirl numbers and recommended employing the Reynolds stress turbulence model (RSM). M.D. Slack et al. [30] used the RSM turbulence model to simulate a gas flow inside the cyclone and revealed that the numerical findings agreed well with the experimental laser Doppler anemometry (LDA) data. G. Gronald and J.J. Derksen [31] applied the large eddy simulation (LES) approach and the two-equation RANS turbulence model to simulate an air flow inside the cyclone and compared the respective CFD results to the experimental findings obtained by LDA. They revealed that the LES model was superior to the RANS model for predicting fluctuating flow rates. However, the LES model requires high-fidelity grids and, therefore, calculations take much time.

Many researchers have emphasized that taking into account the geometric factors is another important problem related to the numerical modeling of flow in cyclones. R.M. Alexander [32] studied some geometric parameters affecting cyclone performance. J. Gimbun et al. [33] investigated the effect of cone tip diameter on pressure drop and cyclone performance.

The literature analysis showed that the problem related to high catalyst consumption due to crushing and abrasion in cyclones has a systemic nature. For cyclones, the most significant factors are the geometric features of the equipment, causing the formation of certain gas-dynamic flow structures affecting the particle–wall interaction. Most studies currently employ numerical simulation of the gas dynamics and particle motion in cyclones using the CFD-DEM model [15,34–37], which allows one to predict cyclone efficiency, hydraulic drag, etc. at different operating parameters [38–41]. It should, however, be mentioned that there is sparse research into erosive wear of the cyclone internal surface.

In general, centrifugal cyclones are the reliable and high-performance equipment that allows one to efficiently separate fine particles from the gas medium. However, cyclones of this type are characterized by low wear resistance and are susceptible to cyclone wall erosion [42,43]. Therefore, designing novel separators characterized by low erosive wear can be a useful solution for the industry. A separator with arc-shaped elements (SAE) is the most promising separation device [44]. An advantage of SAE is that it contains specially designed arc-shaped elements that form non-cyclonic flows. It is reasonable and important from both the fundamental and applied standpoints to investigate erosion in the novel device.

This study aimed to compare erosive wear of the proposed separator with arc-shaped elements and the cyclone for removing solid catalyst particles from gases using numerical CFD-DEM modeling and reveal the features of separation conducted in the novel separator.

2. Experimental Procedure

A standard device for collecting solid particles, the centrifugal cyclone (TsN-15), that is employed for the dehydrogenation of isoparaffins at petrochemical plants (Figure 1), and the novel separator with arc-shaped elements proposed for replacing the conventional cyclone were used to study the erosive wear (Figure 2).

Figure 1. A real-world 3D model of the TsN-15 cyclone: (1) the inlet for the gas contaminated by the catalyst; (2) the outlet for catalyst particles collected during purification; and (3) the outlet for purified gas.

Figure 2. A separator with arc-shaped elements (a cross-sectional view): (1) the inlet for the gas contaminated by the catalyst; (2) rows of arc-shaped elements; (3) the body; (4) the hopper; and (5) the outlet for purified gas.

The operating principle of a cyclone is widely known. The dust-laden flow enters the chamber through the inlet (1). The dust-laden medium then moves with a tornado-like helical flow pattern. Centrifugal forces make particles leave the flow and collide with the internal wall of the cylindrical cyclone body. The particles then fall into the hopper through the orifice (2). The purified gas leaves the cyclone through the outlet of the dipleg (3) submerged into the cyclone to a certain depth (Figure 1).

The separator with arc-shaped elements has a simple design: it consists of several rows of arc-shaped elements (2) enclosed in the body (3) having a rectangular trapezoid shape. The arc-shaped elements inside the SAE are arranged at an angle of 30°, thus ensuring identical flow motion. To prevent loosening of the arc-shaped elements (2) as the gas flow moves, they are attached to the separator body on both sides. In the upper part of the separator, the arc-shaped elements are inserted into special slots in the internal wall of the separator body (3). In the lower part of the device, they are attached to the grid, where the hopper (4) begins (Figure 2).

The operating principle of SAE is that a wavy structure of dust-laden gas flow is generated, giving rise to a high centrifugal field in which the particles are thrown back to the separator walls and gradually settled in the hopper (4).

Numerical simulation of erosive wear in the separator with arc-shaped elements and the cyclone was performed using the ANSYS Fluent software. The k-w SST turbulence model was used for numerical simulation. When solving this model, partial differential equations (the Navier–Stokes equation) were also set:

$$\frac{\partial \vec{v}}{\partial t} = -(\vec{v} \cdot \nabla)\vec{v} + v\Delta\vec{v} - \frac{1}{\rho}\nabla p + \vec{f} \tag{1}$$

where ∇ is the Del operator; Δ is the Laplacian vector operator; t is time, s; v is the coefficient of kinematic viscosity, m^2/s; ρ is the gas density, kg/m^3; p is pressure, Pa; $\vec{v}$ is the velocity vector field; and $\vec{f}$ is the vector field of bulk forces.

The Navier–Stokes equation was supplemented with the continuity equation:

$$\frac{\rho}{\partial t} + \nabla \cdot (\rho\vec{v}) = 0. \tag{2}$$

The discrete phase model (DPM), taking into account the effect of solid particles on the gas flow, was used for modeling. The following equation was applied:

$$\frac{dv_p}{dt} = F_D(v_f - v_p) + g\frac{(\rho_p - \rho)}{\rho_p} + F_x \tag{3}$$

where F_D is the drag force, N_{trajct}; v_f is the gas flow rate, m/s; v_p is the particle speed, m/s; g is the acceleration due to gravity, m/s^2; ρ_p is the density of particles, kg/m^3; and F_x are the additional forces, N_{trajct}.

The erosive wear of the devices was calculated using the equation [45]:

$$R_{erosion} = \sum_{p=1}^{N_{trajct}} \frac{G_p C(dp) f(\alpha) v^n}{A_{face}} \tag{4}$$

where G_p is the mass flow rate of particles, kg/s; $C(dp)$ is the function of particle diameter; $f(\alpha)$ is the function of collision angle; v is the particle impact rate, m/s; n is the number indicating the particle impact rate; and A_{face} is the surface area of a near-wall cell, m^2.

When building the 3D cyclone model, the geometric parameters were assumed to be as follows: cylindrical body diameter, 800 mm; height, 4000 mm; inlet, 500 × 180 mm, outlet diameter; 480 mm; and lower orifice diameter, 240 mm. The cyclone was made of steel of AISI 321 grade (Figure 1).

When building the 3D model of the separator with arc-shaped elements, the geometric parameters were assumed to be as follows: device length, 2500 mm; height, 910 mm; width, 900 mm; thickness of the device walls, 10 mm; the size of arc-shaped elements, 76 × 4 mm; the number of rows of arc-shaped elements, 8; the number of elements per row, 6 (Figure 2). H-shaped elements were studied in refs. [46–48]; however, the hydraulic drag of the separator with this type of elements was higher compared to that of the device with arc-shaped elements.

The boundary and initial conditions were closest to those used in industrial-scale dehydrogenation of paraffin hydrocarbons. The volumetric flow rate of gas = 1.7 m^3/s was set at the device inlet. Pressure of 58,839.9 Pa (0.6 kgf/cm^2) was set at the outlet for purified gas and the outlet orifice for the catalyst entering the hopper. The gas flow temperature was assumed to be 550 °C. The standard aluminum–chromium catalyst particles were used. Particle size varied from 20 to 160 μm; particle density was 1400 kg/m^3. Table 1 shows the mass flow rates of particles depending on their size.

Table 1. Mass flow rate of catalyst particles.

Catalyst Particles, μm	Mass Flow Rate, kg/s
20	0.00818040
25	0.03599376
40	0.03926592
60	0.02977666
100	0.03010387
160	0.00654432

3. Results and Discussion

To provide the empirical foundations for studying erosion and verify the model, we conducted six experiments with varied sizes of catalyst particles (20–160 μm) and flow rates at a varied (assessed) erosive wear rate (0–1.0; 0–5.0; 0–10.0, and 0–15.0 mm/year). The gas flow rate, pressure, and temperature inside the devices was assumed to be constant upon modeling.

The simulated data are shown in Figures 3–9.

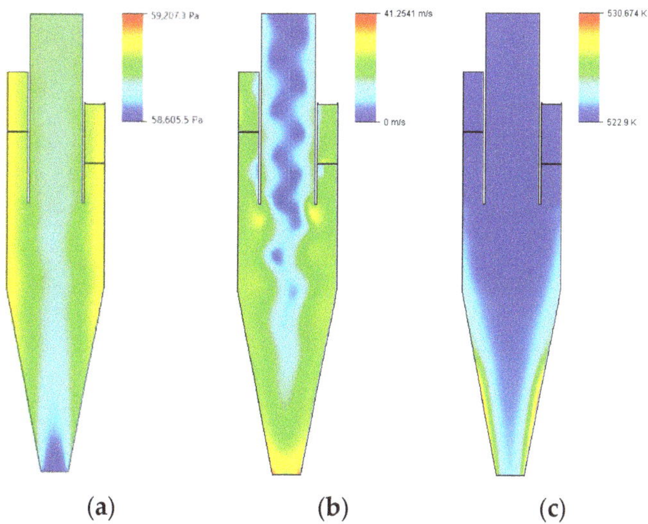

(a) (b) (c)

Figure 3. The data on the (**a**) pressure, (**b**) velocity, and (**c**) temperature profiles for the cyclone.

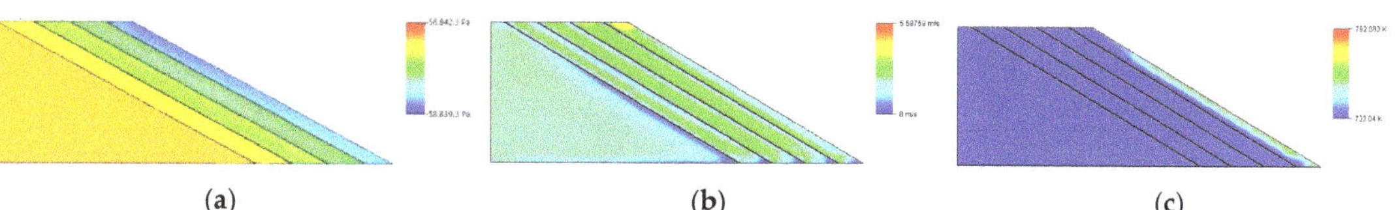

(a) (b) (c)

Figure 4. The data on the (**a**) pressure, (**b**) velocity, and (**c**) temperature profiles for the separator with arc-shaped elements.

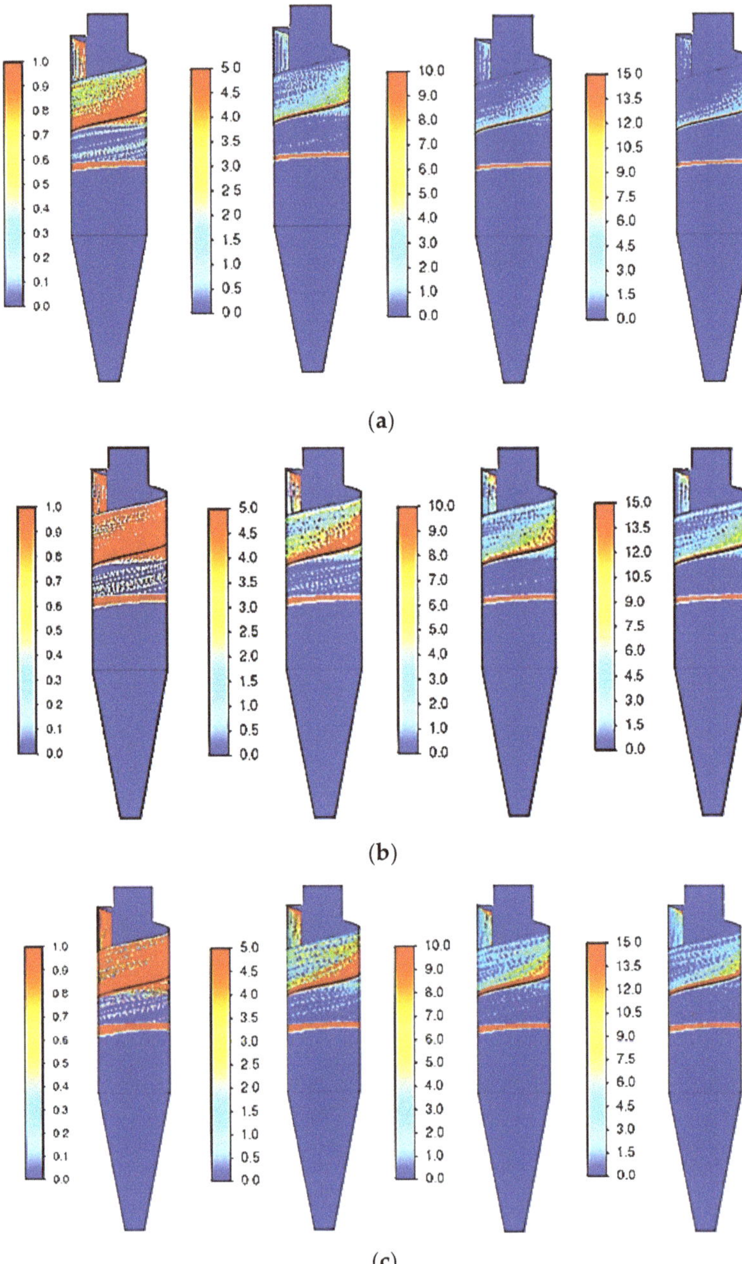

(a)

(b)

(c)

Figure 5. *Cont.*

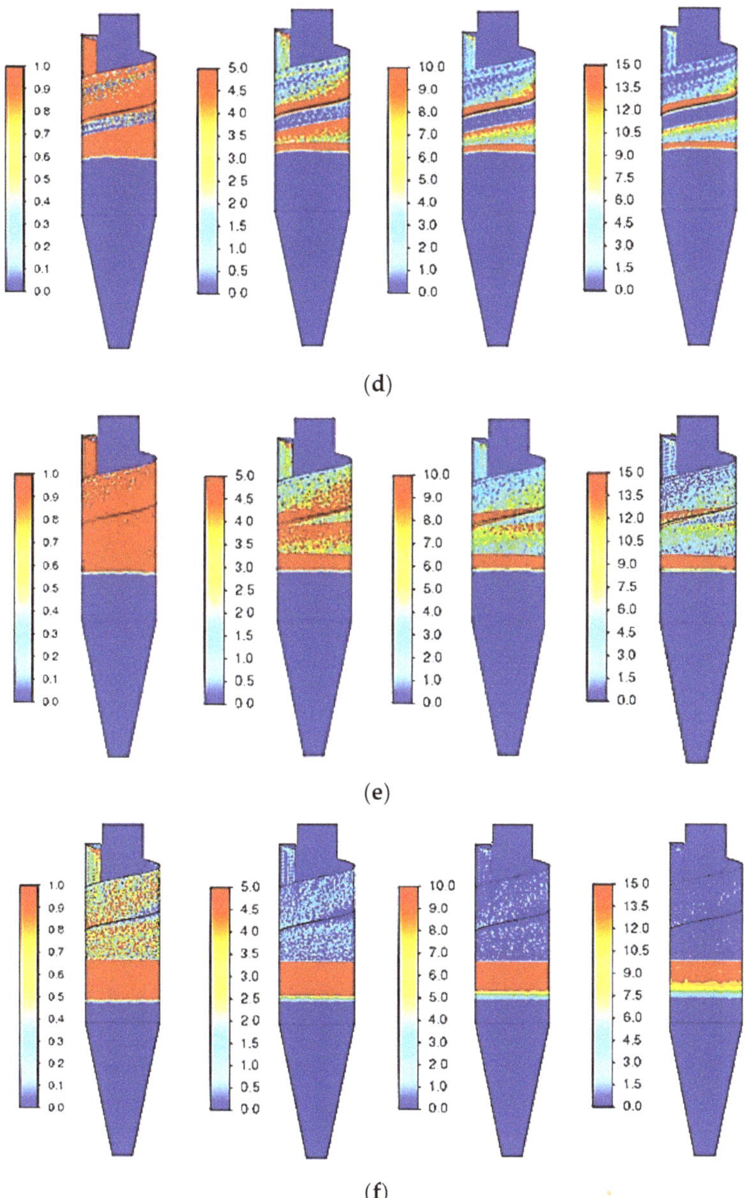

Figure 5. Erosive wear regions in the cyclone at different particle diameters (d_p). The legend shows the numeric value of the erosive wear rate We, mm/year: 0–1.0 mm/year; 0–5.0 mm/year; 0–10.0 mm/year; and 0–15.0 mm/year. (**a**) d_p = 20 μm. (**b**) d_p = 25 μm. (**c**) d_p = 40 μm. (**d**) d_p = 60 μm. (**e**) d_p = 100 μm. (**f**) d_p = 160 μm.

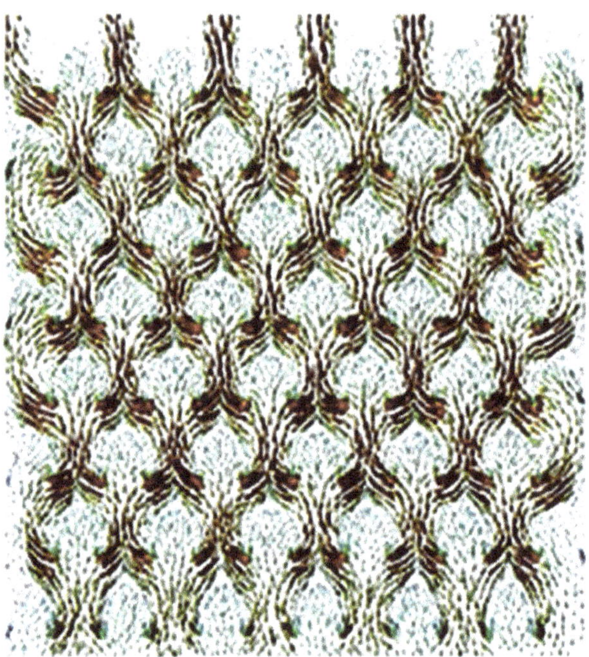

Figure 6. The wave-like structure of the gas flow inside a separator with arc-shaped elements. The vector field of velocities was recorded for the midplane perpendicular to the separator inlet.

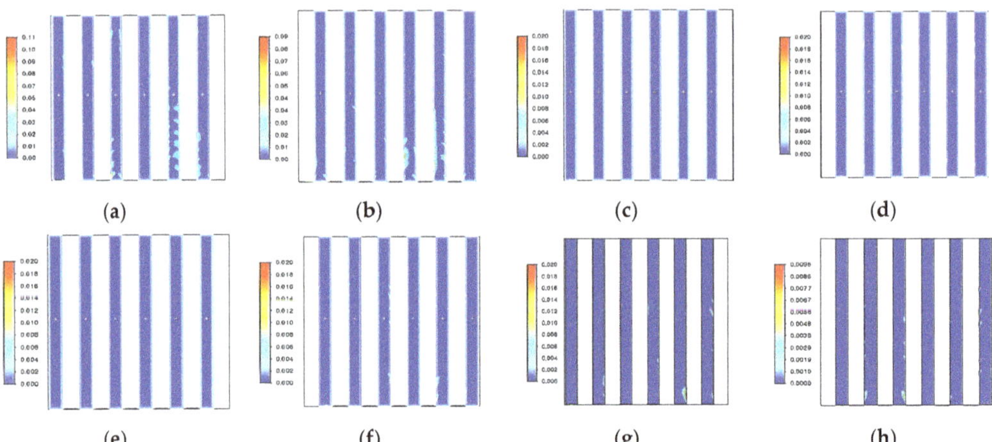

Figure 7. The regions of erosive wear rates W_e, mm/year for different rows of arc-shaped elements in the separator at particle diameter d_p = 20 µm (frontal view, as seen from the inlet). (**a**) row 1. (**b**) row 2. (**c**) row 3. (**d**) row 4. (**e**) row 5. (**f**) row 6. (**g**) row 7. (**h**) row 8.

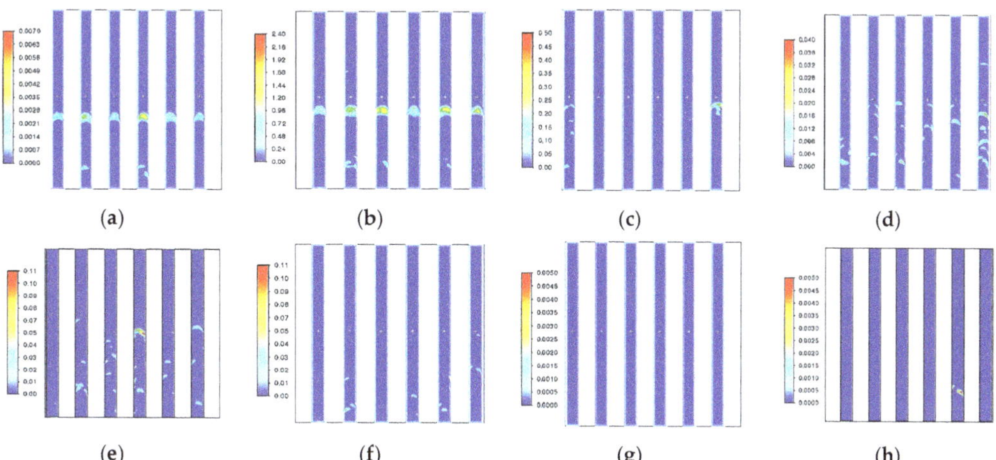

Figure 8. The regions of erosive wear rates W_e, mm/year for different rows of elements in the separator with arc-shaped elements at particle diameter d_p = 160 μm (frontal view, as seen from the inlet). (**a**) row 1. (**b**) row 2. (**c**) row 3. (**d**) row 4. (**e**) row 5. (**f**) row 6. (**g**) row 7. (**h**) row 8.

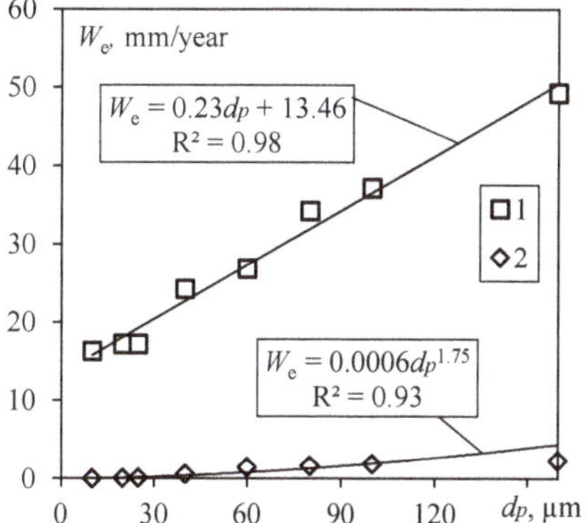

Figure 9. The erosive wear rate W_e, mm/year as a function of particle diameter for different devices: (1) the cyclone and (2) the separator with arc-shaped elements.

3.1. Assessing the Velocity, Pressure, and Temperature Profiles

Figures 3 and 4 show the comparative data on the pressure, velocity, and temperature profiles. According to these data, one can see that increased eddy velocities and pressures are generated at the cyclone margin, thus causing turbulent gas flow and increasing hydraulic drag. Elevated temperatures were also observed in the conical section of the cyclone. Most likely, it occurs due to friction of coarse particles exposed to centrifugal force (Figure 3). Changes in the pressure and velocity profiles for SAE (Figure 4) occur in a regular manner, and no critical deviations were observed. Unlike in the cyclone, the gas flow rate in SAE is stable. Temperature at the device outlet was slightly increased because of centrifugal forces.

Earlier studies have shown that the turbulent gas flow in cyclones increase the probability of fine particle penetration, thus worsening separation efficiency. The high level of fine particle collection is an advantage of the novel device [44,46].

3.2. Numerical Investigation of Erosion for the Cyclone

The results of the studies proved that there is significant erosive wear on the cyclone walls in the component of the technological line of paraffin dehydrogenation (>15 mm/year). As a result, holes appear in certain areas of the cyclones, thus increasing catalyst consumption, as well as causing a pressure drop in the technological line and jeopardizing the environment and health of plant employees. Several cyclones arranged in series in a parallel manner (2, 4, 6, and more cyclones) are most commonly used to increase efficiency of gas purification [48,49]. However, the problem of wear remains as proved by the fact that cyclones are frequently repaired and replaced. The highly loaded first-stage cyclone separators are usually more susceptible to erosive wear in their upper part, starting at the inlet and ending at the lower cylindrical section of the cyclone. The lower part of the cyclone cone is almost not subjected to erosive wear (Figure 5). The reason for that is the scheme of gas flow in the cyclone. After the solid catalyst particles are knocked out of the swirling flow towards the cyclone walls in the upper part of the device, gravity starts playing the key role. As a result, solid particles quickly fall down to the cyclone bottom and almost do not contact the cyclone cone.

Numerical modeling of the separator with arc-shaped elements has shown that, unlike for the cyclone, the separator walls are much less prone to erosive wear ($\leq$2.31 mm/year). The lower annual erosion rate in the proposed separator compared to that in the cyclone can be attributed to differences in the flow structure. However, separation of particles from the gas occurs in both devices mostly due to centrifugal forces. For the wave-like flow structure, particles are thrown to different areas of the separator walls more chaotically, so the erosive wear rate is reduced. The presence of several rows of arc-shaped elements allows one to replace the worn-out elements when necessary.

It was found experimentally that the erosive wear rate depends on particle diameter in both devices. The erosive wear rate goes up as particle diameter is increased from 20 to 160 μm. Variation of catalyst particle size can alter the erosive wear rate in different cyclone regions. Thus, the inlet area and the upper cylindrical part of the cyclone are most susceptible to erosive wear from particles with a diameter of 20–40 μm. As particle size increases from 40 to 160 μm, erosive wear starts to be observed in the lower cylindrical part of the cyclone as well. It is caused by particle coarseness. After being knocked out of the structured swirling flow, particles fall down, but they are constantly and repeatedly thrown to the device walls due to the eddy existing along the entire length of the cylindrical part of the cyclone. Therefore, pressure applied to the lower cylindrical part of the cyclone and the erosive wear rate increase at larger particle sizes (Figure 5).

3.3. Assessment of the Gas Flow Structure for the Novel Separator with Arc-Shaped Elements

Figure 6 shows the wave-like flow structure. The ordered arrangement of the eddy structure without chaotic swirl motion allows one to minimize the pickup of particles previously separated from the flow back into it.

The dust-laden gas flow enters the separator with arc-shaped elements through the inlet (1). The flow runs into the space between the rows of arc-shaped elements (2). The elements are arranged in a checkerboard pattern. Along with the checkerboard pattern of arrangement of arc-shaped element rows (2), there is an important design feature related to their arrangement with respect to each other. In particular, the following condition is met: the elements are arranged in such a manner that it is possible, starting at the central point of the circle (that is equal to two united arc-shaped elements), to draw a circle equal to two of its diameters and running through the central points of the straight outermost margin (the adjacent elements located in the preceding and subsequent rows). The fulfilment of this condition provides the most efficient wave-like structure of gas flow motion, since the centrifugal forces are maximal for this arrangement [41].

The wave-like flow structure is generated as the dust-laden gas goes around the large number of rows of arc-shaped elements (2) arranged in a checkerboard pattern. During the gas motion, there emerge many points of centrifugal forces in each eddy (turn). The high centrifugal forces are caused by the small diameter of semi-circles. Under the centrifugal forces, particles are knocked out of the structured flow towards the walls of the arc-shaped elements (2) and fall down into the hopper (4). The purified gas flow leaves the separator with arc-shaped elements through the outlet (5) (Figure 2).

The absence of rectilinear or nearly rectilinear particle motion towards the device walls and further contact with the walls as the dust-laden flow passes through the separator substantially minimizes the erosive wear.

3.4. Numerical Investigation of Erosion for the Separator with Arc-Shaped Elements

Numerical modeling of erosive wear in the separator with arc-shaped elements showed that the first two rows of elements were most susceptible to erosion. It was presumably caused by the incompletely formed structure of the dust-laden medium in the proposed device. In particular, the structure became fully wave-like in the zone located within the first two rows of the arc-shaped elements (Figure 6). However, some particles did not have enough time to be rearranged along with the gas and inertially move in a rectilinear manner towards a certain area of the walls of the first two rows of arc-shaped elements, thus causing greater erosive wear of these elements compared to the other rows (Figures 7 and 8).

An analysis of the erosive wear patterns shown in Figures 7 and 8 (the catalyst particle diameter (d_p) in the gas flow being 20 and 160 μm, respectively) demonstrates that the erosive wear area on the walls of the arc-shaped elements increased with rising particle diameter. It is clearly seen by examining the first two rows of arc-shaped elements. At a particle diameter of 20 μm, the peripheral regions of arc-shaped elements were more susceptible to erosion. At 160 μm, erosion was observed in certain areas of the arc-shaped elements along the entire perimeter. It is also evident by comparing rows 3–5 of the arc-shaped elements at a particle diameter of 20 and 160 μm (Figures 7 and 8).

The maximum erosive wear rates of the rows of elements in the separator with arc-shaped elements at particle diameters ranging from 20 to 160 μm were digitized during numerical modeling. In Table 2, the first two columns correspond to the two rows of arc-shaped elements that had undergone the maximum erosion at each analyzed catalyst particle diameter. One can see that in this range of particle diameters (20–160 μm), the first two rows of arc-shaped elements are characterized by the maximum erosion rate, which proves the statement given above. For rows 7 and 8, there is almost no wear of the arc-shaped elements, which may indicate that particle concentration in the gas was relatively low (Table 2).

Table 2. The maximum annual erosive wear rate of the rows of elements in the separator with arc-shaped elements (mm/year).

	Row	1	2	3	4	5	6	7	8
						W_e, **mm/Year**			
	20	0.09	0.07	0.016	0.015	0.017	0.02	0.018	0.0096
	25	0.08	0.13	0.06	0.03	0.019	0.03	0.018	0.015
d_p, μm	40	0.14	0.62	0.08	0.008	0.025	0.041	0.002	0.002
	60	0.18	1.47	0.11	0.009	0.041	0.052	0.001	≈ 0
	100	0.47	1.92	0.23	0.02	0.091	0.089	0.003	0.003
	160	0.67	2.31	0.42	0.04	0.099	0.097	0.003	0.001

Figures 4, 7 and 8 show that the erosive wear rate of the cyclone was significantly higher than that for separators with arc-shaped elements. Figure 9 shows it in a more systematized manner. As particle diameter increases from 20 to 160 μm, the erosive wear rate of the cyclone is most accurately described by linear function with the mean square deviation of 0.98:

$$W_e \left(\text{TsN} - 15 \right) = 0.23 d_p + 13.46. \tag{5}$$

The erosive wear rate of the separator with arc-shaped elements is most accurately described by the power-law function with the mean square deviation of 0.93:

$$W_e\ (\mathrm{SAE}) = 0.0006 d_p^{1.75}. \tag{6}$$

Numerical experiments demonstrate that the continuous impact of the high-speed flow of solid catalyst particles on the walls of the devices after the particles were knocked out of the dust-laden medium leads to gradual formation of micro- and macro-defects, which can manifest themselves as roughness, pits, cavities, holes, and other wall defects. As mentioned previously, it increases the catalyst consumption and causes a pressure drop in the technological line and other negative effects.

The study has predicted that using the cyclone for one year for collecting catalyst particles sized 20–160 μm will most likely cause the formation of holes, since erosive wear is more than 15 mm/year. For the wave-like flow structure in the separator with arc-shaped elements, at the volumetric flow rate of 1.7 m^3/s, the erosive wear rate was no higher than 2.31 mm. For cyclone walls thicker than 3 mm, it is highly likely that arc-shaped elements will remain undamaged.

In the separator with arc-shaped elements, erosive wear affects the elements rather than the separator body; therefore, the probability of erosive wear of the body of the proposed separator and excessive catalyst losses is minimized.

4. Conclusions

Modeling of the separation of catalyst particles from the gas using two devices (a cyclone and a separator with arc-shaped elements) revealed that the proposed dust collector allows one to reduce the erosive wear by several times (6.5-fold) for identical operating modes and operating parameters (temperature, pressure, particle size and speed, etc.). The lower erosive wear in the novel separator compared to that in a cyclone is caused by the ordered wave-like flow structure at which particles are thrown into different regions of SAE elements with a greater intensity during eddy formation, so the equipment-damaging side processes are mitigated. A unique feature of the SAE compared to other centrifugal separators is that higher-intensity centrifugal forces are generated in it. Many points of eddy formation appear, with eddy radius being relatively small, as the gas passes around the arc-shaped elements, making particle separation from gas flows more efficient.

It has been demonstrated that the wall erosion rate rises with increasing average size of catalyst particles (>40 μm) for both analyzed devices. Whereas in the presence of particles sized up to 40 μm only the upper part of the cyclone was mainly susceptible to erosive wear, at larger particle size, wall erosion starts to be observed both in the upper part and the lower cylindrical part of the cyclone due to particle coarseness and their chaotic collisions with the device body during settling.

In the separation device with arc-shaped elements, the first two rows of elements are most susceptible to erosion. This is caused by the incomplete formation of the structure of the dust-laden medium in the front part of the device. As a result of rectilinear and inertial motion, particles enter into the same zones within the first two rows of the arc-shaped elements. For the proposed separator, arc-shaped elements undergo erosive wear, thus preventing the formation of holes in the device body and catalyst loss.

The functions showing the dependence between the erosive wear rate and particle diameter for the cyclone and the separator with arc-shaped elements have been obtained; these functions allow one to predict wear of the devices. Speaking about the future outlook, the separator with arc-shaped elements can be recommended to be used at petrochemical plants, and in such processes as fluidized-bed dehydrogenation of C$_4$–C$_5$ isoparaffins to isoolefins in particular.

Author Contributions: All authors listed have made a substantial, direct and intellectual contribution to the work, and approved it for publication: Conceptualization: E.I.S., V.E.Z., A.V.D. and I.I.S.; Investigation: E.I.S. and V.E.Z.; Writing: E.I.S.., V.E.Z. and A.V.D.; writing—review and editing E.I.S., V.E.Z., A.V.D. and I.I.S.; data curation, E.I.S. and V.E.Z.; visualization—V.E.Z.; project administration—I.I.S. All authors have read and agreed to the published version of the manuscript.

Funding: This research received no external funding.

Data Availability Statement: Data are available at request from the authors.

Conflicts of Interest: The authors declare no conflict of interest.

References

1. Zinurov, V.E.; Dmitriev, A.V.; Kharkov, V.V. Design of High-Efficiency Device for Gas Cleaning from Fine Solid Particles. In Proceedings of the 6th International Conference on Industrial Engineering, Delhi, India, 18–19 June 2021; pp. 378–385. [CrossRef]
2. Lim, J.-H.; Oh, S.-H.; Kang, S.; Lee, K.-J.; Yook, S.-J. Development of cutoff size adjustable omnidirectional inlet cyclone separator. *Sep. Purif. Technol.* **2021**, *276*, 1–9. [CrossRef]
3. Fu, P.; Zhu, J.; Li, Q.; Cheng, T.; Zhang, F.; Huang, Y.; Ma, L.; Xiu, G.; Wang, H. DPM simulation of particle revolution and high-speed self-rotation in different pre-self-rotation cyclones. *Powder Technol.* **2021**, *394*, 290–299. [CrossRef]
4. Yu, G.; Dong, S.; Yang, L.; Yan, D.; Dong, K.; Wei, Y.; Wang, B. Experimental and numerical studies on a new double-stage tandem nesting cyclone. *Chem. Eng. Sci.* **2021**, *236*, 1–14. [CrossRef]
5. Farzad, P.; Seyyed, H.H.; Khairy, E.; Goodarz, A. Numerical investigation of effects of inner cone on flow field, performance and erosion rate of cyclone separators. *Sep. Purif. Technol.* **2018**, *201*, 223–237.
6. Celis, G.E.O.; Loureiro, J.B.R.; Lage, P.L.C.; Freire, A.P.S. The effects of swirl vanes and a vortex stabilizer on the dynamic flow field in a cyclonic separator. *Chem. Eng. Sci.* **2022**, *248*, 1–45. [CrossRef]
7. Haig, C.W.; Hursthouse, A.; Mcilwain, S.; Sykes, D. An empirical investigation into the influence of pressure drop on particle behaviour in small scale reverse-flow cyclones. *Powder Technol.* **2015**, *275*, 172–181. [CrossRef]
8. Haake, J.; Oggian, T.; Utzig, J.; Rosa, L.M.; Meier, H.F. Investigation of the pressure drop increase in a square free-vortex cyclonic separator operating at low particle concentration. *Powder Technol.* **2020**, *374*, 1–21. [CrossRef]
9. Zhang, Y.; Jiang, Y.; Xin, R.; Yu, G.; Jin, R.; Dong, K.; Wang, B. Effect of particle hydrophilicity on the separation performance of a novel cyclone. *Sep. Purif. Technol.* **2020**, *237*, 1–11. [CrossRef]
10. Chang, Y.-L.; Jiang, X.; Li, J.-P.; Fu, P.-B.; Yuan, W.; Xin, R.-K.; Huang, Y.; Wang, H.-L. Inlet particle-sorting cyclones configured along a spiral channel for the enhancement of $PM_{2.5}$ separation. *Sep. Purif. Technol.* **2021**, *257*, 1–13. [CrossRef]
11. Seyed Masoud, V.; Farzad, P.; Kamali, M.; Jebeli, H.J. Numerical Investigation of the Impact of Inlet Channel Numbers on the Flow Pattern, Performance, and Erosion of Gas-particle Cyclone. *Chem. Eng. Iran. J. Oil Gas Sci. Technol.* **2018**, *7*, 59–78.
12. Basaran, M.; Erpul, G.; Uzun, O.; Gabriels, D. Comparative efficiency testing for a newly designed cyclone type sediment trap for wind erosion measurements. *Geomorphology* **2011**, *130*, 343–351. [CrossRef]
13. Fulchini, F. Particle Attrition in Circulating Fluidised Bed Systems. Ph.D. Dissertation, University of Leeds, Leeds, UK, 2020; p. 298.
14. Welt, J.; Lee, W.; Krambeck, F.J. Catalyst attrition and deactivation in fluid catalytic cracking system. *Chem. Eng. Sci.* **1977**, *32*, 1211–1218. [CrossRef]
15. Fu, P.; Yu, H.; Li, Q.; Cheng, T.; Zhang, F.; Huang, Y.; Lv, W.; Xiu, G.; Wang, H. CFD-DEM simulation of particle revolution and high-speed self-rotation in cyclones with different structural and operating parameters. *Chem. Eng. J. Adv.* **2021**, *8*, 1–13. [CrossRef]
16. Li, Q.; Wang, J.; Xu, W.; Zhang, M. Investigation on separation performance and structural optimization of a two-stage series cyclone using CPFD and RSM. *Adv. Powder Technol.* **2020**, *31*, 3706–3714. [CrossRef]
17. Azri, M.; Nor, M.; Shahrul, K.; Alemu, L.T. Numerical investigation of API 31 cyclone separator for mechanical seal piping plan for rotating machineries. *Alex. Eng. J.* **2022**, *61*, 1597–1606. [CrossRef]
18. Werther, J.; Reppenhagen, J. Catalyst Attrition in Fluidized-Bed Systems. *AIChE J.* **1999**, *45*, 2001–2010. [CrossRef]
19. Werther, J.; Xi, W. Jet attrition of catalyst particles in gas fluidized beds. *Powder Technol.* **1993**, *76*, 39–46. [CrossRef]
20. Reppenhagen, J.; Werther, J. Catalyst attrition in cyclones. *Powder Technol.* **2000**, *113*, 55–69. [CrossRef]
21. Werther, J.; Reppenhagen, J. Attrition. In *Handbook of Fluidization and Fluid-Particle Systems*; Yang, W., Ed.; CRC Press: Boca Raton, FL, USA, 2003.
22. Ghadiri, M.; Cleaver, J.A.S.; Tuponogov, V.G.; Werther, J. Attrition of FCC powder in the jetting region of a fluidized bed. *Powder Technol.* **1994**, *80*, 175–178. [CrossRef]
23. Haig, C.W.; Hursthouse, A.; McIlwain, S.; Sykes, D. The effect of particle agglomeration and attrition on the separation efficiency of a Stairmand cyclone. *Powder Technol.* **2014**, *258*, 110–124. [CrossRef]
24. Griffiths, W.D.; Boysan, F. Computational fluid dynamics (CFD) and empirical modelling of the performance of a number of cyclone samplers. *J. Aerosol Sci.* **1996**, *27*, 281–304. [CrossRef]
25. Gimbun, J.; Chuah, T.G.; Choong, T.S.Y.; Fakhru'l-Razi, A. A CFD study on the prediction of cyclone collection efficiency. *Int. J. Comp. Meth-Sing.* **2005**, *6*, 161–168. [CrossRef]

26. Gimbun, J.; Chuah, T.G.; Fakhru'l-Razi, A.; Choong, T.S.Y. The influence of temperature and inlet velocity on cyclone pressure drop: A CFD study. *Chem. Eng. Process* **2005**, *44*, 7–12. [CrossRef]
27. Elsayed, K.; Lacor, C. Optimization of the cyclone separator geometry for minimum pressure drop using mathematical models and CFD simulations. *Chem. Eng. Sci.* **2010**, *65*, 6048–6058. [CrossRef]
28. Park, D.; Go, S.J. Design of Cyclone Separator Critical Diameter Model Based on Machine Learning and CFD. *Processes* **2020**, *8*, 1521. [CrossRef]
29. Hoekstra, A.J.; Derksen, J.J.; Van, H.E.A.; Akker, D. An experimental and numerical study of turbulent swirling flow in gas cyclones. *Chem. Eng. Sci.* **1999**, *54*, 2055–2065. [CrossRef]
30. Slack, M.D.; Prasad, R.O.; Bakker, A.; Boysan, F. Advances in cyclone modelling using unstructured grids. *Chem. Eng. Res. Des.* **2000**, *78*, 1098–1104. [CrossRef]
31. Gronald, G.; Derksen, J.J. Simulating turbulent swirling flow in a gas cyclone: A comparison of various modeling approaches. *Powder Technol.* **2011**, *205*, 160–171. [CrossRef]
32. Alexander, R.M. Fundamentals of cyclone design and operation. *Proc. Aust. Inst. Miner. Met.* **1949**, *152*, 152–153.
33. Gimbun, J.; Chuah, T.G.; Choong, T.S.Y.; Fakhru'l-Razi, A. Prediction of the effects of cone tip diameter on the cyclone performance. *J. Aerosol Sci.* **2005**, *36*, 1056–1065. [CrossRef]
34. El-Emam, M.A.; Zhou, L.; Shi, W.; Chen, H. Performance evaluation of standard cyclone separators by using CFD-DEM simulation with realistic bio-particulate matter. *Powder Technol.* **2021**, *385*, 357–374. [CrossRef]
35. Li, H.; Wang, L.; Du, C.; Hong, W. CFD-DEM investigation into flow characteristics in mixed pulsed fluidized bed under electrostatic effects. *Particuology* **2021**, *65*, 1–38. [CrossRef]
36. Nakhaei, M.; Lu, B.; Tian, Y.; Wang, W.; Kim, D.-J.; Wu, H. CFD Modeling of Gas–Solid Cyclone Separators at Ambient and Elevated Temperatures. *Processes* **2019**, *8*, 228. [CrossRef]
37. Li, Z.; Tong, Z.; Yu, A.; Miao, H.; Chu, K.; Zhang, H.; Guo, G.; Chen, J. Numerical investigation of separation efficiency of the cyclone with supercritical fluid-solid flow. *Particuology* **2022**, *62*, 36–46. [CrossRef]
38. Hamed, S.; Mohammad, R.; Dariush, A. Numerical study of flow field in new design cyclones with different wall temperature profiles: Comparison with conventional ones. *Adv. Powder Technol.* **2021**, *32*, 3268–3277. [CrossRef]
39. Chen, J.; Jiang, Z.; Yang, B.; Wang, Y.; Zeg, F. Effect of inlet area on the performance of a two-stage cyclone separator. *Chin. J. Chem. Eng.* **2021**, *36*, 1–35. [CrossRef]
40. Zhang, Z.; Dong, S.; Dong, K.; Hou, L.; Wang, W.; Wei, Y.; Wang, B. Experimental and numerical study of a gas cyclone with a central filter. *Particuology* **2021**, *65*, 1–46. [CrossRef]
41. Mazyan, W.I.; Ahmadi, A.; Brinkerhoff, J.; Ahmed, H.; Hoorfar, M. Enhancement of cyclone solid particle separation performance based on geometrical modification: Numerical analysis. *Sep. Purif. Technol.* **2018**, *191*, 276–285. [CrossRef]
42. Reddy Karri, S.B.; Ray, C.; Knowlton, T. Erosion in Second Stage Cyclones: Effects of Cyclone Length and Outlet Gas Velocity. In Proceedings of the 10th International Conference on Circulating Fluidized Beds and Fluidization Technology-CFB-10, Sun River, OR, USA, 1–5 May 2013. Available online: http://dc.engconfintl.org/cfb10/40 (accessed on 5 October 2022).
43. Peukert, W.; Wadenpohl, C. Industrial separation of fine particles with difficult dust properties. *Powder Technol.* **2001**, *118*, 136–148. [CrossRef]
44. Salakhova, E.I.; Dmitriev, A.V.; Zinurov, V.E.; Nabiullin, I.R.; Salakhov, I.I. Dust Collector for Paraffin Dehydrogenation Units with a Fluidized Catalyst Bed. *Catal. Ind.* **2022**, *14*, 369–375. [CrossRef]
45. Edwards, J.K.; McLaury, B.S.; Shirazi, S.A. Evaluation of Alternative Pipe Bend Fittings in Erosive Service. In Proceedings of the ASME FEDSM'00: ASME 2000 Fluids Engineering Division Summer Meeting, Boston, MA, USA, 11–15 June 2000.
46. Dmitriev, A.V.; Zinurov, V.E.; Dmitrieva, O.S. Collecting of finely dispersed particles by means of a separator with the arc-shaped elements. In Proceedings of the International Conference on Modern Trends in Manufacturing Technologies and Equipment (ICMTMTE 2019), Sevastopol, Russia, 9 September 2019; Volume 126, p. 00007. [CrossRef]
47. Zinurov, V.E.; Kharkov, V.V.; Salakhova, E.I.; Vakhitov, M.R.; Kuznetsov, M.G. Numerical simulation of collection efficiency in separator with inclined double-T elements. *IOP Conf. Ser. Earth Environ. Sci.* **2022**, *981*, 42024. [CrossRef]
48. Zinurov, V.; Dmitriev, A.; Kharkov, V. Influence of process parameters on capturing efficiency of rectangular separator. In Proceedings of the 2020 International Conference on Information Technology and Nanotechnology (ITNT), Samara, Russia, 26–29 May 2020; pp. 1–4. [CrossRef]
49. Fu, P.; Jiang, X.; Ma, L.; Yang, Q.; Bai, Z.; Yang, X.; Lv, W. Enhancement of $PM_{2.5}$ cyclone separation by droplet capture and particle sorting. *Environ. Sci. Technol.* **2018**, *52*, 11652–11659. [CrossRef] [PubMed]

Article

Energy-Saving and Low-Carbon Gear Blank Dimension Design Based on Business Compass

Yongmao Xiao [1,2,3], Jincheng Zhou [1,2,3], Ruping Wang [4,*], Xiaoyong Zhu [5] and Hao Zhang [6]

1 School of Computer and Information, Qiannan Normal University for Nationalities, Duyun 558000, China
2 Key Laboratory of Complex Systems and Intelligent Optimization of Guizhou Province, Duyun 558000, China
3 Key Laboratory of Complex Systems and Intelligent Optimization of Qiannan, Duyun 558000, China
4 Management School, China West Normal University, Nanchong 637002, China
5 School of Economics & Management, Shaoyang University, Shaoyang 422099, China
6 College of Mechanical and Electrical Engineering, Xinjiang Agricultural University, Wulumuqi 830052, China
* Correspondence: wrpqyi@163.com; Tel.: +86-13-016-410-611

Abstract: Sustainable blank dimension design is the key to the implementation of green industrial development. However, blank dimension design only considers the blank production factor of the blank dimension design stage, which cannot guarantee the blank production stage and the use stage's overall goal. In this paper, based on the guiding thinking of a business compass, a low-carbon and low-energy consumption blank dimension optimization design model was proposed. Taking the process parameters of the production and the use of the blank as the variables, the grey wolf optimization algorithm was adopted to solve the problem. Taking the gear blanks dimension as an example, the optimized blank dimension is 98.6, compared with the standard blank dimension of 100, 105, the energy consumption is 95.7% and 93.1%, the carbon emission is 92.6% and 90.2%, and the material consumption is 96.5% and 87.5%, respectively. The sustainable blank dimension design has obvious advantages in terms of low energy consumption and low carbon, and it can save a lot of materials; it can also promote product sustainability.

Keywords: sustainable blank dimension design; energy-saving; low-carbon; business compass; grey wolf algorithm

Citation: Xiao, Y.; Zhou, J.; Wang, R.; Zhu, X.; Zhang, H. Energy-Saving and Low-Carbon Gear Blank Dimension Design Based on Business Compass. *Processes* **2022**, *10*, 1859. https://doi.org/10.3390/pr10091859

Academic Editor: Sergey Y. Yurish

Received: 22 August 2022
Accepted: 9 September 2022
Published: 15 September 2022

Publisher's Note: MDPI stays neutral with regard to jurisdictional claims in published maps and institutional affiliations.

1. Introduction

Modern industrial civilization has created unprecedented wealth for humankind and greatly liberated human productivity [1]. However, excessive industrialization and overdraft of resources have seriously damaged the natural environment on which human beings depend for survival, leading to the worsening of global climate change and environmental pollution, as well as the increasing consumption of resources. It poses a threat to the survival of human beings and triggers global thinking on resource use [2,3]. Blank dimension is a key element which to a great degree can decide blank production and use processes' energy consumption and carbon emission. The blank design stage is very important, and more than 80% of the blank process energy consumption can be determined [4–6]. The dimension design of the blank is of great significance to the energy saving and green development of the industry.

Many scholars have studied the blank dimension design. Wan et al. used the forge, a forging simulation software, to conduct finite element analysis of rolling of flange rings, and gave the optimal dimension of the ring blank before the workpiece was rolled [7]. Li et al. designed the blank dimension according to the forging process characteristics, the work hardening characteristics of the material, the mechanical properties of the blank, the wall thickness before and after the blank forging, and the relationship between the inner and outer diameters, it is verified by the case of automobile swaging shaft [8]. Xu et al. established an intelligent derivation model of rolling ring blank dimension based on the

BP neural network algorithm. The reliability of the intelligent derivation scheme was verified by finite element simulation and physical experiments [9]. Zhang et al. proposed a node mesh mapping method to design a blank and optimize the shape of the blank according to the mapping relationship between the formed part and the target part and the flow trend of the material point. The method was validated through the forming process of square box parts and automotive pillars [10]. Akinnuli et al. formed a global model based on the deep-drawn shell model to predict the optimal blank diameter required for each determined geometry. The model was validated by comparing the results of the four models with those corresponding to the same model [11]. Xiao et al. proposed the concept of step element based on the processing characteristics of complex box blanks and optimized the process route. The method is verified by the processing of emulsion box blank [12]. Gharehchahi et al. used an iterative numerical simulation method to optimize the boundary shape of the blank through continuous iteration [13].

In terms of machining parameter optimization, Li et al. proposed an energy-saving optimization method for NC turning batch machining process parameters considering tool wear. They used a multi-objective simulated annealing algorithm to optimize. The effectiveness of the method was verified compared with empirical schemes [14]. Liu et al. proposed a process parameter optimization decision-making method based on processing process sample prediction and multi-objective genetic optimization algorithm, which was solved by an improved multi-objective genetic algorithm and genetic-backpropagation (GABP) neural network [15]. Tian et al. established a comprehensive optimization model based on tool selection and cutting parameters and took carbon emissions as the optimization objective. The validity of the model was verified by a turbo machining case. [16]. Li et al. proposed a method to comprehensively optimize process planning and cutting parameters to reduce the total energy consumption of machining and balance the machine workload in the workshop; particle swarm optimization was used to solve it [17]. Khan et al. investigated a holistic analysis of surface quality, energy, cost, and carbon emissions and used a Haynes 25 alloy external cylindrical turning process to analyze trade-offs among multiple objectives [18]. Josh et al. conducted a comparative study on the performance of NSGA-II, MOALO, and MODA in micromachining applications [19]. Zhou et al. proposed a cutting parameter optimization method with a carbon emissions objective. They developed NG-NSGA-II to solve the model and used cylindrical turning as an example to verify the method [20]. Xiao et al. proposed a knowledge-driven energy efficient turning optimization method to minimize specific energy consumption and machining time. Compared with the three algorithms of GA, ACA, and PSO, it showed the superiority of this method. [21]. Shin et al. proposed a toolpath-based energy modeling method to optimize energy consumption online and in real-time, which was validated by milling [22].

The above studies have been well verified. However, these studies only consider the optimization of the forming dimension in the blank production stage or the parameter optimization in the blank use stage, respectively. The blank dimension design needs to consider the production and use process as a whole; the optimality of a single stage cannot guarantee the overall optimality. This paper used the business compass as the top-level design model of enterprise sustainable development to guide the blank dimension design and effectively combined design and management. It comprehensively considered the production and use stages of the blank and constructed a five-factor framework of "environment, human, method, material, and machine" for sustainable blank design based on the business compass. And the "method" in the five elements was the focus of research, a blank dimension optimization design model based on energy-saving and low-carbon was established, and the grey wolf algorithm was used to solve it. It was verified with the example of gear blank to enrich the theoretical research on the sustainable design of blank dimension. This method can take into account the energy consumption and carbon emissions of the blank at each stage of its life cycle and help enterprises quickly design the blank dimension that meets the requirements, which is most conducive to the sustainable development of society.

2. Sustainable Blank Dimension Design Method under the Guidance of the Business Compass

2.1. The Guidance of the Business Compass to Business Operations

The business compass is a multidisciplinary and interdisciplinary theoretical framework for the top-level design of enterprise sustainable development. Business compass organically integrates Chinese philosophy and Western management science based on the people-oriented development concept, refer to the five elements "wood, fire, earth, metal and water" in Chinese Taoist philosophy [23], as shown in Figure 1, it proposes five core elements of "trend, path, skill, tool and profit" for the sustainable development of the enterprise, and progressively constructs fifteen sections, including policy, industry, market, mission, vision, value, strategy, tactics, organization, technology, product, service, internal benefit, external benefit, and society, as well as 119 corporate sustainable development management modules, including economic policy, industrial trends, corporate responsibility, strategic goals, corporate ecology, technological innovation, product development, blank procurement, quality management, company achievements, and entrepreneurship, which is a new multidisciplinary cross-theoretical framework for the top-level design of enterprise sustainable development [24]. The business compass guides all aspects of enterprise operation and has important guiding significance for personnel operation management, equipment operation management, financial operation management, product operation management, etc.

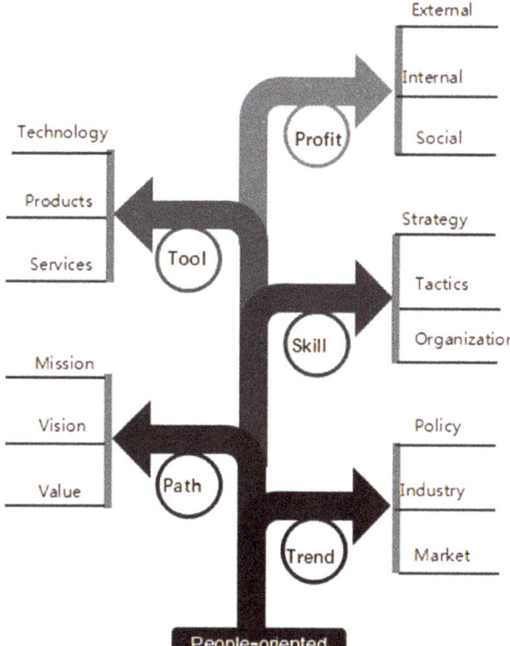

Figure 1. Business compass mind map.

2.2. The Blank Dimension Sustainable Design Framework Based on Business Compass

The guidance of the business compass for product design management can guide the adjustment and optimization of process elements. The essence of sustainable design of blank dimension is to make full use of various scientific and technological design methods, economic management, and other aspects of knowledge according to people's objective needs [25,26], market forecasts, and multi-channel information materials. Through the creative thinking of technical personnel, after repeated deliberation and judgment, the management will make decisions to convert various design models into prototypes, thereby further obtaining blank products with good performance, high economic benefits, creativity, and sustainable environmental protection [27–29].

The business compass can provide guidance for the sustainable development of enterprises. In terms of blank dimension sustainable design, the "trend" is the environment in which a company conducts blank design. In recent years, the manufacturing industry has been developing in the direction of low-carbon and green manufacturing, and enterprise production must keep up with the development trend. The "path" is the purpose of the blank design, the goal of the enterprise is to produce a product, and the blank should be designed according to the product characteristics. The "technology" is choosing the appropriate blank dimension, which needs to be judged according to the product and the current technical level. The "tool" is production equipment, and the performance, parameters, and processing links of production equipment must be considered. The "benefit" is the profit of the product. The product benefits that the blank can bring and the problems it can solve. The business compass is a new way of integrating Chinese and Western management ideas to guide the top-level design of an enterprise. Its five elements of "trend, path, skill, tool, and profit" correspond to the five elements of blank dimension development "environment, human, method, material, and machine." According to the business compass thinking—the concept of sustainable development of blank dimension—the internal connection of circular economy, the framework of sustainable design of blank dimension has been established, as shown in Figure 2.

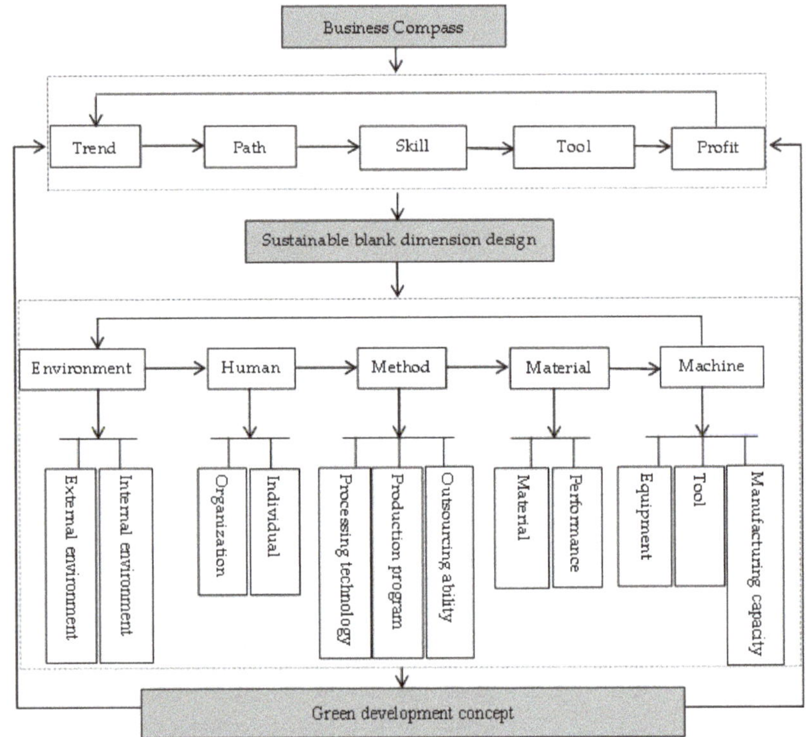

Figure 2. The framework for sustainable design of blank dimension.

The framework of sustainable design of blank dimension based on business compass thinking is a theoretical framework of sustainable development of blank dimension that fully reflects the concept of circular economy development according to the strategic goal of sustainable human development. Its basic idea is to make full use of modern scientific and technological means during the entire life cycle of the blank dimension, whether it is the government, society, enterprises, or individuals within a certain period and a certain range. While meeting the performance of the blank products, natural resources and energy should be positively, actively, and efficiently used, with no pollution or less pollution to the ecological environment, friendly to manufacturers and users, and the final discarded products should be able to be recycled. The wastes that cannot be reused should be properly disposed of to achieve the purpose of cherishing precious natural resources and protecting the ecological environment, thereby promoting the development of a circular economy, and realizing sustainable human development.

More than 80% of the energy consumption and carbon emissions of product processing can be determined in the product design stage. Product design has become an important basis for sustainable design and an important prerequisite for realizing a circular economy. Product design has become an urgent question needed to solve the development of the circular economy. To solve this problem, this paper takes the sustainable design of gears as an example and proposes a blank dimension optimization design model based on low carbon and energy saving.

3. Blank Dimension Optimal Design Model Based on Low-Carbon and Low-Energy Consumption

The blank dimension appears at the last moment of rolling and begins to disappear during the use of the blank, which is a procedural element and has a short existence time. The blank dimension is the sum of back cutting depth and the workpiece dimension. The workpiece dimension is generally known, and the back-cutting depth is a variable in the process and is related to the processing process. Once the machining is completed, the back-cutting depth can be calculated. Therefore, the main work of the research is to optimize the model of energy consumption and carbon emission. The energy consumption and carbon emission models are as follows [30–32].

3.1. Energy Consumption Calculation Model Based on Blank Production and Processing

The manufacturing process includes the stages of energy manufacturing, auxiliary material manufacturing, raw material manufacturing, and product manufacturing. The whole process goes through steel making, pouring, undercutting, forging, machining, and other processes, with the assistance of energy (primary energy, secondary energy, electrical energy), cutting fluid, lubricating oil, fixtures, etc., to finally form the raw material of the gear blank. The production of blanks must go through processes such as raw material mining and transportation, blast furnace ironmaking, steelmaking, and steel rolling. Among them, the mining and transportation of materials, blast furnace ironmaking, steelmaking, and other processes have little impact on the design of the blank dimension. Therefore, the effect of the rolling process energy consumption on the blank dimension is only considered here. According to the billet production and use process, the energy consumption can be divided into rolling energy consumption during blank rolling and cutting energy consumption when the blank is cut.

The rolling energy consumption can be expressed as:

$$E_Z = \sum_{t=1}^{n} q_t = \sum_{t=1}^{n} M_t v_t \tau_t / D_t \tag{1}$$

where t represents the rolling pass, M_t, v_t, τ_t are expressed as the rolling moment, the rolling speed, the rolling time when t pass . q_t represents the t pass rolling energy consumption. D_t represents the roll working diameter, a, n are represented the coefficient, and the rolling pass, respectively. Rolling pass n is a variable, connected with total elongation coefficient u_z and the average elongation coefficient u_p, the expression is:

$$n = \frac{\log(u_z)}{\log(u_p)} \tag{2}$$

$$u_z = \frac{F_0}{F_n} = \frac{(a_t A_t)^2 - 0.86(0.1 a_t A_t)^2}{\pi(D a_{tt})^2 / 4} \tag{3}$$

where a_t, A_t, D are the thermal expansion coefficient, billet side length (mm), and blank diameter (mm), respectively. F_0 and F_n are the red billet section area (mm^2), and the finished product and thermal state section area (mm^2).

The energy consumption in blank processing is primarily the electricity consumption of the machine tool equipment, which will directly affect the shape, dimension, and precision of the blank. In addition, it also includes the lighting energy consumption when the blank is used and the product transportation energy consumption.

The electricity consumption of the machine tool equipment in the blank processing is:

$$E_{D,i} = E_B + E_I + E_C \tag{4}$$

where $E_{D,i}$ is the machine tool total electrical energy consumption (kWh) in the i step in the part process route, E_B is the machine tool basic energy consumption, which is the energy consumption for the machine tool to keep running when the workpiece is loaded, unloaded, positioned, and clamped on the machine tool (kWh), E_I is the machine tool idle energy consumption, it is the spindle running energy consumption when the tool feeds, retracts and adjusts the tool (kWh), E_C is the energy consumed by the tool to cut parts or change the accuracy and surface quality of the parts (kWh);

$$E_C = V * \delta * t_c \tag{5}$$

V is the material removed volume in the process step (mm^3), δ is the specific energy consumption (w.s/mm^3), and t_c is the process step working time.

The energy consumption of lighting and product transportation is

$$E_A = P_L * t_L + \sum_{m=1}^{n} P_T * t_T \tag{6}$$

E_A is the lighting energy consumption and product transportation energy consumption, P_L is the lighting equipment power (kWh/s), t_L is the workpiece average lighting time (s); P_T is the mth transportation equipment power (kWh/s), and t_T is the average transportation time and cutting time of the workpiece of the m-th transportation equipment.

In conclusion, for a part produced by a process route consisting of i steps, the average power consumption of each workpiece can be expressed as,

$$E_J = E_Z + \sum_{i=1}^{l} E_{D,i} + E_A \tag{7}$$

3.2. Carbon Emission Calculation Model Based on Blank Production and Processing

3.2.1. Carbon Emissions from the Blank Production Process

The production of blank must go through the mining and transportation of raw materials, iron smelting, steelmaking, rolling, and other processes. Among them, only the rolling process has a greater impact on the blank dimension design. Therefore, only the impact of rolling carbon emissions on blank production is considered here. The energy consumed is mainly electricity in the hot rolling process, and due to the requirements of product quality, some billets need to wait for a period of time to reduce the temperature after they are released from the heating furnace; that is, temperature control can be performed before rolling. Therefore, the carbon emissions from billet hot rolling can be divided into two parts: the electricity produced carbon emissions in the rolling mill and the temperature drop caused carbon emissions during the waiting process of the rolling. They can be expressed as follows:

$$C_Z = \alpha E_Z = \theta \sum_{t=1}^{n} M_t v_t \tau_t / D_t \tag{8}$$

$$C_T = \varphi \sum_{i=0}^{m-1} [x_{i+1} - (x_i + t_i)] \tag{9}$$

where, C_Z is the rolling mill carbon emission generated by using electricity, C_T is the temperature drop carbon emission caused during the rolling waiting process, m is the total number of steel blanks rolled in the steel plant, θ is the energy consumption carbon emission coefficient, φ is the temperature drop carbon emission coefficient, x_i represents the time when the billet i starts rolling, t_i is the rolling required time of the blank i.

3.2.2. Material Consumption, Energy Consumption, and Waste Generation in Blank Processing

In the process of gear processing, the reaction of raw materials of gear blanks with auxiliary raw materials, electricity, gas, etc., will produce massive carbon emissions. The carbon emission boundary of gear processing can be set as the entire process from the input of gear raw materials to the output of gear products, including the carbon emissions of materials, energy, and wastes consumed in the gear processing. According to production needs, all kinds of materials $i(i = 1, 2, \ldots, I)$ and energy $k(k = 1, 2,\ldots, K)$ enter the gear blank preparation and gear cutting workshop, and after forging, heat treatment, machining, etc., the finished product is obtained, and waste $l(l = 1, 2, \ldots, L)$ is discharged from each workshop.

The materials, energy consumption, and waste generated by the gear passing through workshop m ($m = 1, 2, \ldots, M$) are expressed as $S(i,m)$, $E(k,m)$, and $W(l,m)$, respectively. In the processing time T, the total amount of materials consumption, energy consumption, and waste discharged in the gear processing process are respectively expressed as,

$$M_T = \sum_{m=1}^{M} \sum_{i=1}^{I} S(i,m) \tag{10}$$

$$E_T = \sum_{m=1}^{M} \sum_{k=1}^{K} E(k,m) \tag{11}$$

$$W_T = \sum_{m=1}^{M} \sum_{l=1}^{L} W(l,m) \tag{12}$$

According to the material, energy consumption, and waste generation in the process of gear machining, the processing carbon emissions in the process of quantification is as follows:

$$C_{em} = C_M + C_E + C_W \tag{13}$$

where C_{em}, C_M, C_E, C_W represent the total carbon emission, carbon emission of material consumption, carbon emission of energy consumption, carbon emission of waste treatment. Carbon emission can represent as the product of energy consumption (converted to standard coal amount) and carbon emission factor S_E ($S_E = e_C / t_E$, where e_C is carbon emission (t), t_E is standard coal amount (t)) by the emission factor method.

3.2.3. Material Carbon Emissions Calculation

In gear processing, the indirect carbon emission $C_M^{(i,m)}$ generated by the consumption of the ith material in the processing workshop m is:

$$C_M^{(i,m)} = \sum_{c=1}^{N} S(i,m) J_c^i S_{E_c} \tag{14}$$

J_c^i is the standard coal amount of the c-th energy consumed by the production unit material i, and S_{E_c} is the energy c carbon emission factor.

The total indirect carbon emission of material i consumed by gear processing is

$$C_M^i = \sum_{m=1}^{M} C_M^{(i,m)} = \sum_{m=1}^{M} \sum_{c=1}^{N} S(i,m) J_c^i S_{E_c} \tag{15}$$

The total indirect carbon emission of all materials consumed in gear machining is

$$C_M = \sum_{m=1}^{M} \sum_{i=1}^{I} C_M^{(i,m)} = \sum_{m=1}^{M} \sum_{i=1}^{I} \sum_{c=1}^{N} S(i,m) J_c^i S_{E_k} \tag{16}$$

3.2.4. Energy Carbon Emission Calculation

There are two types of energy consumption carbon emissions: one is the indirect carbon emission C_{IE} generated in the energy preparation process, and the other is the direct carbon emission C_{DE}, which is caused by the fossil energy consumption in the gear processing process, so $C_E = C_{IE} + C_{DE}$.

(1)   Indirect carbon emissions

In gear processing, the indirectly emitted carbon emissions generated by the kth energy consumption in the processing workshop m are:

$$C_{IE}^{(k,m)} = \sum_{n=1}^{N} E(k,m) J_n^k S_{E_n} \tag{17}$$

J_n^k is the standard coal amount of the nth energy consumed by producing unit energy k, and S_{E_n} is the energy n carbon emission factor.

The gear processing process total indirect carbon emissions from kth energy consumed are:

$$C_{IE}^k = \sum_{m=1}^{M} C_{IE}^{(k,m)} = \sum_{m=1}^{M} \sum_{n=1}^{N} E(k,m) J_n^k S_{E_n} \tag{18}$$

The gear processing process total indirect carbon emissions from all energy consumed are:

$$C_{IE} = \sum_{m=1}^{M} \sum_{k=1}^{K} C_{IE}^{(k,m)} = \sum_{m=1}^{M} \sum_{k=1}^{K} \sum_{n=1}^{N} E(k,m) J_n^k S_{E_n} \tag{19}$$

(2)   Direct carbon emissions

In gear processing, the direct emissions of carbon caused by the consumption of the kth fossil energy in the processing workshop m are:

$$C_{DE}^{(k,m)} = E(k,m) P_k S_{E_k} \tag{20}$$

P_k is the standard coal coefficient of energy k, and S_{E_k} is the carbon emission factor of energy.

The total direct carbon emissions of kth energy consumed in the gear processing process are:

$$C_{DE}^k = \sum_{m=1}^{M} C_{DE}^{(k,m)} = \sum_{m=1}^{M} E(k,m) P_k S_{E_k} \tag{21}$$

The total direct carbon emission of energy consumed in the gear processing process are:

$$C_{DE} = \sum_{m=1}^{M} \sum_{k=1}^{K} \sum_{n=1}^{N} E(k,m) J_n^k S_{E_k} \tag{22}$$

3.2.5. Waste Disposal Carbon Emissions

The treatment of wastewater and waste gas generated in the gear processing process consumes electricity and fossil energy. The carbon emission of the waste disposal process generated by the waste l discharged from the manufacturing workshop m are:

$$C_W^{(l,m)} = \sum_{q=1}^{N} W(l,m) J_q^l S_{E_q} \tag{23}$$

where J_q^l is the standard coal amount of the qth energy consumed by the processing unit waste l, and S_{E_q} is the energy q carbon emission factor.

The total carbon emissions from waste l in the gear processing process are:

$$C_W^l = \sum_{m=1}^{M} C_W^{(l,m)} = \sum_{m=1}^{M} \sum_{q=1}^{N} W(l,m) J_q^l S_{E_q} \tag{24}$$

The total carbon emissions of all waste products in gear processing are:

$$C_W = \sum_{m=1}^{M} \sum_{l=1}^{L} C_W^{(l,m)} = \sum_{m=1}^{M} \sum_{l=1}^{L} \sum_{q=1}^{N} W(l,m) J_q^l S_{E_q} \tag{25}$$

4. Optimization Model Solution Method

There are many types of algorithms for solving multi-objective problems. Different algorithms have different characteristics. After comparison, we selected the gray wolf optimizer as the solving method of the optimization model.

4.1. Grey Wolf Algorithm Description

Grey Wolf Optimizer (GWO) is a swarm intelligence optimization algorithm that evolved from the ecological predation habit of the grey wolf. Construct a simple and effective optimization mechanism by using wolves of different social classes to jointly guide the wolves to locate the target and realize the predation process of finding prey, surrounding prey, tracking prey, and capturing prey. Relative to the particle swarm algorithm, cuckoo algorithm, and other swarm intelligence algorithms, the gray wolf algorithm has a stronger global search ability. This paper redesigns the update operator of the algorithm. It adds crossover and mutation operations to solve the model so that the optimization of energy consumption, carbon emission, and cost can be realized in the iterative optimization process [33–36].

It strictly replicates the wolves' internal hierarchy management system. The design of GWO is to build a social hierarchy pyramid of gray wolves, the first to third layers of the social hierarchy pyramid in the wolf pack, namely the best, second best, and third best wolves are denoted as α, β, γ, respectively, and the remaining wolves at the bottom of the pyramid are referred to as δ [37,38]. In GWO, α, β, γ guide the wolves to complete the entire hunting (optimization) process, and the δ wolf must obey the three wolves. This population hierarchy ensures that the process of killing prey is efficient. First, all members of the wolf pack look for, chase, and approach the prey together, and encircle the prey from different angles and directions. As the encirclement circle shrinks, the β and γ wolves that are closest to the prey will attack the prey in the wolf pack led by the α wolf. When the prey escapes, the δ will be supplemented to ensure that the wolves continue to tackle the prey in each direction, and finally, catch prey smoothly [39].

First, the gray wolf forms a prey encirclement, and the individual's distance from the prey is computed as follows:

$$D = \left| C{\cdot}X_p(t) - X_w(t) \right|. \tag{26}$$

$$X_w(t+1) = X_p(t) - A. \tag{27}$$

$X_p(t)$ is the position coordinate of the prey, $X_w(t)$ is the gray wolf's location coordinates, and t is the number of iterations. A and C are synergy coefficient vectors, which are the convergence factor and the swing factor, respectively.

$$A = 2a{\cdot}r_2 - a \tag{28}$$

$$C = 2{\cdot}r_1 \tag{29}$$

r_1 and r_2 are random vectors in the [0, 1] interval, as the number of iterations increases, a gradually decreases linearly from 2 to 0.

Since the position of the prey is unknown in the abstract search space, three wolves (α, β, γ) with the best results are saved in each iteration, and other wolves are guided to approach the prey according to their position information. The update method is as follows:

$$D_\alpha = \left| C_1{\cdot}X_\alpha(t) - X_w(t) \right| \tag{30}$$

$$D_\beta = \left| C_2{\cdot}X_\beta(t) - X_w(t) \right| \tag{31}$$

$$D_\gamma = \left| C_3{\cdot}X_\gamma(t) - X_w(t) \right| \tag{32}$$

$$X_1 = X_\alpha - A_1{\cdot}D_\alpha \tag{33}$$

$$X_2 = X_\beta - A_2{\cdot}D_\beta \tag{34}$$

$$X_3 = X_\gamma - A_3{\cdot}D_\gamma \tag{35}$$

$$X_p(t+1) = (X_1 + X_2 + X_3)/3 \tag{36}$$

The search process of GWO is mainly composed of two parts: prey positioning and gray wolves' movement. First, in the solution space, the gray wolf population is produced randomly, starting from the random population of gray wolves, and then (α, β, γ) guides other individuals to complete the prey capture. A is a random number between $[-a, a]$ because as the number of repetitions increases, a reduces linearly. A has a higher value in the early period of evolution and a lower value in the later period.

When $|A| > 1$, the gray wolves will expand the encirclement and suppression circle to increase the search range so that the wolves can search for prey in a wider range and ensure the global search ability of the algorithm. When $|A| < 1$, the gray wolves will shrink the encircling circle to attack and capture the prey. Finally, when the stopping condition is satisfied, it will capture the prey and output the optimal solution [40–42].

4.2. Algorithm Flow

The flow chart of GWO is shown in Figure 3, including the following steps.

Step 1. Code the process route, equipment, and tool.

Step 2. Population initialization and parameter setting.

Step 3. Sort the solution set.

Step 4. Determine whether the termination condition is reached, output the result if it is reached, and go to Step 5 if it is not reached.

Step 5. Four social classes of wolves were established, the wolves hunt under the leadership of the alpha wolf, and the solution set is obtained by cross-mutation.

Step 6. Merge the new wolf group obtained in Step 5 with the previous old wolf group.

Step 7. Update the population position.

Step 8. Output the optimal solution, and the algorithm ends.

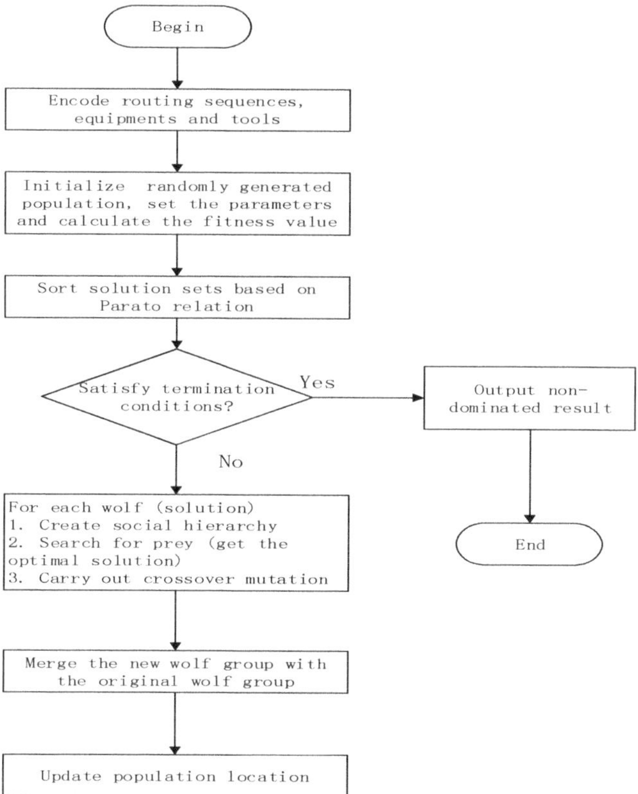

Figure 3. The flow of Grey Wolf Optimizer.

4.3. Encoding and Decoding

In order to solve the problem of process route optimization, the selection of equipment and tools and the sequence of processes need to be reasonably reflected in the part coding method in the algorithm. Each individual in the population has three substrings, including the workpiece process route sequence, equipment, and tools; the number of steps in the component is equal to the extent of the three substrings, as shown in Figure 4. The order substring represents the operation order of part processing in a continuous list, and its genes should take into account the constraints of processing priority. The equipment substring consists of the equipment number that has been assigned to each operation, and the jth gene on the substring represents the equipment used to complete operation j. The meaning of the tool substring is similar to that of the equipment substring.

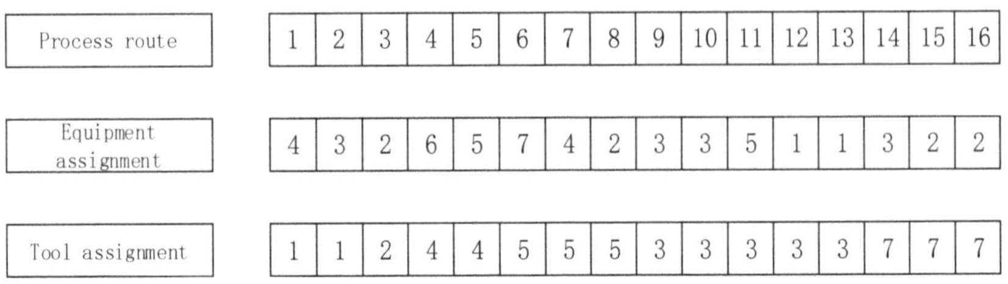

Figure 4. Individual coding method.

4.4. Population Initialization and Fitness Function

According to the above coding method, each wolf (solution) is a sequence, and the initial solution is obtained by random initialization. Individuals in the wolf pack are classified according to their fitness level using the fitness function. During the position update process of the gray wolf, the wolves with higher fitness are reserved, and the gray wolves with lower fitness are assigned to search for and capture the prey. According to the cutting parameters of the process, the corresponding fitness value is calculated. In this paper, two objective functions, f_1 (energy consumption in the manufacturing process) and f_2 (carbon emissions in the manufacturing process), are integrated into one fitness function by the weight method. The weight is set as ω, the fitness function is as follows,

$$min\ fitness_i = min\left(\omega_1 \frac{f_{1i} - f_{1min}}{f_{1max} - f_{1min}} + \omega_2 \frac{f_{2i} - f_{2min}}{f_{2max} - f_{2min}}\right) \tag{37}$$

$$f_1 = min\left(E_Z + \sum_{i=1}^{l} E_{D,i} + E_A + \sum_{i=1}^{l} E_{M,i}\right) \tag{38}$$

$$f_2 = min(C_Z + C_T + C_M + C_E + C_W) \tag{39}$$

f_{1i} and f_{2i} are the ith wolf's fitness values, respectively. f_{1max} and f_{1min} represent the maximum and minimum energy consumption, respectively, when the only optimization goal is to reduce the energy consumption of the manufacturing process. f_{2max} and f_{2min} respectively represent the maximum and minimum carbon emissions obtained when the carbon emission of the manufacturing process is taken as the single optimization objective, ω_1, ω_2 represent the weight of the corresponding item, and satisfy $\omega_1 + \omega_2 = 1$. The analytic hierarchy process, the fuzzy evaluation approach, the expert scoring method, the group decision-making method, and other methods can be used to determine the weight.

4.5. Constraints

The selection of cutting parameters for blank development is affected by many aspects. The rolling process is constrained by the bite angle and the stability conditions of the rolling stock in the pass. The cutting parameters of the process are subject to the constraints of machine tool speed, feed, cutting force, machine tool cutting power and roughness, etc., which can be expressed as follows,

(1) Biting condition.

The actual bite angle α of the rolling piece α should be less than the maximum bite angle α_{max} of the rolling mill pass to make the rolling piece enter the roll smoothly,

$$\alpha \leq \alpha_{max} \tag{40}$$

(2) Stability condition of the rolling piece in the pass.

$$\beta_{min} < \beta < \beta_{max} \tag{41}$$

where, β, β_{min}, and β_{max} are the rolling ratio of the non-equiaxed section and the minimum and maximum allowable axial ratios, respectively.

(3)   Machine tool speed constraint.

The cutting speed directly determines the spindle speed, and the spindle speed will directly affect the tool life and the processing effect of the workpiece. After the machine tool is determined, the spindle speed of the machine tool should be within the spindle speed range specified by the allowable nameplate of the machine tool,

$$n_{min} \leq n \leq n_{max} \tag{42}$$

where n_{min} and n_{max} indicate the machine tool spindle's minimum and maximum speeds, respectively.

(4)   Feed limit constraint.

In order to achieve the machining accuracy criteria, the value of the machine feed or feed speed selected for the process must be within the range of the feed specified by the allowable nameplate,

$$f_{min} \leq f \leq f_{max} \tag{43}$$

$$f_{v_{min}} \leq z f_z \leq f_{v_{max}} \tag{44}$$

where f_{min} and f_{max} are the feed rates of the minimum and maximum level of the lathe respectively. $f_{v_{min}}$ and $f_{v_{max}}$ are the feed rates of the minimum and maximum level of the milling machine, respectively.

(5)   Cutting force constraint.

The dimension of the cutting force will affect the life of the machining resources and machining accuracy. During the cutting process, the overall cutting force, which includes the main cutting force, rear force, and feed force, cannot exceed the maximum cutting force $F_{k_{max}}$ allowed by the machine tool.

$$F_i = \sqrt{F_{c_i}^2 + F_{f_i}^2 + F_{p_i}^2} \leq F_{k_{max}} \tag{45}$$

$$\begin{cases} F_c = C_{F_c} a_p^{x_{F_c}} f^{y_{F_c}} v_c^{n_{F_c}} k_{F_c} \\ F_f = C_{F_f} a_p^{x_{F_f}} f^{y_{F_f}} v_c^{n_{F_f}} k_{F_f} \\ F_p = C_{F_p} a_p^{x_{F_p}} f^{y_{F_p}} v_c^{n_{F_p}} k_{F_p} \end{cases} \tag{46}$$

where F_c, F_f, and F_p are the main cutting force, rear force, and feed force in the turning process, respectively. Taking turning as an example, C_{F_c}, x_{F_c}, y_{F_c}, n_{F_c}, k_{F_c} are the relevant coefficients of turning main cutting force, C_{F_f}, x_{F_f}, y_{F_f}, n_{F_f}, k_{F_f} are the correlation coefficients of turning feed force, C_{F_p}, x_{F_p}, y_{F_p}, n_{F_p}, k_{F_p} are the correlation coefficients of turning back force. The above coefficients are related to the processing conditions and workpiece material and can be obtained by consulting the cutting allowance manual. At the same time, the feed force of the machine tool must be within the allowable feed force range,

$$F_f = C_{F_f} a_p^{x_{F_f}} f^{y_{F_f}} v_c^{n_{F_f}} k_{F_f} \leq F_{f_{max}} \tag{47}$$

(6)   Machine tool power constraint.

During the machining process, the machine tool's output power is connected to the main cutting force F_c, the cutting speed v_c, and the effective coefficient τ of the selected machine tool power. The output power of the machine tool must also be controlled within the range of the effective cutting power P_{max} allowed by the machine tool.

$$\frac{F_c v_c}{\tau} \leq P_{max} \tag{48}$$

(7)   Feature roughness constraint.

Each machining feature of the part has machining quality requirements, the selection of cutting parameters in process i must ensure the surface quality of the machining feature, and the roughness must meet the Equation (50),

$$R_a = \frac{0.0312 f^2}{r} \leq R_{a_{max}} \tag{49}$$

where r is the tool nose arc radius, the maximum allowable surface roughness is $R_{a_{max}}$.

4.6. Establishing Population Classes and Location Updates

Since the search process of the original GWO algorithm is guided by three optimal solutions, but the Pareto solution set is obtained in the multi-objective optimization process, so the population is divided into multiple levels according to the dominance relationship, the non-dominated solutions are denoted by α, β and γ, respectively. When the number of non-dominated levels of the population is 1, α, β, γ are randomly generated by the population; when the number of non-dominated levels is 2, α is derived from the first level, and β, γ are derived from the second level; When the number of non-dominant level is 3 and above, α, β, and γ are obtained from the above three ranks, respectively [43,44].

Since the gray wolf algorithm was originally designed to handle continuous optimization issues, so as to ensure that the solutions generated by each evolution are feasible for the process planning problem, a transformation method is used to improve the update operator to achieve neighborhood changes and select one of them as the offspring with equal probability, as shown below.

$$X(\pi)_i^{t+1} = \begin{cases} shift\left(X(\pi)_i^t, C \cdot \left(X(\pi)_\alpha^t - X(\pi)_i^t\right)\right) \; if \; 0 \leq rand \leq \frac{1}{3} \\ shift\left(X(\pi)_i^t, C \cdot \left(X(\pi)_\beta^t - X(\pi)_i^t\right)\right) \; if \; \frac{1}{3} \leq rand \leq \frac{2}{3} \\ shift\left(X(\pi)_i^t, C \cdot \left(X(\pi)_\gamma^t - X(\pi)_i^t\right)\right) \; if \; \frac{2}{3} \leq rand \leq 1 \end{cases} \tag{50}$$

where, $shift\left(\vec{x}, \vec{d}\right)$ indicates that the individual $X(\pi)_i^t$ can move left or right, and $\vec{d}$ indicates the distance that the element moves $|d|$, *rand* is a random number in $[0, 1]$, and C takes 1 in this paper. The shift function can help the wolves to update their positions.

4.7. Genetic Operations

Genetic operations generally fall into two categories: crossover and mutation, the genetic operations are different for each substring. For the sake of algorithm efficiency, the two-point intersection is adopted for the equipment substring and the tool substring. The mutation operation of equipment substrings and tool substrings is performed by selecting points to replace any optional equipment, tools, and machining paths, as shown in Figure 5a. Use an improved two-point intersection method for sequential substrings. The operator will inherit the effective sequence substrings that satisfy the priority constraints and avoid duplication and omission. Two splicing points are randomly selected in the substring. The genes before splicing site 1 and after splicing site 2 in parent P_1 are directly copied to the same site in offspring O_1, remove the existing genes of O_1 in the parental P_2, and copy the remaining genes to the remaining sites of O_1 (and the gene segment between the two splicing sites) according to the order in P_2. For another offspring, O_2, it is generated according to the same principle after swapping P_1 and P_2. This intersection operation can result in an order of operations that does not violate constraints [45–47].

The mutation operation of sequential substrings is as follows: first, a process is randomly selected, and then a potential site is defined on the substring, that is, the site where the selected process can be replaced without violating the priority constraint. In the flexible sequence problem, the selected operation needs to satisfy all its predecessors and takes precedence over all its immediate successors. Therefore, potential sites appear consecutively on the substring. Among the potential insertion sites, one was randomly selected to replace the site that was initially selected, as shown in Figure 5b.

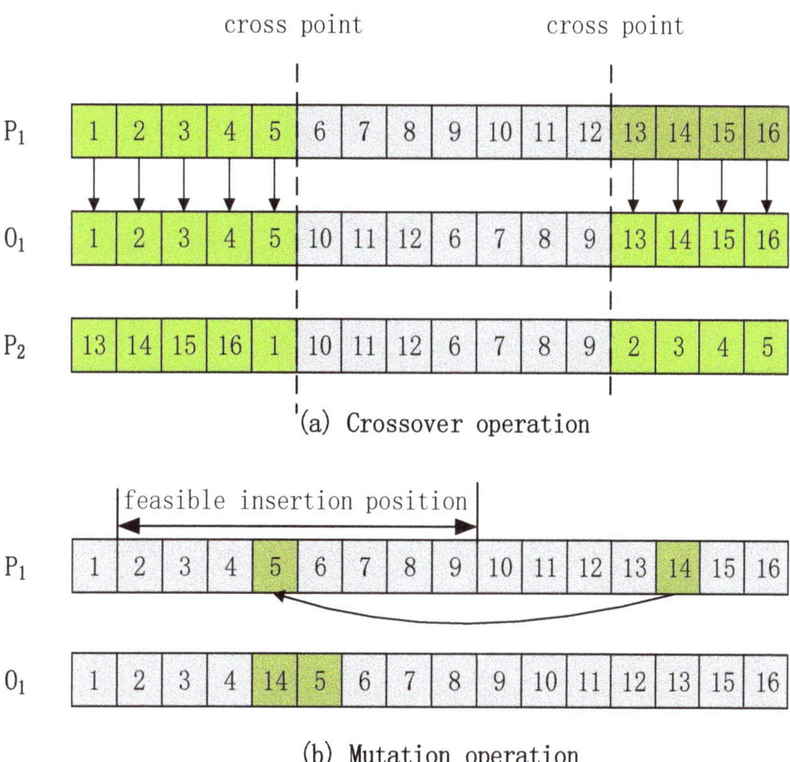

Figure 5. Sequence substring genetic operation.

5. Case Study

5.1. Instance Parameters

A factory received the task of designing the blank of the second gear of the automobile transmission intermediate shaft (20CrMnTiH, the outer diameter is 97.25 mm, the tooth thickness is 15 mm, the modulus is 1.75 mm, the number of teeth is 46, and the weight is about 0.665 kg), the quantity is 500,000 pieces. The manufacturing process must meet the requirements of circular economy development.

The business compass is a model that guides enterprises to design for sustainable development. It can be used for research ranging from enterprise business strategy path selection to small research on product design. Specific to the design of the automobile transmission gear blank, we analyzed the existing situation and production needs of the enterprise. Based on the theoretical framework of the business compass, we can dynamically guide the optimization selection of the objective influencing factors in the design process from the requirements of the blank design. The machining parameters at the time of the obtained target optimal value are reversed to calculate the blank dimension.

5.2. Experimental Equipment and Parameters

The existing rough rolling equipment of the enterprise is: the entire rolling line uses a total of 22 rolling mills, which are divided into roughing, neutralizing, first finishing, and second finishing mills, arranged in the way of 4, 6, 6, and 6, respectively. The billet used is 45 steel, and the length and width are both 165 mm. The process flow of continuous round steel rolling is shown in Figure 6. The heated billet is sent to the rough rolling mills for four passes, cut head by the #1 flying shear, sent to the medium rolling mills for six passes, cut head by the #2 flying shear, and sent to the first finishing rolling mills for six passes, cut head by the #3 flying shear, sent to the second finishing rolling mills for six passes, the ruler is cut by the flying shears, and then sent to the cooling bed for cooling. Finally, it is automatically collected and bundled, and the finished products are collected and put into the warehouse. The company's existing blank processing conditions are: there are two CNC lathes (M1, M2), two CNC milling machines (M3, M4), one drilling machine (M5), and one grinding machine (M6) in the workshop, involving turning and milling, drilling, grinding four processing methods. According to the actual processing conditions, the detailed information of the machine tool is shown in Table 1, Table 2 is the turning tool parameters, and Table 3 is the turning force coefficient.

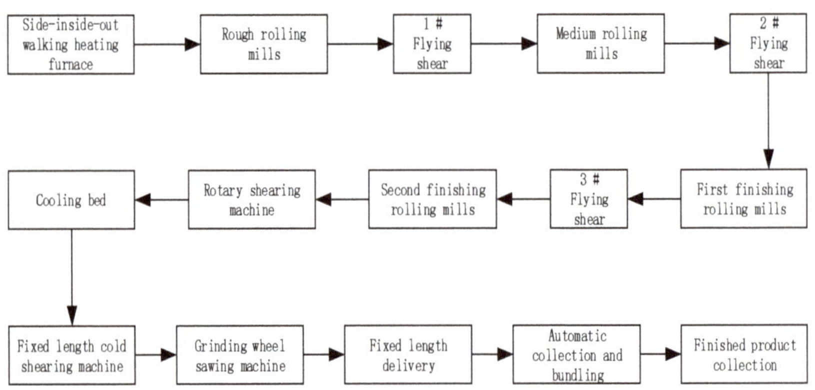

Figure 6. Flow chart of the rolling process.

Table 1. Machine parameters.

Lathe	n_{min}(r/min)	n_{max}(r/min)	f_{min}(mm/r)	f_{max}(mm/r)	F_{max}(N)	P_{max}(kW)	τ
M1	100	1400	0.1	0.25	1700	8.0	0.85
M2	120	1600	0.1	0.35	1700	10	0.8

Table 2. Tool parameters.

Tool	Material	Main Deflection Angle(°)	Rake Angle(°)	Blade Angle(°)	Tip Arc Radius r_θ(mm)
K1	High-speed steel	45°	20°	5°	0.8
K2	Cemented carbide	45°	20°	5°	0.8

Table 3. Cutting force coefficient.

	C_{F_c}	x_{F_c}	y_{F_c}	n_{F_c}	k_{F_c}	C_{F_f}	x_{F_f}	y_{F_f}	n_{F_f}	k_{F_f}
1	1750	0.9	0.75	0	1	580	1.1	0.65	0	1
2	2855	1.0	0.75	−0.1	1	2920	1	0.5	−0.35	1

	C_{F_p}	x_{F_p}	y_{F_p}	n_{F_p}	k_{F_p}
1	1100	0.9	0.65	0	1
2	1930	0.9	0.6	−0.35	1

The carbon emission factors of the material preparation process produced by gear processing are shown in Table 4; Table 5 shows the values of indirect carbon emission factors of energy consumption in the process of energy preparation; Table 6 shows the relevant factor values of direct carbon emissions from the consumption of fossil energy; Table 7 shows the relevant factor values of carbon emissions from waste disposal.

Table 4. Values of carbon emission factors in the material preparation process.

Carbon Emission Category	Material i Consumption	Production Process Consumes Energy c	Energy c Carbon Emission Factor S_{E_c}
Material preparation process carbon emissions C_M	Steel	Raw coal	2.653

Table 5. Values of indirect carbon emission factors in the energy preparation process from energy consumption.

Carbon Emission Category	The nth Energy Type Consumed by Energy k	Production Process Consumes Energy	Energy n Carbon Emission Factor S_{E_n}
Indirect carbon emissions in the energy production process C_{IE}	Electricity	Raw coal	2.565
		Crude	2.221
		Natural gas	1.642
	Coal	Crude	2.221
		Natural gas	1.642
		Electricity	8.220
	Natural gas	Raw coal	2.565
		Crude	2.221
		Natural gas	1.642
	Fuel/Circulating oil/Lubricant	Raw coal	2.565
		Crude	2.221
		Natural gas	1.642

Table 6. Values of direct carbon emission factors from fossil energy consumption.

Carbon Emission Category	Consumption Type of Material k	Energy Carbon Emission Factor S_{E_k}
Processing direct carbon emissions C_{DE}	Coal	0.6764
	Natural gas	0.4593
	Fuel/Circulating oil/Lubricant	0.6878

Table 7. Values of carbon emission factors for waste treatment.

Carbon Emission Category	Waste l Discharge Type	Energy Consumed Type in the Waste Treatment Process	Energy Carbon Emission Factor S_{E_q}
Waste treatment carbon emissions C_W	Wastewater/waste oil	Electricity	8.221
	Scra0ps	Electricity	8.221

According to the energy consumption and carbon emission calculation model proposed above, some required equipment parameter information has been listed, and the GWO was used to optimize and solve the model. The parameters of the algorithm are as follows: population number is 100, maximum generation number is 500, crossover rate is 0.8, the mutation rate is 0.1, and the optimization goal is a low carbon and low energy consumption.

5.3. Optimization Results and Analysis

The optimized blank dimension is 98.6, compared with the optimized blank dimension of the standard dimension of 100 and 105, respectively, the algorithm convergence diagrams of carbon emission and energy consumption are shown in Figures 7–9.

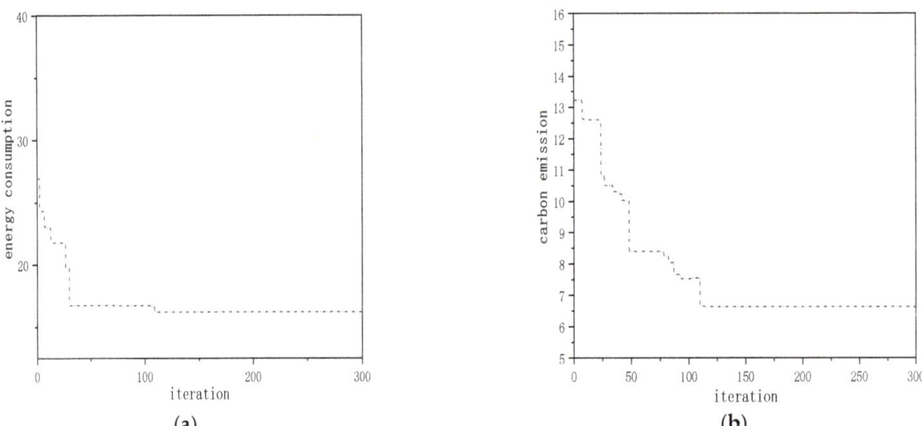

Figure 7. Iterative convergence graph of blank 1. (**a**) Energy consumption (**b**) Carbon emission.

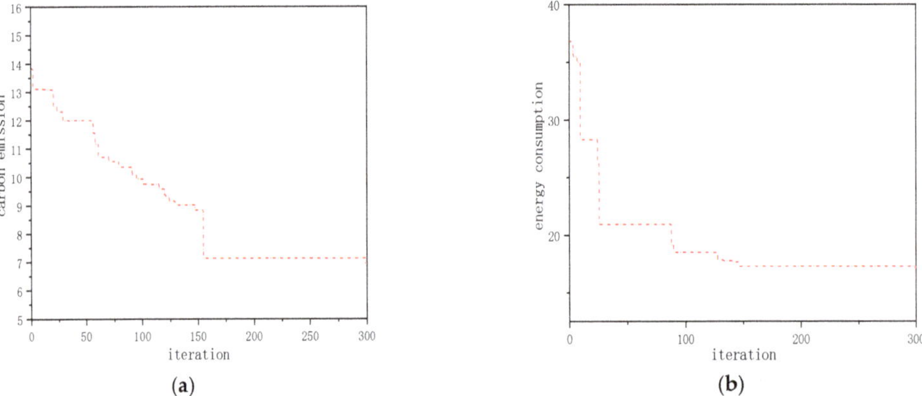

Figure 8. Iterative convergence graph of blank 2. (**a**) Energy consumption (**b**) Carbon emission.

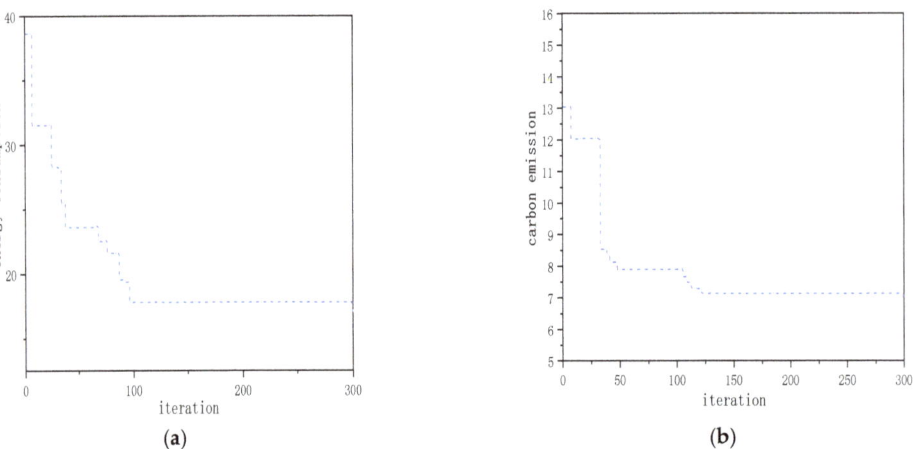

Figure 9. Iterative convergence graph of blank 3. (**a**) Energy consumption (**b**) Carbon emission.

The coordinated optimization of many elements in the production and use of blanks can result in optimal energy consumption and carbon emission. The blank dimension designed according to sustainable thinking is 98.6, the energy consumption per piece is 15.865, the carbon emission is 6.31, and the material consumption is 0.893. The standard blank dimensions of 100 and 105 are selected for comparison. As shown in Table 8, the optimal energy consumption is 16.57 and 17.03, carbon emissions are 6.93 and 6.85, and material consumption is 0.925 and 1.02. Compared with the available selected blank dimensions of 100 and 105, and considering enterprise conditions, the optimal designed blank dimension's energy consumption is only close to 95.7% and 93.1% of the selected blank dimensions, respectively, and the carbon emission is 92.6% and 90.2% of the selected dimensions, respectively. Material consumption is close to 96.5% and 87.5% of the selected dimensions, respectively. This batch of gears requires 500,000 pieces. Compared with the standard dimension 100 and 105 blanks, the optimized blanks reduce energy consumption by 3.525×10^5 and 5.825×10^5 and reduce carbon emissions by 3.1×10^5 and 2.7×10^5, decreasing material consumption by 1.6×10^4 and 6.3×10^4, respectively. The above results showed that the optimized blank dimension could not only save a lot of energy consumption and reduce carbon emissions but also save a lot of production materials. It can be seen that the blank dimension design method has the advantages of saving energy and low carbon and reducing resource consumption. This method can help enterprises design the blank dimension according to the requirements of enterprise energy saving and low carbon and provide suggestions for the sustainable development of enterprises.

Table 8. Comparison of different blank dimensions and optimization objectives.

Scheme	Blank Dimension	Energy Consumption	Carbon Emission	Material Consumption
1	98.6	15.865	6.31	0.893
2	100	16.57	6.93	0.925
3	105	17.03	6.85	1.02

5.4. Comparison with Previous Works

In the first chapter, the literature [7–22] provides many ideas and methods of blank dimension optimization, which effectively achieves the goal. However, the literature [7–13] only considered the blank production stage and optimized the blank shape and forming process. The literature [14–22] only considered the blank use stage and optimized the design of the processing parameters in the use process. The production and use of blanks are a unified whole, and sustainable design blanks need to be considered together. Based on the business compass, this paper comprehensively considered the production and use stages of blanks and constructed a calculation model for energy consumption and carbon emissions during blank production and use was proposed. The improved gray wolf algorithm was used to solve the problem, which was faster.

Compared with references 14–22, the selection of energy consumption and carbon emission as research objects is more in line with the requirements of the times and compared with the optimization methods in references 14–22, compared with the blank selection in references 14–22, the designed blank dimension can save energy consumption and carbon emission by more than 4%.

5.5. Practical Implications and Future Steps

In this paper, a blank dimension design method considering process variation was proposed and verified by a machining case of gear blank. This research can provide guidance for enterprises to design energy-saving and low-carbon blank dimensions and help the manufacturing industry improve the utilization rate of raw materials and energy. Implementing blank dimension optimization design will help governments and business managers develop energy-saving and low-carbon strategies to reduce the harmful environmental impact of their products. However, this paper only considered the impact of blank dimension design on energy-saving and low-carbon objectives, and future research should include consideration of the impact on more objectives, as well as the impact of blank dimension optimization design on other types of process equipment and process elements.

6. Conclusions

1. In order to achieve the goal of minimum energy consumption and carbon emission in the process of blank production and use, the optimization design of the blank dimension is carried out. The factors in the process of blank production and use are coordinated according to the business compass, and an energy-saving and low-carbon blank dimension optimization design method considering the process of dynamic change is proposed. The gray wolf algorithm is used to solve the calculation model. Taking gear blank as an example, the feasibility of the method is verified. Compared with the two standard blank dimensions, 100 and 105, this method can reduce energy consumption by 4.3% and 6.9%, respectively, and reduce carbon emissions by 7.4% and 9.8%, respectively. The results show that this method can effectively help production managers and designers design appropriate blank dimensions to achieve the goal of energy saving and emission reduction in the process of blank production and use.

2. The value of this method mainly lies in: (1) selecting the optimization design of blank dimension, which is less studied at present, to analyze the process of energy consumption and carbon emission change in the process of blank production and processing; (2) taking the business compass as a guide, comprehensively coordinating the factors of the stage of blank production and the stage of using; (3) selecting the optimization objective which synthetically considers the energy-saving and low-carbon composite economic indexes, and establishing the optimization model of processing parameters.

Author Contributions: Conceptualization, Y.X. and R.W.; methodology, Y.X., J.Z. and R.W.; software, Y.X.; validation, J.Z. and R.W.; formal analysis, J.Z. and R.W.; investigation, Y.X. and X.Z.; writingtable—original draft preparation, Y.X. and H.Z.; writing—review and editing, Y.X. and H.Z.; visualization, X.Z. All authors have read and agreed to the published version of the manuscript.

Funding: The authors are grateful for National Science Foundation, China (No.s 61862051, 51975432); Natural Science Foundation of Hunan Province, China (No. 2022JJ50244); China Education Department of Hunan Province (No. 21B0695); Project of Hunan social science achievement evaluation committee in 2022 (No. XSP22YBC081); the Science and Technology Foundation of Guizhou Province under Grant (No. [2019]1299); the Top-notch Talent Program of Guizhou province under Grant (No. KY [2018]080); the Program of Qiannan Normal University for Nationalities under Grant (Nos. QNSY2018JS013, QNSYRC201715).

Institutional Review Board Statement: Not applicable.

Informed Consent Statement: Informed consent was obtained from all subjects involved in the study.

Data Availability Statement: Not applicable.

Conflicts of Interest: The authors declare no conflict of interest.

References

1. Hirschnitz-Garbers, M.; Tan, A.R.; Gradmann, A. Key drivers for unsustainable resource use–categories, effects and policy pointers. *J. Clean. Prod.* **2016**, *132*, 13–31. [CrossRef]
2. Murray, A.; Skene, K.; Haynes, K. The circular economy: An interdisciplinary exploration of the concept and application in a global context. *J. Bus. Ethics* **2017**, *140*, 369–380. [CrossRef]
3. Lin, K.; Zhao, H. The Impact of Green Finance on the Ecologicalization of Urban Industrial Structure-Based on GMM of Dynamic Panel System. *J. Artif. Intell. Technol.* **2022**, *2*, 123–129. [CrossRef]
4. Zheng, J.; Feng, G.; Ren, Z. China's energy consumption and economic activity at the regional level. *Energy* **2022**, *259*, 124948. [CrossRef]
5. Zhou, L.; Li, J.; Li, F. Energy consumption model and energy efficiency of machine tools: A comprehensive literature review. *J. Clean. Prod.* **2016**, *112*, 3721–3734. [CrossRef]

6. Starman, B.; Cafuta, G.; Mole, N. A Method for Simultaneous Optimization of Blank Shape and Forming Tool Geometry in Sheet Metal Forming Simulations. *Metals* **2021**, *11*, 544. [CrossRef]
7. Wan, M.Z.; Zhu, Q.H.; Wu, J. Research and Development of Forming Process for Super Heavy Flange Forgings with Thick Wall. *Hot Work. Technol.* **2020**, *49*, 123–125.
8. Li, D.C.; Lu, X.; Sun, Z.Y. Investigation on workblank size requirements of rotary-swaging shafts of an automobile. *J. Plast. Eng.* **2019**, *26*, 72–76.
9. Xu, L.L.; Deng, J.D.; Hu, Z.L.; Xu, L. Intelligent deduction design method of ring forging rolling blank based on machine learning. *J. Plast. Eng.* **2022**, *29*, 23–29.
10. Zhang, X.Q.; Wang, Y.; Wang, C.; Hu, P. Blank Optimization for Sheet Metal Forming Using Inverse Finite Element Method and Mesh Mapping. *J. Shanghai Jiaotong Univ.* **2019**, *53*, 1389–1394.
11. Akinnuli, B.O.; Olatunji, O.E.; Bodunde, O.P. Symmetrical shell deep drawing material optimal blank diameter prediction and waste control model. *Arctic* **2018**, *71*, 60–69.
12. Xiao, Y.; Yan, W.; Wang, R. Research on Blank Optimization Design Based on Low-Carbon and Low-Cost Blank Process Route Optimization Mode. *Sustainability* **2021**, *13*, 1929. [CrossRef]
13. Gharehchahi, H.; Kazemzadeh-Parsi, M.J.; Afsari, A. Optimum blank shape design in deep drawing process using a new boundary updating formula. *Int. J. Mater. Form.* **2021**, *14*, 1375–1389. [CrossRef]
14. Li, C.B.; Yu, B.S.; Xiao, Q. A Cutting Parameter Energy-saving Optimization Method for CNC Turning Batch Processing Considering Tool Wear. *J. Mech. Eng.* **2021**, *57*, 217–229.
15. Liu, Y.; Yan, C.; Ni, Y.; Mou, Y. Multi-objective Optimization Decision of High-speed Dry Hobbing Process Parameters Based on GABP and Improved NSGA-II. *China Mech. Eng.* **2021**, *32*, 1043–1050.
16. Tian, C.; Zhou, G.; Zhang, J.; Wang, C. Integration optimization of tool selection and cutting parameters based on machining features considering carbon emissions. *Comput. Integr. Manuf. Syst.* **2020**, *26*, 2060–2072.
17. Li, L.; Li, C.; Tang, Y. An integrated approach of process planning and cutting parameter optimization for Energy-aware CNC Machining. *J. Clean. Prod.* **2017**, *162*, 458–473. [CrossRef]
18. Khan, A.M.; Liang, L.; Mia, M. Development of process performance simulator (PPS) and parametric optimization for sustainable machining considering carbon emission, cost and energy aspects. *Renew. Sustain. Energy Rev.* **2021**, *139*, 110738. [CrossRef]
19. Joshi, M.; Ghadai, R.K.; Madhu, S. Comparison of NSGA-II, MOALO and MODA for Multi-Objective Optimization of Micro-Machining Processes. *Materials* **2021**, *14*, 5109. [CrossRef]
20. Zhou, G.; Qi, L.; Xiao, Z. Cutting parameter optimization for machining operations considering carbon emissions. *J. Clean. Prod.* **2018**, *208*, 937–950. [CrossRef]
21. Xiao, Q.; Li, C.; Tang, Y. A knowledge-driven method of adaptively optimizing process parameters for energy efficient turning. *Energy* **2019**, *166*, 142–156. [CrossRef]
22. Shin, S.; Woo, J.; Rachuri, S. Energy efficiency of milling machining: Component modeling and online optimization of cutting parameters. *J. Clean. Prod.* **2017**, *161*, 12–29. [CrossRef]
23. Wang, R.P. *Business Compass*; Science Press China: Beijing, China, 2020.
24. Xiao, Y.; Wang, R.; Yan, W.; Ma, L. Optimum Design of Blank Dimensions Guided by a Business Compass in the Machining Process. *Processes* **2021**, *9*, 1286. [CrossRef]
25. Khan, M.; Idrees, M.; Rauf, M. Green Supply Chain Management Practices' Impact on Operational Performance with the Mediation of Technological Innovation. *Sustainability* **2022**, *14*, 3362. [CrossRef]
26. Bocken, N.M.; Pauw, I.; Bakker, C.; Grinten, B.C. Product design and business model strategies for a circular economy. *J. Ind. Prod. Eng.* **2016**, *33*, 308–320. [CrossRef]
27. Maxwell, D.; Vorst, R. Developing sustainable products and services. *J. Clean. Prod.* **2003**, *11*, 883–895. [CrossRef]
28. Vrchota, J.; Pech, M.; Rolínek, L. Sustainability Outcomes of Green Processes in Relation to Industry 4.0 in Manufacturing: Systematic Review. *Sustainability* **2020**, *12*, 5968. [CrossRef]
29. Xu, B.; Qu, H. Impact of the design industry on carbon emissions in the manufacturing industry in china: A case study of zhejiang province. *Sustainability* **2022**, *14*, 4261. [CrossRef]
30. Hammond, G.P.; Jones, C.I. Embodied Energy and Carbon In Construction Materials. *Constr. Mater.* **2009**, *162*, 1–64. [CrossRef]
31. Li, R.; Liu, Q. Service-oriented Research on Multi-pass Milling Parameters Optimization for Green and High Efficiency. *Chin. J. Mech. Eng.* **2015**, *51*, 89–98. [CrossRef]
32. Xiao, Y.; Zhang, H.; Jiang, Z. An approach for blank dimension design considering energy consumption. *Int. J. Adv. Manuf. Technol.* **2016**, *87*, 1229–1235. [CrossRef]
33. Lu, C.; Gao, L.; Li, X. A hybrid multi-objective grey wolf optimizer for dynamic scheduling in a real-world welding industry. *Eng. Appl. Artif. Intell.* **2017**, *57*, 61–79. [CrossRef]
34. Komaki, G.M.; Kayvanfar, V. Grey Wolf Optimizer algorithm for the two-stage assembly flow shop scheduling problem with release time. *J. Comput. Sci.* **2015**, *8*, 109–120. [CrossRef]
35. Teng, Z.J.; Lv, J.L.; Guo, L.W. An improved hybrid grey wolf optimization algorithm. *Soft Comput.* **2019**, *23*, 6617–6631. [CrossRef]
36. Xiao, Y.M.; Zhang, H. Multiobjective optimization of machining center process route: Tradeoffs between energy and cost. *J. Clean. Prod.* **2021**, *28*, 21–26. [CrossRef]

37. Joshi, H.; Arora, S. Enhanced Grey Wolf Optimization Algorithm for Global Optimization. *Fundam. Inform.* **2017**, *153*, 235–264. [CrossRef]
38. Zhou, B.; Lei, Y. Bi-objective grey wolf optimization algorithm combined Levy flight mechanism for the FMC green scheduling problem. *Appl. Soft Comput.* **2021**, *111*, 107717. [CrossRef]
39. Dong, Y.; Liu, J.; Liu, Y. The Optimization Research of Diesel Cylinder Gasket Parameters Based on Hybrid Neutral Network and Improved Grey Wolf Algorithm. *Math. Probl. Eng.* **2020**, *2020*, 1–16. [CrossRef]
40. Huang, G.; Cai, Y.; Liu, J. A Novel Hybrid Discrete Grey Wolf Optimizer Algorithm for Multi-UAV Path Planning. *J. Intell. Robot. Syst.* **2021**, *103*, 1–18. [CrossRef]
41. Karakoyun, M.; Ozkis, A.; Kodaz, H. A new algorithm based on grey wolf optimizer and shuffled frog leaping algorithm to solve the multi-objective optimization problems. *Appl. Soft Comput.* **2020**, *96*, 106560. [CrossRef]
42. Choubey, C.; Ohri, J. Optimal Trajectory Generation for a 6-DOF Parallel Manipulator Using Grey Wolf Optimization Algorithm. *Robotica* **2020**, *39*, 411–427. [CrossRef]
43. Yin, L.; Zhuang, M.; Jia, J. Energy Saving in Flow-Shop Scheduling Management: An Improved Multiobjective Model Based on Grey Wolf Optimization Algorithm. *Math. Probl. Eng.* **2020**, *2020*, 1–14. [CrossRef]
44. Naserbegi, A.; Aghaie, M.; Zolfaghari, A. Implementation of Grey Wolf Optimization (GWO) algorithm to multi-objective loading pattern optimization of a PWR reactor. *Ann. Nucl. Energy* **2020**, *148*, 107703. [CrossRef]
45. Hudzikowski, A.; Gluszek, A.; Krzempek, K. A compact, spherical mirrors-based dense astigmatic-like pattern multipass cell design aided by a genetic algorithm. *Opt. Express* **2021**, *29*, 26127–26136. [CrossRef]
46. Elasko, D.; Ksiek, W.; Pawiak, P. Transmission Quality Classification with Use of Fusion of Neural Network and Genetic Algorithm in Pay&Require Multi-Agent Managed Network. *Sensors* **2021**, *21*, 4090. [CrossRef]
47. Rezaeipanah, A.; Mojarad, M. Modeling the Scheduling Problem in Cellular Manufacturing Systems Using Genetic Algorithm as an Efficient Meta-Heuristic Approach. *J. Artif. Intell. Technol.* **2021**, *1*, 228–234. [CrossRef]

Article

Energy-Saving and Efficient Equipment Selection for Machining Process Based on Business Compass Model

Yongmao Xiao [1,2,3], **Jincheng Zhou** [1,2,3], **Ruping Wang** [4,*], **Xiaoyong Zhu** [5] **and Hao Zhang** [6]

1   School of Computer and Information, Qiannan Normal University for Nationalities, Duyun 558000, China
2   Key Laboratory of Complex Systems and Intelligent Optimization of Guizhou Province, Duyun 558000, China
3   Key Laboratory of Complex Systems and Intelligent Optimization of Qiannan, Duyun 558000, China
4   Management School, China West Normal University, Nanchong 637002, China
5   School of Economics & Management, Shaoyang University, Shaoyang 422099, China
6   College of Mechanical and Electrical Engineering, Xinjiang Agricultural University, Wulumuqi 830052, China
*   Correspondence: wrpqyi@163.com; Tel.: +86-130-1641-0611

Abstract: The optimal selection of machine equipment can reduce the energy consumption and processing time of the parts processing process in enterprises. The energy consumption and time of using different equipment to process the same product vary greatly. Traditional equipment selection is only through qualitative analysis comparing the process characteristics of using different equipment or optimizing parameters for a single piece of equipment. It does not take into account the dynamics of the production process and does not consider the impact of process factors on production decisions. To solve this problem, we established a production equipment selection model based on the business compass model and proposed a calculation method that considered energy consumption and time objectives in the production process. Quantitative analysis can be performed for different equipment. The energy consumption and processing time of different equipment are calculated by the beetle antennae search (BAS) algorithm. A case study of machining end cap holes was carried out. The results showed that this method can calculate the optimal energy consumption and the optimal time of different equipment for producing the same product, which has good theoretical and practical significance for enterprises and governments to choose energy-saving and efficient production equipment.

Keywords: equipment selection decision; business compass; energy consumption; processing time; beetle antennae search algorithm

Citation: Xiao, Y.; Zhou, J.; Wang, R.; Zhu, X.; Zhang, H. Energy-Saving and Efficient Equipment Selection for Machining Process Based on Business Compass Model. *Processes* **2022**, *10*, 1846. https://doi.org/10.3390/pr10091846

Academic Editor: Sergey Y. Yurish

Received: 12 August 2022
Accepted: 6 September 2022
Published: 13 September 2022

Publisher's Note: MDPI stays neutral with regard to jurisdictional claims in published maps and institutional affiliations.

1. Introduction

Industry plays a very important role in global economic development and has a huge impact on the development of various countries and regions [1]. Energy consumption in the industrial sector accounts for about 70% of the total energy consumption. Energy shortage and serious environmental pollution are the two major problems affecting economic growth and sustainable development [2]. The production model of high consumption, high pollution, and low profit in traditional industries is no longer suitable for the needs of social development. Modern manufacturing is changing to high-quality development and developing towards a green and sustainable direction [3,4]. The optimal selection of machine tool equipment is one of the effective ways to improve the greenness of the enterprise parts processing process. Under the condition that the production process requirements are met, there are often many options for machine tool equipment. The machining quality, time quota, cost, energy consumption, and noise pollution of different machine tools are different. Therefore, the selection of suitable processing equipment and processing technology is of great significance for enterprises to save energy and time [5–7].

Many scholars have studied the selection of processing technology. Liu et al. proposed a multi-objective optimization method for CNC machine tools that integrates the advantages of quality function deployment (QFD), fuzzy linear regression, and 0–1 objective planning. The engineering practicability of this method is verified by the case of the multi-objective decision-making problem of CNC machine tools in the process of building an intelligent manufacturing platform [8]. Zhou et al. proposed a machine tool selection method based on the combination of fuzzy analytic hierarchy process and entropy weight ideal point method and verified the method through the selection of machine tools for processing camshafts [9]. Li et al. established an evaluation model for machine tool equipment selection based on AHP and the ideal point method. The AHP was used to determine the weight of each influencing factor, and the selection model of machine tool equipment was evaluated by the ideal point method (TOPSIS). Combined with the selection of a batch of valve body and valve core stepped hole processing equipment in a factory, the feasibility and practicability of this method are illustrated [10]. Han et al. combined the entropy weight method with the TOPSIS method and established a decision-making model for the green process scheme of machining based on the entropy weight TOPSIS method. The feasibility and practicability of the model are verified by taking three machining process schemes of the lifting beam of the dump truck [11]. Zanuto et al. used a commercial life cycle assessment (LCA) tool to compare the environmental impact of different process operation strategies [12]. Klink et al. study grinding, EDM, laser machining, and milling manufacturing techniques, and different manufacturing processes are compared with each other with examples of machining mold cavities [13]. The era of Industry 4.0 will fundamentally change the production mode of the manufacturing industry, and the application of information technology in the manufacturing industry will be more extensive. Liu et al. proposed a digital twin-driven dynamic evaluation method for machining processes and verified it [14]. Chen et al. proposed an automated process planning approach for hybrid manufacturing processes, capable of automatically setting key parameters such as depth, tool accommodation, angle, and tool selection [15]. Komatsu proposed an automatic method for selecting machine tools by evaluating various machining processes and this method was validated [16]. Koremura et al. proposed a prediction method for process evaluation indicators using a computer-aided process planning (CAPP) system. By using these indicators, the operator can choose the most suitable one from the alternatives. A case study of process planning was carried out through bar machining, and the results showed the application value of the evaluation index [17]. The above literature proposed many methods for the selection of equipment in the production process, which effectively achieve the purpose. However, these methods used qualitative analysis when evaluating the equipment selection scheme, and the objectivity of the data cannot be well reflected.

To optimize machining parameters, Han et al. proposed a variable parameter drilling (VPD) method that can improve the machining efficiency and hole surface quality of porous parts and verified by a combined algorithm (CA) [18]. He et al. proposed an improved method to comprehensively optimize the distribution and parameters of machining allowance and conducted two case studies [19]. Wang et al. established a multi-objective optimization model of CNC turning process parameters based on the second-order regression equation of the response surface and used an improved artificial bee colony algorithm to solve the optimal parameter combination. The comparison with the results obtained by the NSGA-II algorithm showed the superiority of this method [20]. Li et al. constructed and validated the energy consumption and quality model of laser welding [21]. Jia et al. studied the power and energy consumption of the drilling process and established a mathematical model of energy consumption and verified the method through a hole machining case [22]. Xiao et al. proposed a CNC machining center process parameter optimization model that comprehensively considered energy consumption and cost, used a CA to solve the model, and verified the method by plane milling [23]. Ma et al. integrated CAD and CAM applications for virtual machining and process parameter optimization for complex end milling. The method was validated by machining an impeller [24]. Zhang et al. established an optimization model for the sequence of steps with the auxiliary processing time as the objective function and obtained the sequence of steps with the shortest auxiliary processing time through an improved genetic algorithm, which effectively reduced the processing time of parts [25]. The above literature optimized the process parameters of the machining process and compared and validated the data through quantitative analysis. However, these studies were performed on the same equipment and did not consider the effects of different processing equipment options.

There are many types of machine tools, and a certain processing feature of a product can often be realized by different machine tools. Under the condition that the production process requirements are met, the energy consumption and time of the processing process of different equipment vary greatly. The choice of machine tool equipment is a multi-objective and multi-scheme decision-making problem. In the actual design process, designers often rely on qualitative analysis to select processing equipment and cannot provide exact data support for equipment selection. Based on the business compass model, this paper established a production equipment selection model and analyzed the management process of enterprise equipment selection. A unified energy consumption and time objectives calculation model for different equipment was established, which was solved by the beetle antennae search algorithm. The method was verified by two kinds of equipment selection for processing a certain end cap hole, which can provide a reference for the efficient and energy-saving production of enterprises.

2. Production Equipment Selection Model Based on Business Compass Model

2.1. Business Compass Model

The business compass model is a new type of enterprise management and operation model, which organically integrates Chinese and Western management sciences, and summarizes a theoretical system with practical guiding significance from an innovative perspective. The "five dimensions" in the business compass not only come from the experience summary of business management theory but also the results of a large amount of practical research, which are closely related to the ancient Chinese "five elements" theory. A systematic view of Taoist philosophy in ancient China, the "five elements" theory is an important point of view [26–28]. The business compass model is shown in Figure 1. The "five dimensions" in the business compass are trend, path, skill, tool, and benefit, which provide a systematic view of business management. It is not only the five elements of business operations but also the five business capabilities that can introduce the production and operation of enterprises in detail.

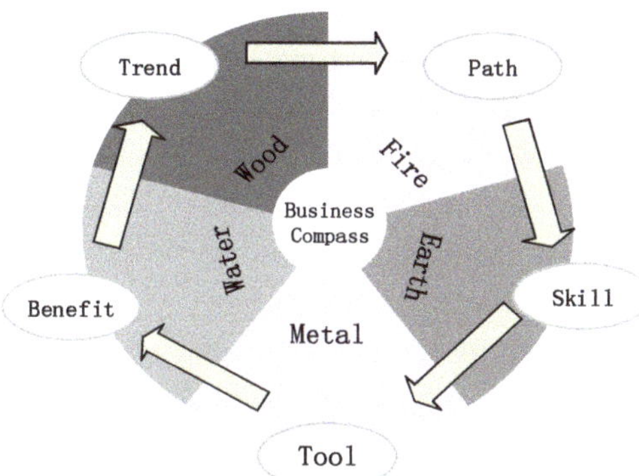

Figure 1. Business compass model.

The trend of the business compass is the environment in which an enterprise operates, including policies, industry prospects, and market demands. An enterprise's grasp of the "trend" often determines whether the enterprise can develop in the long term. The path of the business compass is the guiding ideology of enterprise management and the direction of enterprise development. The skill of the business compass is an enterprise's strategy, tactics, and strategic choice and layout, which determines the specific field in which the enterprise raises the technical threshold and defines the direction of technological innovation, product iteration, and service upgrade. The tool of the business compass is the product and technology of the enterprise, and it is the realization tool of the enterprise's production. The benefit of the business compass refers to the profit. The effective distribution of profit can fully mobilize the enthusiasm of the internal members of the enterprise and can also promote the enterprise to have a good growth environment [29].

2.2. Equipment Selection Model Based on Business Compass

There are often a variety of machine tools that can be selected for processing the same product, The energy consumption and cost of different equipment vary greatly. Choosing the right processing equipment is of great significance for the high efficiency and energy saving of enterprises. As the guiding concept of enterprise management, the business compass can provide guidance for the selection of equipment for processing the same product. The business compass can provide guidance on enterprise equipment selection. "Trend" is the environment in which enterprises make equipment selection. In recent years, the manufacturing industry has been developing in the direction of low-carbon manufacturing and green manufacturing, and enterprise production should keep up with the development trend. "Path" is the purpose of equipment selection. The goal of an enterprise is to produce products, and equipment alternatives should be selected according to product characteristics. "Skill" is how to choose the appropriate production equipment, which needs to be judged according to the product and the current technical level. "Tool" is the production equipment, and the performance, parameters, and processing links of the equipment must be considered. "Benefit" is the profit of the product and what equipment to choose can make the profit of the product the highest. This article summarized the model of enterprise production equipment selection based on the model of the business compass, as shown in Figure 2. This model takes the production and operation of the enterprise as the core. The management concept of the business compass guides the management and operation of the enterprise. The operation department is responsible for the management of the production conditions and technological innovation of the enterprise. The designer initially selects several suitable schemes according to the production conditions of the enterprise and the characteristics of the workpiece to be processed, conducts a comprehensive evaluation of the schemes, and then selects the most suitable production equipment. It can be seen from this that the model has good industrial applicability when selecting equipment for the production of the same product and can help enterprises select suitable processing equipment. The selection of equipment can help enterprises to produce with high efficiency and energy saving, and enterprises can obtain more profit space and promote enterprise development.

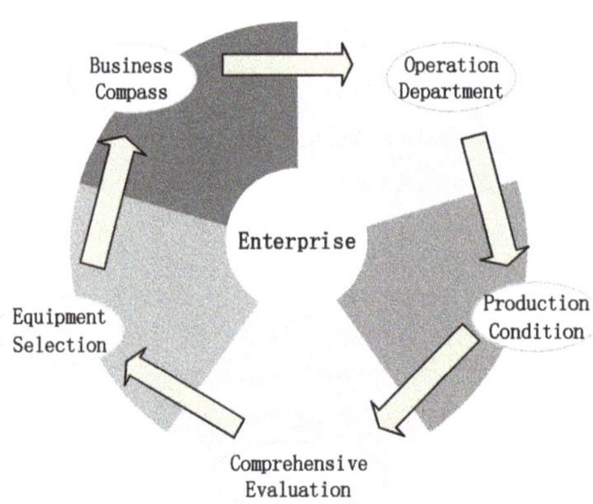

Figure 2. Production equipment selection model.

3. Multi-Objective Unified Computing Model for Different Equipment

The industrial industry pays great attention to the economy of the production process. By optimizing the energy and time objectives of product processing, the optimal energy consumption and time of the production process can be obtained, thereby reducing the investment of resources and promoting the rational use of resources. Energy cost and worker time cost are important components of the total cost. Reducing energy consumption and processing time can also reduce production costs and increase profits. Many types of equipment can be used to produce the same product, and it is important to choose the right one. The establishment of a uniform energy consumption and time objective function model is the basis of equipment selection. The unified energy consumption and time model has better engineering applicability. The calculation process of energy consumption and time is as follows [30–33].

3.1. Energy Consumption Objective

According to the state of the machine tool in the working process, the energy consumption of the machining process can be divided into standby energy consumption E_f, no-load energy consumption E_{air}, actual cutting energy consumption E_c, and additional load energy consumption E_a. During the cutting process, the state of the machine tool has little effect on the energy consumption of the auxiliary systems, such as lighting systems, cooling fans, and lubrication systems, and its power is only related to its own characteristics. The energy consumption calculation process is as follows.

The power of the machine tool with no load Is related to the spindle speed, and the expression is as follows:

$$P_{air} = An^2 + Bn + C \tag{1}$$

where coefficients A, B, C can be obtained by fitting the measured data, P_{air} is the no-load power, and n is the spindle speed.

The no-load energy consumption is

$$E_{air} = (An^2 + Bn + C) \times t_{air} \tag{2}$$

where E_{air} is no-load energy consumption, and t_{air} is machine no-load time.

The actual cutting energy consumption E_c is,

$$E_c = P_c t_c = F_c v_c t_c \tag{3}$$

P_c is the cutting power for machine tool, t_c is cutting time, F_c is cutting force, and v_c is cutting speed.

The milling process of machine tools is accompanied by additional load loss, and the additional loss mechanism is very complicated. It is generally believed that the loss of energy of the additional load is approximately proportional to the cutting energy consumption.

$$E_a = bE_c \tag{4}$$

where b is the correlation coefficient, generally between 0.15 and 0.25.

The energy consumption of the cutting process is

$$E = E_f + E_{air} + E_c + E_a = P_f t_f + (An^2 + Bn + C)t_{air} + (1 + b)F_c v_c t_c \tag{5}$$

where P_f, t_f are the standby power and standby time of the machine tool, respectively; A, B, C are coefficients related to no-load power consumption; n is the spindle speed, t_{air} is machine no-load time; b is the correlation coefficient; and F_c, v_c, t_c are the cutting force, cutting speed, and cutting time, respectively.

3.2. Time Objective

The total time T of the workpiece cutting process mainly includes: standby time t_f, no load time t_{air}, and actual cutting time t_c. The total time T of the cutting process is calculated as follows:

$$T = t_f + t_c + t_{air} \qquad (6)$$

$$t_c = \frac{60L}{nf} \qquad (7)$$

where L is the total length of the toolpath during the machining process, n is the spindle speed of the machine tool, and f is the feed rate during the machining process.

3.3. Constraints

The value of the objective function must meet the machining cutting parameters and quality requirements, and the relevant constraints are such as Formulas (8)–(13).

$$n_{\min} \leq n \leq n_{\max} \qquad (8)$$

where $n_{\min}$ is the minimum spindle, and $n_{\max}$ is the maximum spindle speed.

$$f_{\min} \leq f \leq f_{\max} \qquad (9)$$

where $f_{\min}$ is the lowest feed rate, and $f_{\max}$ is the fastest feed rate.

$$P_c \leq \eta P_{\max} \qquad (10)$$

where $P_{\max}$ is the machine maximum power, and η is the effective coefficient of machine power

$$F \leq F_{\max} \qquad (11)$$

where $F_{\max}$ is the maximum cutting force.

$$T_{\min} \leq T \leq T_{\max} \qquad (12)$$

where $T_{\min}$ is the tool life lower limit, and $T_{\max}$ is the tool life upper limit.

$$R \leq R_{\max} \qquad (13)$$

where $R_{\max}$ is the maximum allowable surface roughness.

4. Beetle Antennae Search Algorithm

4.1. Analysis of Beetle Antenna Search Algorithm

The BAS algorithm was proposed by Jiang et al. in 2017 [34]. The BAS algorithm was proposed based on the foraging behavior of beetles. In the process of beetle foraging, it uses its left and right antennas to detect the food taste concentration. If the food taste concentration detected by the left antenna is greater, it will move to the left; otherwise, it will move to the right. During the whole process of moving, its position is constantly adjusted and changed until it moves to the position of the food. Different from many multi-swarm heuristic algorithms, the BAS algorithm requires only one beetle; therefore, its operation is simple, the amount of calculation is less, and the iteration speed is faster [35,36].

4.2. Beetle Antenna Search Algorithm Flow

The flowchart of the BAS algorithm is shown in Figure 3. The following are the specific steps [37]:

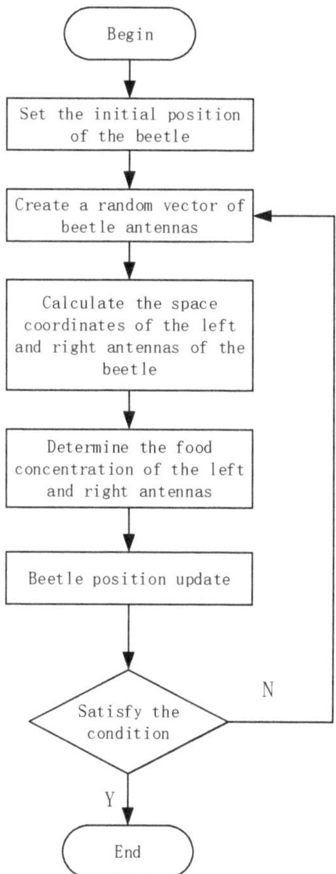

Figure 3. BAS algorithm flow.

1. For an n-dimensional optimization problem, record the centroid of the beetle as $x(x_1, x_2, \cdots, x_n)$; then, the fitness function can be expressed as $f(x_1, x_2, \cdots, x_n)$. Before the beetle searches for food, its initial position needs to be set, that is, the initial value of x.

2. Since the direction of the beetle search is random, it is necessary to establish a random vector of beetles and normalize it.

$$V = \frac{rands(n, 1)}{||rands(n, 1)||} \tag{14}$$

where $rands(n, 1)$ is a randomly generated n-dimensional vector between 0 and 1. At this time, the beetle's left antenna coordinate x_l and right antenna coordinate x_r can be obtained as:

$$x_l = x + d \times V \tag{15}$$

$$x_r = x - d \times V \tag{16}$$

where d is the distance between the antenna of left and right.

3. Compare the food taste concentration and fitness value of the left antenna and the right antenna of beetles.

When $f(x_l) < f(x_r)$, then:

$$x_k = x_{k-1} + step \times \frac{x_l - x_r}{||x_l - x_r||} \tag{17}$$

When $f(x_l) \geq f(x_r)$, then:

$$x_k = x_{k-1} - step \times \frac{x_l - x_r}{||x_l - x_r||} \tag{18}$$

In the above two formulas, *step* represents the moving step of the beetle, and x_k and x_{k-1} represent the value of x at the k and $k-1$ iterations, respectively.

4. Enter the iterative process. When the maximum number of iterations is reached or the fitness value meets the requirements, the iteration stops and the result is output, which is the optimal value at this time.

The algorithm runs on MATLAB 2016b. The dimension n is 2, the initial step size of the beetle *step* is 0.3, the distance between the two whiskers of the beetle d is 2, and the number of iterations is 300.

5. Case Study

An enterprise needs to process a batch of end caps. The size of the workpiece to be processed is shown in Figure 4. The material is Q235, and the thickness is 20 mm, with 100,000 pieces produced. Now, it is necessary to process four holes on the end cap. Available equipment is a lathe and a drill.

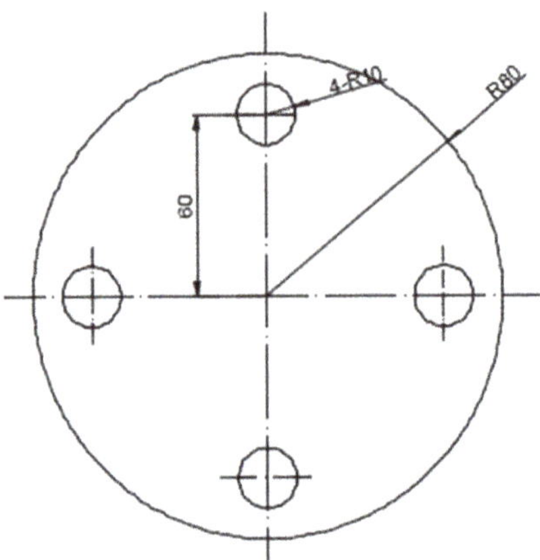

Figure 4. Workpiece dimensions.

5.1. Analysis of Available Equipment Conditions

The company currently has two types of equipment for processing this part, namely, a lathe and a drilling machine. Tables 1–3 show the main parameters of the lathe. Table 1 shows the lathe parameters, Table 2 shows the range of the lathe cutting parameters, and Table 3 shows the relevant parameters of the turning experience model.

Table 1. Lathe parameters.

Model	Spindle Motor Power/kw	Low Gear Speed Regulation Range/(r·min⁻¹)	High Gear Speed Regulation Range/(r·min⁻¹)
C2-6136HK/1	5.5	100–1000	300–2100

Table 2. Selection range of cutting parameters.

Cutting Parameters	Value Range
$f/(\mathrm{mm \cdot r^{-1}})$	0.05–0.3
a_p/mm	0.5–4
$n/(\mathrm{r \cdot min^{-1}})$	50–1000

Table 3. Turning empirical model parameters.

C_{Fc}	x_{Fc}	y_{Fc}	n_{Fc}	k_{Fc}
2795	1	0.75	−0.15	0.778

The model of the drilling machine is the ZXK50 CNC vertical drilling machine. Table 4 shows the relevant parameters of the drilling machine, and Table 5 shows the parameters of the drill bit.

Table 4. Drilling Machine Parameters.

Machine Rated Power p/kw	Spindle Speed Range (r/min)	Feed Speed Range (mm/min)	Machine Efficiency η
3.7	45–2000	20–600	0.8

Table 5. Drill Bit Parameters.

Number of Cutting Edges	Material	Economic Life/*min*
2	YG8	50

5.2. Optimization Results

The application process of the BAS algorithm in the end cap processing includes the following steps:

1. Establish a fitness function according to the energy consumption and time models of the production processes of different equipment.
2. Determine the random vector of beetles. The dimension n is 2, the initial step size of the beetle *step* is 0.3, the distance between the two whiskers of the beetle d is 2, and the number of iterations is 300.
3. The coordinates of the left and right antennas of the beetle are calculated, and the corresponding fitness values are calculated.
4. Update the location of beetles.
5. Determine whether the number of iterations is satisfied and output the result if it is satisfied. If not, go back to Step 2.

The optimal energy consumption and processing time of different equipment when processing the same product can be obtained through the BAS algorithm. Figure 5 shows the energy consumption and iteration times of lathe processing, and Figure 6 shows the energy consumption and iteration times of drilling machine processing. Figures 7 and 8 separately show the time and iteration times of lathe and drilling processing. Through the optimization calculation, the optimal objectives when different equipment process the same product can be obtained.

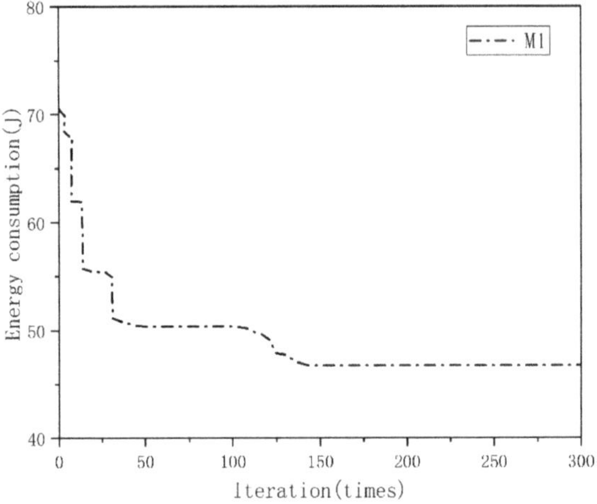

Figure 5. Energy consumption iterative curve of lathe.

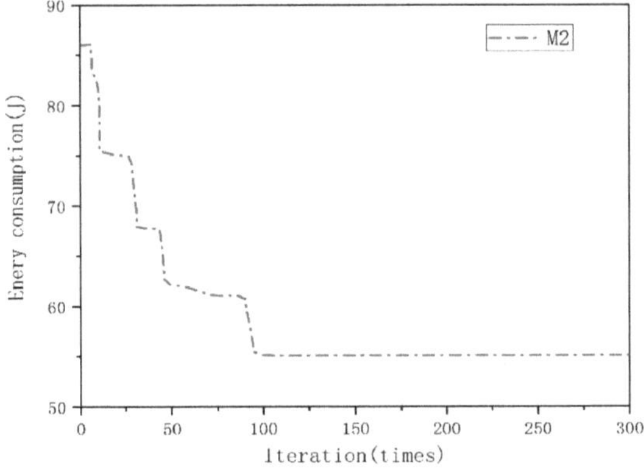

Figure 6. Energy consumption iterative curve of drilling machine.

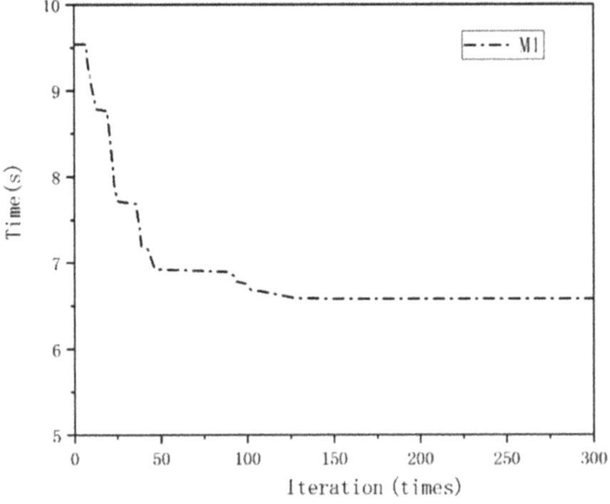

Figure 7. Time iterative curve of lathe.

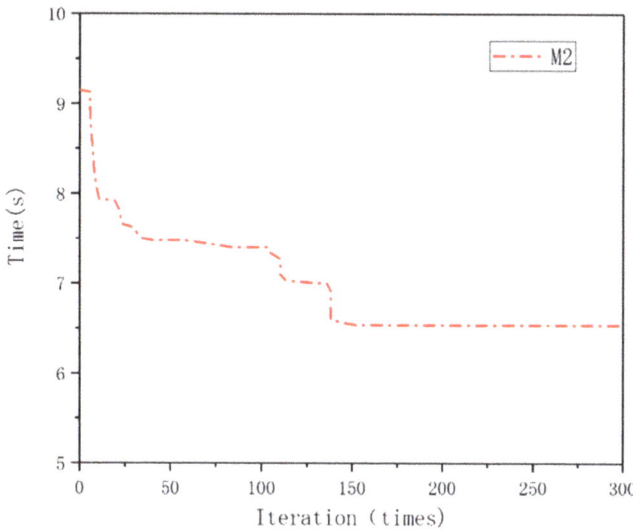

Figure 8. Time iterative curve of drilling machine.

5.3. Results and Discussion

5.3.1. Comparison

The optimization results of processing the same product with different equipment are shown in Table 6. As can be seen from the table, the energy consumption of lathe processing is reduced by 15.2%, and the processing time is reduced by 0.61% compared with drilling machine processing. Less energy and time are used when machining this part with a lathe, so a lathe is chosen as the machining equipment. The optimal values of energy consumption and time are 46.73 and 6.54, respectively. This method improves the profit of the enterprise by choosing a way of less energy consumption and processing time.

Table 6. Comparison of Different Equipment Optimization Results.

Equipment	Energy Consumption	Time
Lathe	46.73	6.54
Drilling machine	55.12	6.58

5.3.2. Compared with Previous Works

In order to achieve energy-saving and high efficiency in the production process, scholars at home and abroad have studied some equipment selection optimization methods and algorithms and have effectively achieved the goal. However, there are also some problems. The comparison between the traditional equipment selection method and the method proposed in this paper is shown in Figure 9. The calculation of different schemes in [8–17] was static, and only qualitative analysis was carried out for different equipment. Qualitative analysis in the new era is difficult for enterprises to make the appropriate choice based on the actual situation. Refs. [18–25] carried out parameter optimization for the same equipment without considering the influence of different processing equipment for the same product. In [22], Jia et al. proposed a method for obtaining and saving power and energy consumption during drilling and established a mathematical model of energy consumption. The method is verified by a case of hole machining, and the optimal energy consumption is obtained. If the turning method is used, the energy consumption can be reduced by more than 5%.

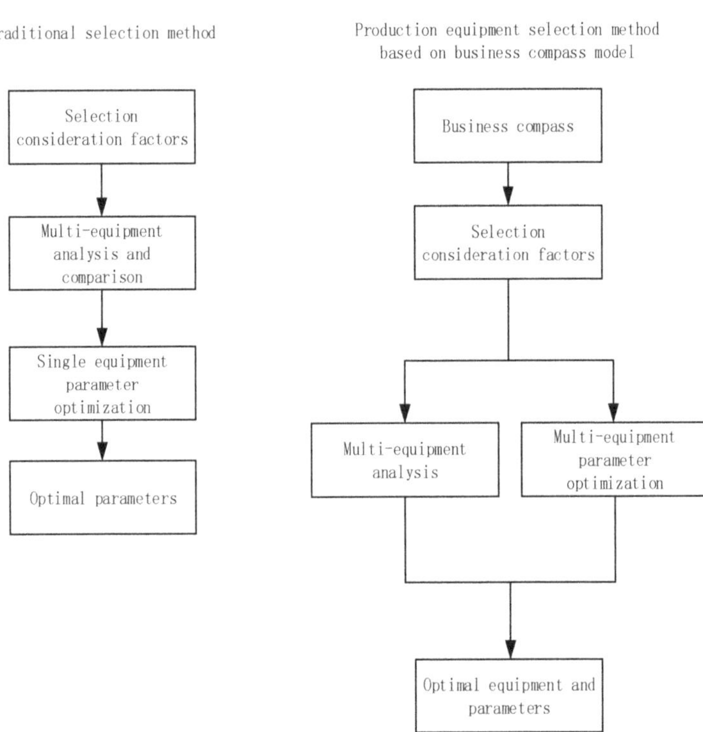

Figure 9. Comparison between traditional equipment selection method and production equipment selection method based on business compass.

Based on the model of the business compass, this article combined enterprise management with production equipment selection and established a production equipment selection model. The unified energy consumption and time model of different equipment were established and used the beetle search algorithm to calculate. Compared with other algorithms, the beetle antennae search algorithm only needs one body, and the amount of computation is greatly reduced. It is simple in principle, uses fewer parameters, requires less computation, and is faster to solve. The result proves the necessity and significance of the method.

5.4. Practical Implications and Future Steps

This article proposed an equipment selection model based on the business compass, and a unified calculation method for energy consumption and time of different devices, which was verified by an example. This research can help enterprises choose the best production equipment and provide a reference for energy-saving and efficient production of enterprises. However, this article only considers the comparison of different equipment for processing the same product. In the future, the influence of workshop workers, equipment status, material information, and other factors on the equipment selection will be analyzed.

6. Conclusions

Equipment selection is an important part of product production decision-making and is of great significance to product production and operation management. Choosing the right production equipment can save resources, improve production efficiency, help the planning and implementation of production and operation management activities, and promote the realization of efficient, flexible, punctual, and clean production and operation management goals. The selection of production equipment is a complex issue, which can promote the sustainable development of society. Based on the enterprise management model of the business compass, this article established a model of enterprise production equipment selection and proposed a unified method to calculate the energy consumption and processing time of various equipment. This model can help enterprises to produce energy-saving and efficient production.

1. Based on the model of the business compass, this article established a model of enterprise production equipment selection. This model combines enterprise operation management with production equipment selection and analyzes the enterprise management process to realize equipment selection.
2. A unified energy consumption and time calculation model to produce the same product with multiple equipment was established and the model was verified by the case of machining end cap holes.
3. The BAS algorithm was used to optimize and calculate the energy consumption and time of multiple equipment processing.

The research result showed that this method can quantitatively analyze the energy consumption and time when different equipment processes the same product. It has important significance for enterprise production. From the perspective of enterprise management, a method for selecting production equipment is proposed, which can provide guidance for enterprises to choose energy-saving and efficient equipment during production and can also provide advice for the government to save energy. However, in this paper, only the energy consumption and processing time of the production process were studied. More optimization objectives such as the economy and carbon emission will be considered in the future. This paper only considers the comparison of different equipment for processing the same product and will analyze the influence of workshop workers, equipment status, materials, and other factors on equipment selection in the future.

Author Contributions: Conceptualization, Y.X. and J.Z.; methodology, Y.X., J.Z. and R.W.; software, Y.X.; validation, J.Z., R.W. and X.Z.; formal analysis, Y.X. and R.W.; investigation, Y.X. and X.Z.; writing—original draft preparation, Y.X. and H.Z.; writing—review and editing, Y.X. and H.Z.; visualization, X.Z. All authors have read and agreed to the published version of the manuscript.

Funding: The authors are grateful to the National Science Foundation, China (Nos. 61862051, 51975432); the Natural Science Foundation of Hunan Province, China (No. 2022JJ50244); the China Education Department of Hunan Province (No. 21B0695); the Project of Hunan Social Science Achievement Evaluation Committee in 2022 (No. XSP22YBC081); the Science and Technology Foundation of Guizhou Province under Grant (No. [2019]1299); the Top-Notch Talent Program of Guizhou Province under Grant (No. KY [2018]080); and the Program of Qiannan Normal University for Nationalities under Grant (Nos. QNSY2018JS013, QNSYRC201715).

Institutional Review Board Statement: Not applicable.

Informed Consent Statement: Informed consent was obtained from all subjects involved in the study.

Data Availability Statement: Not applicable.

Conflicts of Interest: The authors declare no conflict of interest.

References

1. Zhu, Q.; Li, X.; Li, F. Analyzing the sustainability of China's industrial sectors: A data-driven approach with total energy consumption constraint. *Ecol. Indic.* **2021**, *122*, 107235. [CrossRef]
2. Brodny, J.; Tutak, M. Analysis of the efficiency and structure of energy consumption in the industrial sector in the European Union countries between 1995 and 2019. *Sci. Total Environ.* **2022**, *808*, 152052. [CrossRef] [PubMed]
3. Liu, H.; Fan, L.; Shao, Z.X. Threshold effects of energy consumption, technological innovation, and supply chain management on enterprise performance in China's manufacturing industry. *J. Environ. Manag.* **2021**, *300*, 113687. [CrossRef] [PubMed]
4. Kahraman, A.; Kantardzic, M.; Kahraman, M.M. A Data-Driven Multi-Regime Approach for Predicting Energy Consumption. *Energies* **2021**, *14*, 6763. [CrossRef]
5. Chen, Y.; Luca, G. Technologies Supporting Artificial Intelligence and Robotics Application Development. *J. Artif. Intell. Technol.* **2021**, *1*, 1–8. [CrossRef]

6. Lin, K.; Zhao, H. The Impact of Green Finance on the Ecologicalization of Urban Industrial Structure-Based on GMM of Dynamic Panel System. *J. Artif. Intell. Technol.* **2022**, *2*, 123–129. [CrossRef]

7. Li, H.; Wang, W.; Fan, L. A novel hybrid MCDM model for machine tool selection using fuzzy DEMATEL, entropy weighting and later defuzzification VIKOR. *Appl. Soft Comput.* **2020**, *91*, 106207. [CrossRef]

8. Liu, S.H.; Du, Y.B.; Yao, K.H.; Tang, D.B. Multi-objective Optimum Seeking Method of Intelligent Manufacturing Oriented CNC Machine Tool. *Trans. Chin. Soc. Agric. Mach.* **2017**, *48*, 396–404.

9. Zhou, L.X.; Yuan, B.; Wang, Y.C. The Optimization Selection of Machine Tool for Green Manufacturing. *Modul. Mach. Tool Autom. Manuf. Tech.* **2018**, *02*, 157–160.

10. Li, L.; Wang, Y.C.; Liu, X.C.; Tang, Y.; Ye, L. Machine Tool Equipment Selection Based on TOPSIS-AHP. *Modul. Mach. Tool Autom. Manuf. Tech.* **2016**, *12*, 129–132.

11. Han, Z.Q.; Shan, W.B.; He, T.F. Decision on green process scheme of mechanical processing based on entropy weight and TOPSIS. *Mod. Manuf. Eng.* **2020**, *08*, 87–91.

12. Zanuto, R.; Hassui, A.; Lima, F. Environmental impacts-based milling process planning using a life cycle assessment tool. *J. Clean. Prod.* **2019**, *206*, 349–355. [CrossRef]

13. Klink, A.; Arntz, K.; Johannsen, L. Technology-based assessment of subtractive machining processes for mold manufacture. *Procedia CIRP* **2018**, *71*, 401–406. [CrossRef]

14. Liu, J.F.; Zhao, P.; Zhou, H.G. Digital twin-driven machining process evaluation method. *Comput. Integr. Manuf. Syst.* **2019**, *25*, 1600–1610.

15. Chen, N.; Barnawal, P.; Frank, M.C. Automated post machining process planning for a new hybrid manufacturing method of additive manufacturing and rapid machining. *Rapid Prototyp. J.* **2018**, *24*, 1077–1090. [CrossRef]

16. Komatsu, W.; Nakamoto, K. Machining process analysis for machine tool selection based on form-shaping motions. *Precis. Eng.* **2021**, *67*, 199–211. [CrossRef]

17. Koremura, K.; Inoue, Y.; Nakamoto, K. Machining Process Evaluation Indices for Developing a Computer Aided Process Planning System. *Int. J. Autom. Technol.* **2017**, *11*, 242–250. [CrossRef]

18. Han, C.; Luo, M.; Zhang, D. Optimization of varying-parameter drilling for multi-hole parts using metaheuristic algorithm coupled with self-adaptive penalty method. *Appl. Soft Comput.* **2020**, *95*, 106489. [CrossRef]

19. He, K.; Hong, H.; Tang, R. Analysis of Multi-Objective Optimization of Machining Allowance Distribution and Parameters for Energy Saving Strategy. *Sustainability* **2020**, *12*, 638. [CrossRef]

20. Wang, Q.L.; Wei, P.; Duan, X.H. Multi-objective Optimization of Computer Numerical Control Turning Process Parameters Based on Response Surface Method and Artificial Bee Colony Algorithm. *Ind. Eng. Manag.* **2022**, *27*, 117–126.

21. Li, Y.F.; Xiong, M.; He, Y.; Xiong, J.J.; Tian, X.C.; Mativenga, P. Multi-objective optimization of laser welding process parameters: The trade-offs between energy consumption and welding quality. *Opt. Laser Technol.* **2022**, *149*, 107861. [CrossRef]

22. Jia, S.; Yuan, H.; Ren, D. Power and Energy Consumption Acquisition and Energy-Saving Control Method of CNC Machine Tool Drilling Process. China Patent 108037734A, 15 May 2018.

23. Xiao, Y.; Jiang, Z.; Gu, Q.; Yan, W.; Wang, R. A novel approach to CNC machining center processing parameters optimization considering energy-saving and low-cost. *J. Manuf. Syst.* **2021**, *59*, 535–548. [CrossRef]

24. Ma, H.; Liu, W.; Zhou, X. An effective and automatic approach for parameters optimization of complex end milling process based on virtual machining. *J. Intell. Manuf.* **2020**, *31*, 967–984. [CrossRef]

25. Zhang, S.B.; Zhang, S. Auxiliary Process Time Modeling and Machining Step Sequencing Optimization of Box Parts. *Modul. Mach. Tool Autom. Manuf. Tech.* **2022**, *7*, 90–94, 98.

26. Wang, R.P.; Yi, J. Characteristics and Mission of Management Thinking with Chinese Characteristics Based on the Business Compass Perspective. In Proceedings of the 8th International Symposium on Project Management (ISPM 2020), Beijing, China, 4–5 July 2020.

27. Chen, X. Introduction to Taoist philosophy. *Chin. Philos.* **2019**, *106*, 14.

28. Sun, X.P. Taoism Naturalistic Technology Views. *Stud. Dialectics Nat.* **2022**, *38*, 42–47.

29. Wang, R.P. *Business Compass*; Science Press China: Beijing, China, 2020.

30. Deng, W.; Luo, Y.; Luo, J. Optimization of milling parameters based on low energy consumption. *Chin. J. Constr. Mach.* **2019**, *17*, 397–400.

31. Xiao, Y.; Zhao, R.; Yan, W.; Zhu, X. Analysis and Evaluation of Energy Consumption and Carbon Emission Levels of Products Produced by Different Kinds of Equipment Based on Green Development Concept. *Sustainability* **2022**, *14*, 7631. [CrossRef]

32. Xie, J.; Ma, J.H.; Luo, X. Study on Energy Consumption Model and Cutting Parameter Decision Method of Machine Tool Oriented to Energy Saving Optimization. *J. Chongqing Univ. Technol. (Nat. Sci.)* **2020**, *34*, 77–86.

33. Xiao, Y.; Wang, R.; Yan, W.; Ma, L. Optimum Design of Blank Dimensions Guided by a Business Compass in the Machining Process. *Processes* **2021**, *9*, 1286. [CrossRef]

34. Jiang, X.; Li, S. BAS: Beetle Antennae Search Algorithm for Optimization Problems. *Int. J. Robot. Control* **2018**, *1*, 1–5. [CrossRef]

35. Hao, J.; Huang, J.; Zhang, A. Optimal coordinated control of hybrid AC/VSC-HVDC system integrated with DFIG via cooperative beetle antennae search algorithm. *PLoS ONE* **2020**, *15*, e0242316. [CrossRef] [PubMed]

36. Jie, Q.; Ping, W.; Cpa, B. Joint application of multi-object beetle antennae search algorithm and BAS-BP fuel cost forecast network on optimal active power dispatch problems. *Knowl.-Based Syst.* **2021**, *226*, 107149. [CrossRef]

37. Zhang, J.; Wang, L. Fault Diagnosis of Hydraulic Pumps Based on Support Vector Machine Optimized by Beetle Antennae Search. *Noise Vib. Control* **2022**, *42*, 5.

 processes

Article

An Agent-Based Approach for Make-To-Order Master Production Scheduling

Faezeh Bagheri [1,*], **Melissa Demartini** [2], **Alessandra Arezza** [3], **Flavio Tonelli** [1], **Massimo Pacella** [4] **and Gabriele Papadia** [4]

[1] Department of Mechanical, Energy, Management and Transportation Engineering (DIME) Polytechnic School, University of Genoa, 16145 Genoa, Italy; flavio.tonelli@unige.it
[2] SDU—Center for Sustainable Supply Chain Engineering, Department of Technology and Innovation, University of Southern Denmark, Campusvej 55, 5230 Odense, Denmark; med@iti.sdu.dk
[3] DGS SPA, Via Paolo di Dono, 73, 00142 Roma, Italy; alessandra.arezza@dgsspa.com
[4] Department of Engineering for Innovation, University of Salento, 73100 Lecce, Italy; massimo.pacella@unisalento.it (M.P.); gabriele.papadia@unisalento.it (G.P.)
* Correspondence: faezeh.baagheri@edu.unige.it; Tel.: +39-010-33-52888

Abstract: In recent decades, manufacturers' intense competitiveness to suit consumer expectations has compelled them to abandon the conventional workflow in favour of a more flexible one. This new trend increased the importance of master production schedule and make-to-order (MTO) strategy concepts. The former improves overall planning and controls complexity. The latter enables the production businesses to reinforce their flexibility and produce customized products. In a production setting, fluctuating resource capacity restricts production line performance, and ignoring this fact renders planning inapplicable. The current research work addresses the MPS problem in the context of the MTO production environment. The objective is to resolve Rough-Cut Capacity Planning by considering resource capacity fluctuation to schedule the customer's order with the minimum cost imposed by the company and customer side. Consequently, this study is an initial attempt to propose a mathematical programming approach, which provides the optimum result for small and medium-size problems. Regarding the combinatorial intrinsic of this kind of problem, the mathematical programming approach can no longer reach the optimum solution for a large-scale problem. To overcome this, an innovative agent-based heuristic has been proposed. Computational experiments on variously sized problems confirm the efficiency of the agent-based approach.

Keywords: master production scheduling; make-to-order; mathematical programming; agent-based; overtime; earliness; tardiness

Citation: Bagheri, F.; Demartini, M.; Arezza, A.; Tonelli, F.; Pacella, M.; Papadia, G. An Agent-Based Approach for Make-To-Order Master Production Scheduling. *Processes* **2022**, *10*, 921. https://doi.org/10.3390/pr10050921

Academic Editor: Sergey Y. Yurish

Received: 1 April 2022
Accepted: 3 May 2022
Published: 6 May 2022

1. Introduction

In the last few decades, the interest of industry and academia has led to a stable increase in the production planning (PPC) area, as this is considered one of the most central and relevant choices faced by firms [1,2]. A traditional PPC problem originates with the details of a customer demand that must be covered with a specific production plan by managing different resources and constraints (i.e., demand, process, and supply) while minimizing costs [3]. Mula et al. (2006) identified five main PPC areas: (i) Master Production Schedule (MPS), (ii) Material Requirement Planning and Manufacturing Resources Planning (MRP); (iii) Supply Chain Planning; (iv) Aggregate Production Planning and (v) Hierarchical Production Planning [4]. MPS defines the optimal production plan by meeting customer demand and minimizing holding and set-up costs. MRP accomplishes PPC components by using a bill of materials (BOM) and the MPS outputs. An additional component, the Rough-Cut Capacity Plan (RCCP) module, can be used to check the feasibility of the MPS plan. However, due to the supply chain complexity and the increasing need for integration and coordination among supply chain players, Supply Chain Planning modules have been introduced in the last decades to manage multisite PPC. Finally, Hierarchical Production Planning is used to distinguish between several planning levels, and Aggregate Production Planning to establish production, inventory, and workforce levels. Manufacturing businesses encounter complicated production planning problems due to a growing focus on customer requirement and service and a rise in product complexity. A high degree of manufacturing flexibility is also necessitated due to short product life cycles, ongoing market volatility, and unpredictable demand. Understanding operational interdependence and reacting effectively to market or demand shifts is even more crucial in an ever-complex environment [5]. In today's competitive market, manufacturing companies are striving to modify their production strategy to enhance their market share by responding to a broader variety of client needs. One supporting strategy in this regard is Make-To-Order (MTO). MTO enables business owners to produce customized products according to what is desirable for customers, and can be facilitated through the implementation of Industry 4.0 technologies [6]. The higher the customer satisfaction and the more needs are met, the more successful the company will be in attracting higher number of customers and increasing its market share [7–9]. According to the capability of MTO to confront market competitiveness and MPS to handle intrinsic uncertainty in production, their accompaniment could play a noteworthy role for manufacturers in the direction of overcoming the above-mentioned challenges. Therefore, the goal of this paper is to focus on an MPS/RCCP problem similar to that proposed in [10], where the authors developed a decision support framework to improve the MPS process in a MTO environment. They extended the RCCP functionalities of Microsoft Navision by implementing an Extended-RCCP based on a Genetic Algorithm (GA). In this paper, instead of using the GA, we present an innovative agent-based heuristic, and we compare its results versus a mathematical programming approach. The paper is organized as follows; the next section depicts the state of art related to MPS, MTO problems, the focus of this research, and the main characteristics of the agent-based heuristic. Section 3 describes the model formulation and the solution approach. Section 4 presents the comparison between results obtained by the mathematical programming approach and agent-based heuristics applied to some problem instances in 'real' industrial cases. Finally, Section 5 presents conclusions and future outlook.

2. Literature Review

The MPS process defines production plans for product families or products regarding fluctuating demands [11]. The results of demand planning and forecasting influence the MPS as it aims to balance the demand and available capacities. In return, the resulting plans determine purchased parts for the MRP and the production volume for the lot sizing. In the literature for the MPS, various mathematical optimization models can be found, most of which are used in linear programming, integer linear programming, and mixed-integer linear programming [12]. Basic MPS models consider a single-period single-stage case [13]. However, there are many extensions as practical examples; there are usually many different stages, with planning horizons spanning from a few months up to one year. Therefore, multi-period multi-stage models are considered contributions. Additionally, the insertion of capacity restrictions is the state of the art for MPS models. In this scenario, to allow demand to be met, the number of stored products needs to be determined [14]. Different models consider further additional adjustments for uncertainties, such as quality issues, rework, and uncertain demand. An example of this can be found in Taşkın and Ünal (2009) [15]. They describe an MPS model applied within the glass industry. They come up against the problem of inconsistent product qualities and downgrade substitution to meet demand. The reworking of rejects is integrated into the MPS by Inderfurth, Lindner and Rachaniotis (2005) [16]. At present, many models focus on existing uncertainties in tactical and operational planning, though at the forefront are those focusing on demand. Researchers can broadly be split into two groups: those who consider non-cost-based objective functions like flow time, job tardiness, job earliness and schedule makespan [17], and those who consider combined cost-based and non-cost-based objective functions [18,19]. Among the latter group, some recent studies [20,21] indicate that smoothed series of production volumes, through the minimization of (i) 'total variations in production volumes'; (ii) the total cost; and (iii) other objectives; are sought by production managers. Driven by the complexity of the automotive industry, Mansouri, Golmohammadi, and Miller (2019), in their paper, first examine how the throughput of complex job shop systems can be forecasted based on problem characteristics and different MPS methods [22]. Next, they analyse how different MPS approaches balance the relationship between problem characteristics and throughput. A mixed-effects model based on operational characteristics and the MPS development method was established to obtain these objectives and predict the system's throughput. The analyses are based on a real case study taken from the automotive industry and two complex job shop systems in the literature. The experimental results indicate that the throughput of job shop systems can be predicted with a high level of accuracy. Golmohammadi (2013) developed a neural network model focused on detailed scheduling for analysis of job shop scheduling. Instead of a simulation model, which is a costly and complex approach for scheduling, the output of the model proposed by the author helps managers estimate the throughput based on historical data with a trained neural network model [23]. The main shortcoming of the research is that the prediction results may not be accurate due to the training data set potentially not comprising new problem characteristics. The paper by Guillaume, Thierry and Zieliński (2017) focuses on the tactical level by examining the MPS and MRP planning processes; specifically, they work on the CLSP (for the MPS process) and MLCLSP (for the MRP process), both with back-ordering, with uncertain cumulative demand [24]. An essential difficulty for the production planning systems is the issue of tactical production and capacity planning under uncertainty in demand. In the paper, the authors cover (i) the MPS (CLSP) under small uncertainty in the cumulative demand; (ii) the MRP (MLCLSP) problems under uncertainty in the cumulative demand. The model of uncertainty in the cumulative demand enables us to take into account simultaneously the imprecision on order quantities and dates. For both problems, linear programming models, including back-ordering and the cumulative demand, have been presented. Efficient methods for evaluating the impact of uncertainty on production plans and linear programming for computing optimal robust production plans for MPS and MRP problems in the cumulative demand are proposed. The authors prove that the computational complexity of optimization processes, with the min-max criterion, is not significantly increased when introducing uncertainty in the cumulative demand, compared with the deterministic counterparts–they remain polynomial solvable. Therefore, they can be applied in the industrial context, namely in the manufacturing planning tools using linear programming solvers. Sahin, Powell Robinson and Gao (2008) explored the MPS problem in a MTO environment [25]. They developed a simulation model to analyse cost and schedule metrics and design specific advanced order commitment policies. They conducted full-factorial experiments to define the main drivers of MPS policy cost and schedule stability. Powell Robinson, Sahin, and Gao (2008) also develop a two-stage rolling schedule environment with a particular focus on the policy related to the schedule flexibility in the non-frozen time intervals [26]. Sawik (2007) explored the same topic by developing multi-objective, long-term production scheduling in an MTO environment and a lexicographic approach with a hierarchy of integer programming formulations [27]. The goal was to assign customer orders with different due dates to minimize tardiness and maximize the input and output inventory. The same topic is explored by Nedaei and Mahlooji (2014), who developed a multi-objective MPS and rolling schedule policies in a two-stage MTO environment [28]. Finally, very recent articles address MPS and MTO problems in connection with Industry 4.0. Indeed, Yin, Stecke and Li (2018) explained that due to the relevance of the mass-customization trend and due to the difficulties companies are experiencing with the current manufacturing systems, Industry 4.0 technologies can enable and facilitate a MTO environment as a typical strategy of customization [29]. Gu and Koren (2022) stated that internet of things, cyber physical systems, machine learning and deep learning technologies should be integrated to develop a mass-individualisation

MTO [30]. Mladineo et al. (2022) stated that companies need to advance their product configurators to satisfy customer needs and to keep the whole process economical and efficient. In this regard, data integration is defined as a fundamental requirement for MTO, especially in the context of horizontal and vertical integration of the value chain [9]. Kundu, Rossini and Portioli-Staudacher (2018) discussed the importance of Industry 4.0 technologies in workload control for MTO companies. They stated that the first condition is to implement production automation to improve flexibility, then they highlighted the importance of cyber physical systems as a representation of physical components. To this end, sensors are considered one of the core technologies to enable cyber physical systems. Finally, they emphasised the importance of a proper communication network within the factory to improve data sharing between manufacturing processes and production orders [31]. Lee et al. (2019) implemented an MTO strategy in the context of Industry 4.0. They developed a product configuration system to match customer demands and manufacturer orders [32]. Rahman, Janardhanan and Nielsen (2019) focused on real-time order acceptance and scheduling as key concepts of MTO in an Industry 4.0 environment. They developed a real time system capable of accepting orders and scheduling decisions through a hybrid genetic algorithm and particle swarm optimization model [33]. Micieta et al. (2019) designed an innovative approach to products segmentation in an MTO environment based on Industry 4.0 concepts. This approach allowed companies to reduce work in progress lead time and increase efficiency [34]. Woschanka, Dallasega and Kapeller (2020) aimed to enhance logistics performance in a MTO environment by using real data and analysing several planning granularity levels. The approach has been validated through a discrete event simulation model [35].

2.1. Main Focus of the Proposed Research

This study aims to closen the MPS problem in the context of MTO to the real production environment. Various limitations that frequently occur in the production factories have been imposed to realize this goal. For example, limited resource capacity is one of the most frequent challenges that restrict the production rate and affect customer order delivery date. There are also other kinds of constraints that make the production setting more complicated. More details and clarification have been explained in the rest of the paper. This extension and modification enable a manufacturer to achieve more practical results. To achieve these results, two solution approaches, including Mathematical Programming (MP_RCCP) and an Agent-Based approach (AB_RCCP), have been presented. Garey and Johnson have proven that when capacity and setup limitations are imposed, the MPS problem is taken into account as an NP-hard problem. In practice, this implies that the processing time required to answer such issues will swiftly and tremendously increase as the problem size rises [36]. Following this, achieving the optimal solution through mathematical programming with increasing instances' size is quite complicated (MP_RCCP). To overcome this challenge, an AB_RCCP has been developed to perform an acceptable solution in a reasonable time.

The 'conceptual' generated problem instance is described as follows:

- *Customer Order (CO):* this is the customer's request. It is managed in an MTO environment. Each customer order presents a due date and a specific quantity that needs to be satisfied.
- *Bill of material (BOM):* this lists the raw materials, parts, and components needed to make a product. It presents different levels according to the customer order.
- *Bill of the process (BOP):* this is comprised of detailed plans explaining the manufacturing processes for a particular product. These plans contain in-depth information on machinery, plant resources, equipment layout, configurations, tools, and instructions.
- *Resources:* these can be physical (work centers, tooling, process materials) or skills. Each resource presents its capacity and feature.

2.2. The Main Characteristics of the Proposed Approach

The Cooperative Supply Chain (Coop SC) framework is a prototype capable of supporting a decision-making process through adequate, modern, and flexible tools, capable of hosting algorithms from third parties [37]. CoopSC includes an RCCP engine to optimize multi-site contexts using a modified multi-agent architecture by adopting three different levels of supervision: demand, production sites, and resources.

Starting from the highest planning levels (Figure 1), the Demand Management and Inventory Management modules generate medium and long-term forecasts for macro-families and/or product types related to a CO. The Inventory Management is a collection of inventory planning tools and applications that can optimize the mix and quantity of planned inventory. They provide a validation of the input required to satisfy the CO for a set of KPI. The MPS component is, therefore, able to provide decision-making support for the resolution of typical business constraints, generated by conflicts between the company needs and the constraints of a multi-plant/multi-supplier context. The requests, processed by CoopSC, can be defined manually or imported by an external forecasting module [38] capable of forecasting the sales and use of the products (appropriately divided into production lots by the Inventory Management level) so that they can be purchased or made in adequate quantities in advance. The production and procurement commitments, which are required by the estimate of future requirements, are simulated according to the capacity of the resources, the planning cycles, and the actual processing bills, and considering the supply and transport constraints (RCCP module). The CoopSC MPS/RCCP component is based on the Service Lanes Planning (SLP) concept [39]. It articulates the evaluation of customer service levels by identifying service lanes based on assignment criteria and heuristics that verify inventory availability and production plans in the processing, production and supply stages. The SLP develops a proposal by considering a multi-constrained context as depicted in Figure 2, the goal of which is to provide a feasible solution that takes into consideration the available stock, the CO, the production constraints, and the SL.

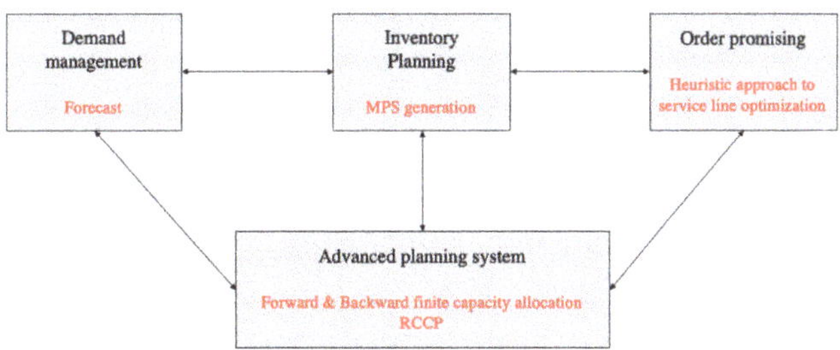

Figure 1. The MPS/RCCP component of the Cooperative supply chain framework.

The CoopSC MPS component provides two analyses:

1. Allocate CO by using an infinite capacity approach;
2. Execute a master planning activity with a finite capacity and develop several proposals to satisfy CO and several production constraints.

In the first case, CoopSC MPS can satisfy the SL, but it doesn't take into account the capacity of resources. In the second case, the tool splits up the proposals defining the allocation rules that make it possible to move the individual planning lots to respect the finished capacity, always according to the service level. Therefore, CoopSC MPS divides the proposal into minimal lots and, by early and tardy times, attempts to change the allocation of the lots to meet the SL and the capacity of resources.

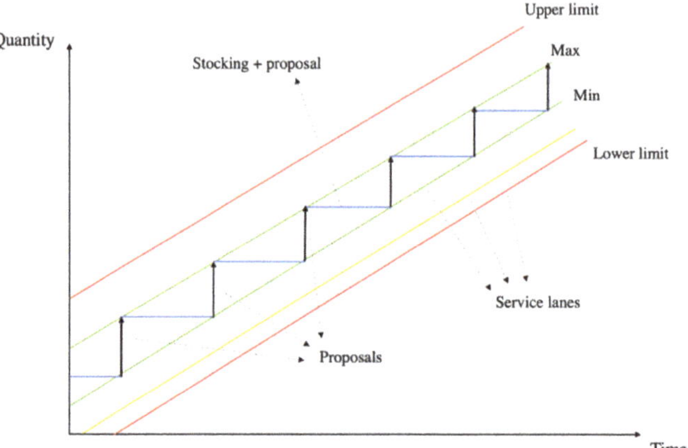

Figure 2. The Service Lane Planning concept–Note Adapted from Bernocco et al. (2003).

Therefore, the goal of the CoopSC MPS is to minimize capacity overflows by using several levers, or the user can force a proposal to make the allocation feasible. If there is a capacity overflow, the system allocates the CO to the next bucket and executes a new backward assignment of that CO.

The backward assignment is based on an AB_RCCP heuristics where a set of agents can solve the planning problem by simulating the resources allocation to meet the objective function.

Here, two solution approaches, namely Mathematical Programming (MP_RCCP) and Agent-Based approach (AB_RCCP), have been described in detail in Sections 2 and 3, respectively.

3. Problem Description and Solution Approaches

The production environment characterized by the MTO strategy possesses a high level of flexibility concerning satisfying the customer requirement. This specification supports business owners to cover a wide range of customers' needs by customizing their requirements into the product features and enabling them to maintain and raise their market share in today's competitive market [7].

In the make-to-store (MTS) strategy, the factory could produce the products in advance and store them in the warehouse. This feature enables more efficiently handling of demand fluctuations. Establishing a balance between resources and production line within MTS is more achievable [40], while in the MTO strategy the factory will proceed with production after receiving the customer order. It is obvious, regarding the limitation of the resources, that if fluctuation occurs in the customer order, the production process will encounter the challenge of instituting balance in the factory's resources [8]. In this regard, the main concentration of current research is presenting an efficient solution approach for coping with this kind of challenge in the MTO production setting. Sections 2.1 and 2.2 have been dedicated to the description of MP_RCCP and AB_RCCP, respectively.

3.1. The Mathematical Programming Approach (MP_RCCP)

In this section, the MILP model presented by [41] has been extended according to the MTO production strategy. In the context of MTO, each customer could customize their order. Each customer order (CO) consists of the sequence of operations and due date. The resources have a limited capacity for processing the operations in the real production environment. Concerning this fact, this model aims to plan/schedule the orders to minimize tardiness and overtime cost.

To apply the approach, it is required to formulate a mathematical model of the production environment. In the current study, the first requirement has been satisfied in Section 2. Depending on the managers' preference, whether this preference can originate from the customer side or the production environment, or both, the goal of mathematical model is adjusted accordingly and production line features are introduced as the series of constraints. At this stage, the optimal solution is achieved by relying on the optimizing software and coding the formulated mathematical model.

Other assumptions of the proposed model are as follows:

- There is no priority for each customer order.
- Each resource could process only one operation at a time.
- Each operation is eligible to operate with only one resource.
- Resumption of operations is possible.
- The orders are processed in the batch.
- Resources have limited capacity.
- Resources are available from time zero.

The index, parameters, and variables of the model are as follows:

Index	
Customer	$i \in \{1, \dots, C\}$
Resource	$j \in \{1, \dots, M\}$
Operation	$o \in \{1, \dots, O\}$
Time horizon	$t \in \{1, \dots, T\}$

Parameters		
Customer	D_i	The quantity of customer order i
	DD_i	The due date for customer order i
	$TardiC_i$	Tardiness cost for customer order i
Resource	CAP_j^t	The capacity of resource j in time bucket t
	$OvertC_j$	The overtime cost of resource j
	$ACAP_j^t$	Equal to 1 if overtime is allowed, otherwise 0
Operation	PT_{io}	Processing time operation o of customer order i
	ER_{io}^j	Equal to 1 if resource j is eligible to process operation o of customer order i, otherwise 0
other	Dur	The length of each time bucket
	N	A very large number

Variables	
lo_{io}^t	The length of the time that operation o of customer order i is performed in time bucket t
α_{io}^t	Equal 1, if operation o of customer order i is begun in time bucket t
$\beta_{io\acute{i}\acute{o}}^{tj}$	Equal 1 if operation o of customer order i is started in time bucket t before operation $\acute{o}$ of customer order $\acute{i}$ and $ER_{io}^j == ER_{\acute{i}\acute{o}}^j$
st_{io}^t	Start time operation o of customer order i in time bucket t
ct_{io}^t	Completion time operation o of customer order i in time bucket t
lst_{io}	A lower bound for starting time of operation o of customer order i
uct_{io}	A upper bound of completion time of operation o of customer order i
l_i	The amount of lateness related to customer order i
ot_j^t	The amount of overtime related to resource j in time bucket t

The proposed model is as follows:

$$Min\ Z = \sum_i TardiC_i * l_i + \sum_t \sum_j OvertC_j * ot_j^t \qquad (1)$$

$$\sum_t lo_{io}^t = D_i * PT_{io} \qquad \forall\ i,\ o \qquad (2)$$

$$lo_{io}^t \leq N * \alpha_{io}^t \qquad \forall\ i,\ o,\ t \qquad (3)$$

$$ct_{io}^t = st_{io}^t + lo_{io}^t \qquad \forall\ i,\ o,\ t \qquad (4)$$

$$uct_{io} \geq ct_{io}^t - \left(1 - \alpha_{io}^t\right) * N \qquad \forall\ i,\ o,\ t \qquad (5)$$

$$lst_{io} \leq st_{io}^t - \left(1 - \alpha_{io}^t\right) * N \qquad \forall\ i,\ o,\ t \qquad (6)$$

$$lst_{i(o+1)} \geq uct_{io} \qquad \forall\ i,\ o \qquad (7)$$

$$\beta_{ioí\acute{o}}^{tj} + \beta_{í\acute{o}io}^{tj} \geq \alpha_{io}^t + \alpha_{í\acute{o}}^t - 1 \qquad \forall i,\ o,\ í,\acute{o},\ j,\ t \qquad (8)$$

$$ct_{io}^t \leq st_{í\acute{o}}^t + \left(1 - \beta_{ioí\acute{o}}^{tj}\right) * N \qquad \forall i,\ o,\ í,\acute{o},\ j,\ t \qquad (9)$$

$$st_{io}^t \geq (t - 1) * Dur \qquad \forall\ i,\ o,\ t \qquad (10)$$

$$f_{jh}^t \leq t * Dur \qquad \forall\ i,\ o,\ t \qquad (11)$$

$$uct_{iO} \geq DD_i + l_i \qquad \forall\ i,\ O : \text{last operation} \qquad (12)$$

$$\sum_i \sum_o lo_{io}^t * ER_{io}^j \leq CAP_{jt} + ot_j^t \qquad \forall\ j,\ t \qquad (13)$$

$$ot_j^t \leq \left(Dur - CAP_j^t\right) * ACAP_j^t \qquad \forall\ j,\ t \qquad (14)$$

$$lo_{io}^t \geq 0,\ st_{io}^t \geq 0,\ ct_{io}^t \geq 0,\ lst_{io} \geq 0,\ uct_{io} \geq 0,\ l_i \geq 0,\ ot_j^t \geq 0 \qquad \forall\ i,\ o,\ j,\ t \qquad (15)$$

$$\alpha_{io}^t \in \{0,1\},\ \beta_{ioí\acute{o}}^{tj} \in \{0,1\} \qquad \forall\ i,\ o,\ í,\acute{o},\ j,\ t \qquad (16)$$

The objective function (1) involves the cost related to tardiness and overtime and tries to minimize it with respect to the constraints. Constraint (2) ensures that the summation of processing time in the all-time bucket should satisfy the whole quantity of customer orders. Through constraint (3), the value for the variable α_{io}^t is determined. This variable is used to avoid overlapping in scheduling. Constraint (4) calculates the completion time of each operation in each time bucket. Constraints (5) and (6), define the upper bound for completion time and lower bound for starting time of operation o in time bucket t. Constraint (7) ensures the sequence of operations. It means that the next operation cannot be started unless the previous operation is completed. Constraints (8) and (9) jointly arrange the sequence of operations from different customer orders on the same machine. Constraints (10) and (11) define the start and completion time boundaries. Constraint (12) computes customer order lateness. Constraint (13) ensures that resource loading does not exceed the available capacity. Constraint (14) limits the amount of resource overloading. Constraints (15) and (16) define the non-negative and binary variables.

3.2. The Agent Based Approach (AB_RCCP)

The AB_RCCP aggregates the individual proposals into minimum planning batches and, according to previously defined lead times or delays, tries to shift the allocation of production batches to provide the required level of service and ensure compliance with the finite capacity.

The proposed multi-agent heuristics is distinguished by efficient interactivity and computational convergence thanks to the optimized simulation engine, and the ability to work in 'back and forth' allocation to time.

The planning process, developed through the backward loading of the production sites and the macro resources associated with them, is carried out by two supervising agents (plant and resource). Their goal is to choose the production site where the demand will be allocated and then decide which production resources will be destined for the batch to be produced according to the CO and their feature and capacity. This process is depicted in Figure 3.

Each agent has information related to (i) the production system and (ii) the option for exchanging data with other agents to start the negotiations. They are also equipped with a decision-making system to determine the planning of tasks based on the information defined above.

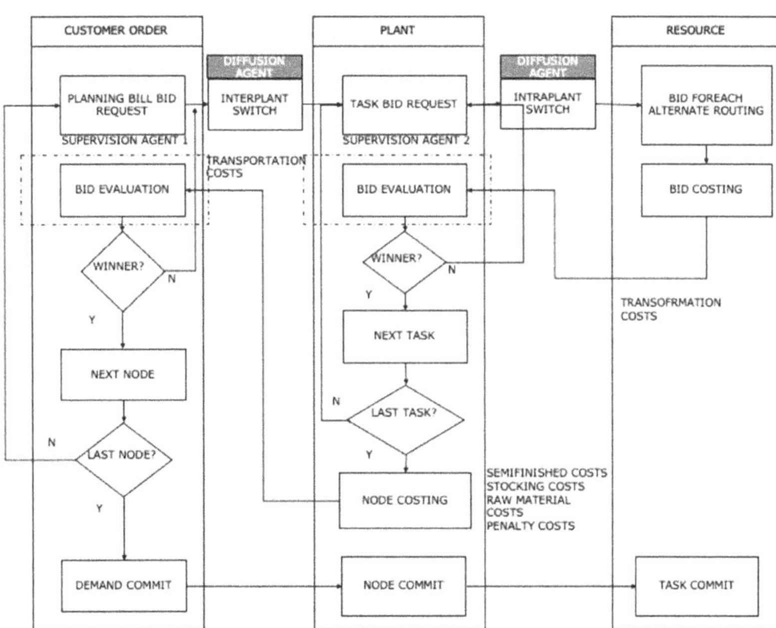

Figure 3. Multi-agent planning architecture (AB_RCCP)–Note Adapted from Bernocco et al. (2003).

The same logic is applied at the hierarchical level of resources. The final goal is to define which resources will perform which operations in which buckets. The supervisory agents (plants and resources) engage in negotiations with their level agents to satisfy the CO and optimize a specific target objective function. Therefore, the plant and resource agents sell a service to satisfy their load profile.

Accordingly to these mechanisms, the AB_RCCP of CoopSC calculates different "scenarios" by applying several planning heuristics parameters. Then, allowing for the comparison of the plans through a set of performance indicators.

4. Computational Results

This section evaluates the performance and efficiency of the MP and AB approaches. The MP approach was coded in General Algebraic Modeling System (GAMS) software, and the CPLEX solver was employed. The AB approach has been developed on Net (C#). Both approaches are executed on a laptop characterized by Intel(R) Core(TM) i7-10750H CPU @ 2.60 GHz processor and 16 GB memory to achieve comparable results.

In order to quantify this evaluation, two measurement indexes, including "Service level" and "Average Resource Overtime" have been formulated. The former, presented by Equation (17), assesses how well the company is able to deliver customer orders on time. The latter one, presented by Equation (18), focuses on resource utilization and calculates how much additional resource capacity is used on average.

$$Service\ Level_i = \frac{Due\ date_i}{Due\ date_i + lateness_i} \tag{17}$$

$$Average\ Resource\ overtime = \frac{\left(\sum_j \frac{Imposed\ Overtime_j}{Max\ Overtime_j}\right)}{Number\ resource} \tag{18}$$

The "Service level" value would be (0, 1]. The higher the service level value, the more successful the factory has been in delivering the customer's orders on time.

The "Average Resource Overtime" value would be [0, 1]. The less the average Resource overtime receives, the less load will be applied to the resources. Consequently, the factory is more successful in establishing balance on the production line.

The proposed MP approach tries to find the best solution for the resource capacity constraint while trying to minimize tardiness and overtime. These two objective functions simultaneously seek to satisfy the customer's expectations and improve resource usage.

The AB approach enables anticipating the production planning by considering earliness besides tardiness and overtime. This approach by imposing earliness is more flexible. The earliness is activated within the AB approach when the factory has enough space for holding the final products until their due date.

4.1. Systematic Generation of Test Instances

In this section, various random instances, according to Table 1, have been generated to evaluate the performance of the MP and AP approaches. To make these two approaches comparable, it is assumed that the earliness is diactive in the AB approach.

Table 1. Parameter adjustment.

Parameters	Notation	The Function of Generation
The quantity of customer orders i	D_i	The uniform distribution between [1, 10]
The due date for customer order i	DD_i	The uniform distribution between [5, 25]
Tardiness cost for customer order i	$TardiC_i$	The uniform distribution between [0.625, 2.5]
The overtime cost of resource j	$OvertC_j$	The uniform distribution between [2.5, 10]

In this study, the size of instances is defined through the number of customers. By increasing the number of customers, the size of the problem will increase from small to medium and large. In this regard, ten instances will be generated in three different sizes.

4.2. Results for Systematically Generated Test Instances

In this section, the two proposed approaches are evaluated through executing instances generated in Section 3.2. Two measurement indexes presented in Section 3 have been calculated to provide a clear evaluation. These quantities enable managers to have an explicit vision of the planning. While Tables 2–4 provide details of the calculations, Figures 4–6 illustrate an overview of how these two approaches work.

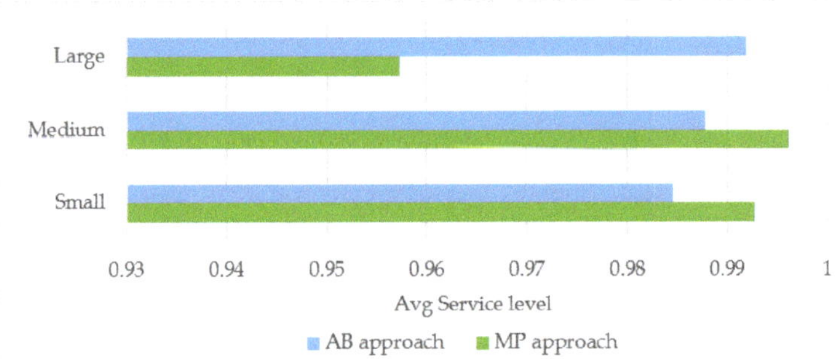

Figure 4. Average service level for small, medium, and large instances for MP and AP approach.

Table 2. Result of small size problem (EC PART I).

INS	Service Level		Average Resource Overtime		Time (s)	
	MP	AB	MP	AB	MP	AB
01	0.9976	0.9937	0.0080	0.0161	15	9
02	0.9950	0.9891	0.0009	0.0000	18	9
03	1.0000	0.9947	0.0000	0.0000	17	9
04	0.9979	0.9882	0.0066	0.0000	15	9
05	1.0000	0.9903	0.0057	0.0123	15	9
06	0.9953	0.9896	0.0038	0.0000	14	10
07	0.9855	0.9585	0.0099	0.0204	18	10
08	0.9988	0.9944	0.0071	0.0118	18	9
09	0.9933	0.9760	0.0038	0.0000	14	9
10	0.9938	0.9706	0.0090	0.0000	18	11
Avg	0.9927	0.9845	0.0055	0.0061	16.2	9.4

Table 3. Result of medium size problem (EC PART II).

INS	Service Level		Average Resource Overtime		Time (s)	
	MP	AB	MP	AB	MP	AB
01	0.9975	0.9896	0.0208	0.0771	1004	15
02	0.9969	0.9921	0.0885	0.1084	1004	14
03	0.9899	0.9883	0.1354	0.1382	1007	13
04	0.9891	0.9867	0.1283	0.1387	1025	13
05	0.9963	0.9781	0.0918	0.0946	1004	11
06	0.9950	0.9791	0.0430	0.0454	1004	10
07	0.9996	0.9917	0.0724	0.0809	1007	9
08	0.9980	0.9865	0.0080	0.0108	1004	10
09	0.9994	0.9878	0.0184	0.0213	1005	8
10	0.9986	0.9983	0.0123	0.0179	1003	10
Avg	0.9960	0.9878	0.0618	0.0733	1006.7	11.3

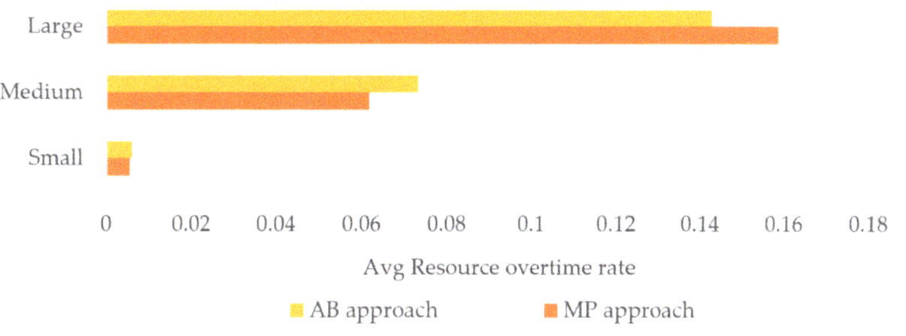

Figure 5. Average Resource overtime for small, medium, and large instances for MP and AP approach.

Table 4. Result of large size problem (EC PART III).

INS	Service Level		Resource Overtime		Time (Second)	
	MP	**AB**	**MP**	**AB**	**MP ***	**AB**
01	0.9403	0.9928	0.0971	0.0724	More than 3600	16
02	0.9566	0.9895	0.1458	0.1501	More than 3600	19
03	0.9296	0.9865	0.1941	0.1828	More than 3600	20
04	0.9510	0.9937	0.1847	0.2093	More than 3600	17
05	0.9651	0.9940	0.1226	0.1321	More than 3600	18
06	0.9417	0.9927	0.1117	0.1387	More than 3600	16
07	0.9550	0.9924	0.1960	0.2334	More than 3600	15
08	0.9854	0.9909	0.1875	0.0180	More than 3600	17
09	0.9745	0.9938	0.1728	0.1198	More than 3600	16
10	0.9642	0.9929	0.1747	0.1705	More than 3600	15
Avg	0.9573	0.9919	0.1587	0.1427	> 3600	16.9

* The Mp approach requires more than 3600 s to provide an optimum result. The best result at 3600 s has been presented in this column.

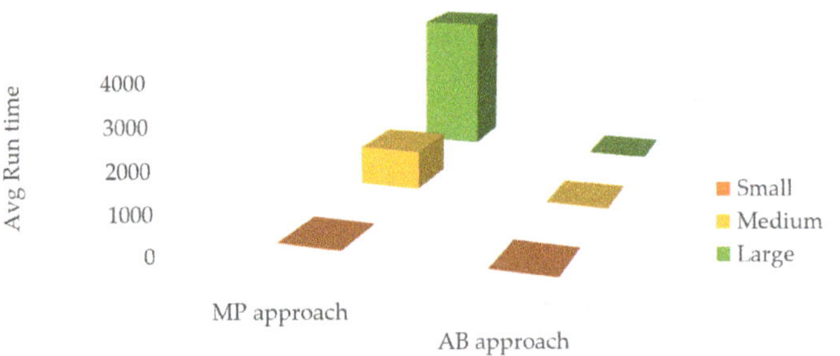

Figure 6. Average Run time for small, medium, and large instances for MP and AP approach.

4.3. Sensitivity Analysis of MP Approach

The proposed MP approach in this study aims to schedule the customer orders in a way that tries to satisfy the demands according to the company's beneficiaries. The company's beneficiary is specified through the penalty value for tardiness and overtime. The MP approach behaves differently with respect to the various value of penalties to establish the best balance. As the analysis of medium-sized instances has shown in Figures 7–10 by raising the penalty value for tardiness, the mathematical model struggles to compensate by applying more overtime. This conduct is also valid for overtime penalty value.

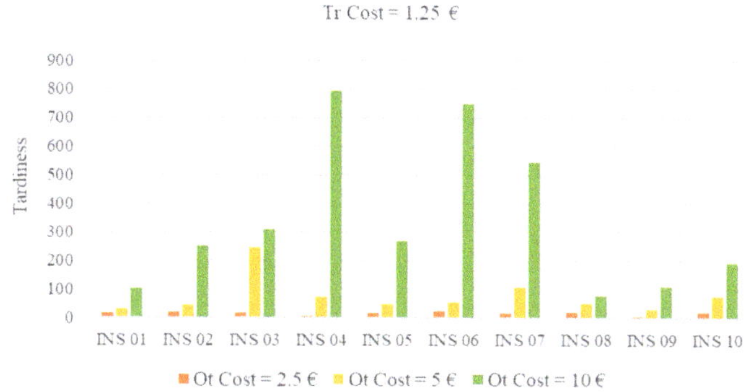

Figure 7. The amount of imposed "Tardiness" with respect to various overtime penalties.

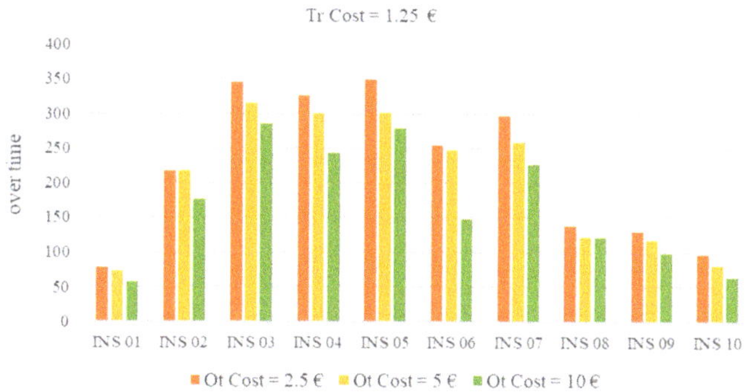

Figure 8. The amount of imposed "Overtime" with respect to various overtime penalties.

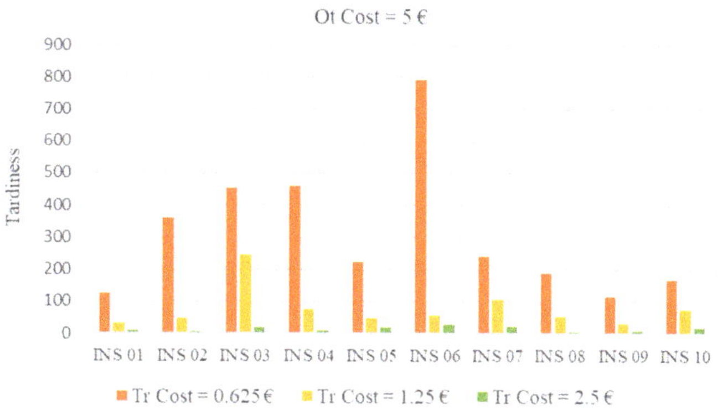

Figure 9. The amount of imposed "Tardiness" with respect to various tardiness penalties.

Figure 10. The amount of imposed "Overtime" with respect to various tardiness penalties.

4.4. Case Study and Sensitivity Analysis of AB Approach

According to the revealed information in Section 3.2, by increasing the dimension of instances, the required time for discovering an optimum solution by the MP approach will increase impressively. Since, in the real production environment, the number and the diversity of customer orders are far from what has been supposed previously, the case study was scrutinized in this section to unfold how the AB approach provides the solution for real-world production planning problems under various scenarios stemming from manager's priorities.

Here, the case of production planning for the "Table production factory" for a one-year time horizon has been reviewed. This company accepts up to 35 customer orders in each time bucket (each time bucket is equal to the length of the month). The production procedure of this company is split into several main operations, which could be attributed to fabricating legs and top, constructing frames, and assembling and colouring the main body. The design features of each piece could be customized according to the customer's desire. Figure 11 shows the Table production diagram.

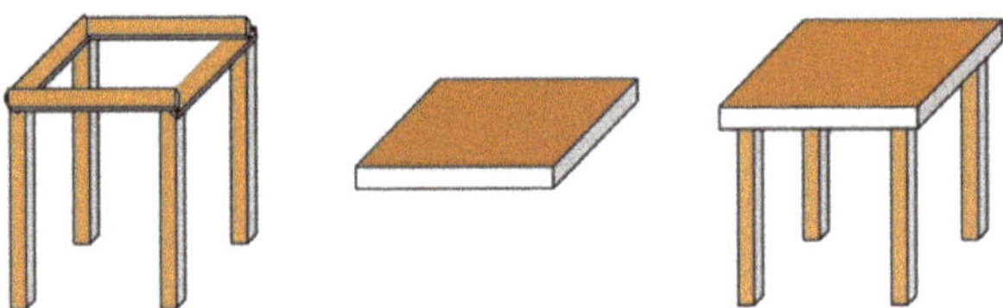

Figure 11. Production process diagram of "Table".

In order to assess the performance of the AB approach for the real problem, five different scenarios have been developed according to combinations of management's priorities to discover how the AB approach reacts to various circumstances. These scenarios have been presented in Table 5. One notable feature of the AB approach is that the earliness strategy could be initiated if the production factory contains enough storage space. Following this, the proposed solution by the AB approach could easily make itself compatible with the nature of the production environment.

Table 5. Scenarios for case study.

Scenario		Over Time		Tardiness		Earliness	
		High Priority	**Less Priority**	**High Priority**	**Less Priority**	**High Priority**	**Less Priority**
A	A1		✓	✓		is not activated	
	A2	✓			✓	is not activated	
B	B1		✓	✓		✓	
	B2	✓			✓	✓	
	B3	✓		✓			✓

In Table 5, scenario type A supposes that the factory does not contain enough storage. Hence, producing the customers' orders in advance is not possible, while strategy B assumes enough space for storing, and the company could anticipate production orders by imposing earliness. Strategy A and B include two and three subcategories, respectively. In each combination, one out of three criteria, including "Overtime", "Tardiness", and "Earliness," has a high level of priority for the manager. In contrast, the rest of them have less priority. Inside each combination, the AB approach tries to find the proper solution by imposing the planning burden on low-level priority criteria and avoiding high-level priority criterion occurrence.

Table 6 reports the computational result for the "case study" under various scenarios. As has been exhibited in this table and Figures 12 and 13, the reaction of the AB approach is adjusted according to what is desirable for managers.

Table 6. Computational results for "Case study".

	Scenario	Over Time (h)	Tardiness (h)	Earliness (h)	Run Time (s)
A	A1	2822	0	is not activated	19
	A2	660	2843	is not activated	20
B	B1	2822	0	308	19
	B2	563	2920	232	22
	B3	437	0	3655	23

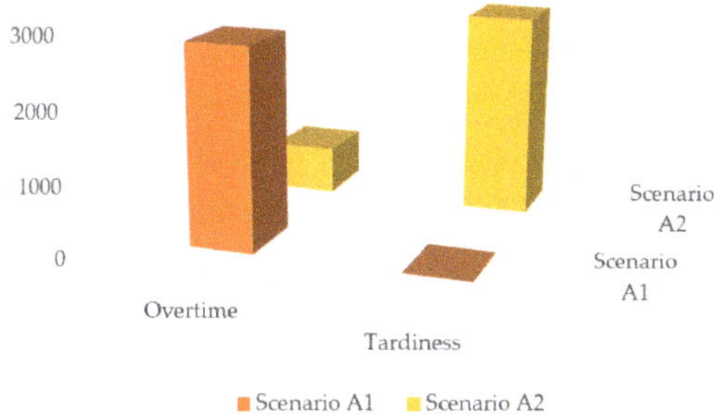

Figure 12. The comparison between Scenario A1 and A2.

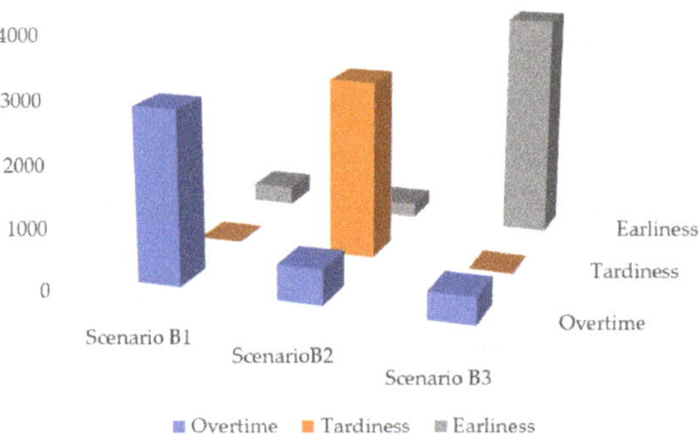

Figure 13. The comparison among Scenario B1, B2, and B3.

5. Conclusions

In today's technologically advanced world, the variety of products available to the market is rising. As a result, customers' expectations are always shifting and evolving. In this environment, if the manufacturers want to maintain and enhance their market share, they need to align their production strategies with customers' expectations. Among various production strategies with their own special pros and cons, MTO's production strategy strives to cover a wide range of customers' expectations by relying on customization. The proper solution for its MPS should be provided to make this alignment successful.

In addition to the complex nature of MTO, the fluctuation in resource availability in the real production environment, which has been ignored by previous studies, restricts the production activities and makes MPS more complicated, and the previous result not applicable. Following this, the main concentration of this study is proposing a proper approach to respond to this requirement.

To this end, the mixed-integer linear programming model was formulated, which is capable of presenting the optimum solution for small and medium-sized factories that receive limited number of orders, while for the large-sized factories, where the amount and diversity of client orders are far from the rest, the MP approach is no longer capable of providing the optimum solution reasonably. To overcome this weakness, an innovative AB heuristic has been developed which provides an effective and flexible MPS planning. The striking feature that distinguishes the AB approach from MP is that it could anticipate the production planning by considering earliness when the factory has enough space to store the final products until their due date. Computational experiments on variously sized problems confirm the proposed AB approach's efficiency, even for increasingly complex industrial instances in terms of flexibility and running time.

One outstanding point of this study is that a MPS which is defined through AB and MP approaches not only enables the businesses owner to maintain and raise their market share by maximizing customers' service level, but also improves resource utilization. Both of these fulfillments which address the managers' external and internal desires are accomplished by setting the goal of AB and MP approaches for MPS planning in the direction of minimizing tardiness and overtime cost.

This current study could be extended in various aspects. One limitation of this study is ignoring the operator availability for imposing the overtime on the resource. In some manufacturing environments, the production procedure is operators-dependent and overlooking this fact would make the results not applicable.

From the manufacturing perspective, this study could be extended for those factories that prepare a series of semi-finished products in advanced and assemble them according to customers' orders.

From the supply chain viewpoint, an inevitable crucial requirement for establishing successful production is supplying the raw materials at the right time and in the sufficient quantity. This prerequisite could be combined with the current problem in future research studies, and broader planning could be provided for the manufacturers.

Author Contributions: Conceptualization, F.B.; methodology, F.B. and M.D.; software, F.B.; validation, F.B. and A.A.; investigation, M.D.; writing—original draft preparation, F.B. and M.D.; writing—review and editing, F.B. and M.D.; supervision, F.T., M.P. and G.P.; funding acquisition, F.T. All authors have read and agreed to the published version of the manuscript.

Funding: This research was funded by Regione Puglia: Project 'Cooperative Supply Chain' (CoopSC).

Conflicts of Interest: The authors declare no conflict of interest.

References

1. Jans, R.; Degraeve, Z. Modeling Industrial Lot Sizing Problems: A Review. *Int. J. Prod. Res.* **2008**, *46*, 1619–1643. [CrossRef]
2. Karimi, B.; Fatemi Ghomi, S.M.T.; Wilson, J.M. The Capacitated Lot Sizing Problem: A Review of Models and Algorithms. *Omega* **2003**, *31*, 365–378. [CrossRef]
3. Mula, J.; Poler, R.; Garcia-Sabater, J.P. Capacity and Material Requirement Planning Modelling by Comparing Deterministic and Fuzzy Models. *Int. J. Prod. Res.* **2008**, *46*, 5589–5606. [CrossRef]
4. Mula, J.; Poler, R.; García-Sabater, J.P.; Lario, F.C. Models for Production Planning under Uncertainty: A Review. *Int. J. Prod. Econ.* **2006**, *103*, 271–285. [CrossRef]
5. Gansterer, M. Aggregate planning and forecasting in make-to-order production systems. *Int. J. Prod. Econ.* **2015**, *170*, 521–528. [CrossRef]
6. Demartini, M.; Evans, S.; Tonelli, F. Digitalization Technologies for Industrial Sustainability. *Procedia Manuf.* **2019**, *33*, 264–271. [CrossRef]
7. Rajagopalan, S. Make to order or make to stock: Model and application. *Manag. Sci.* **2002**, *48*, 241–256. [CrossRef]
8. Stevenson, M.; Hendry, L.C.; Kingsman, B.G. A review of production planning and control: The applicability of key concepts to the make-to-order industry. *Int. J. Prod. Res.* **2005**, *43*, 869–898. [CrossRef]
9. Mahdavisharif, M.; Cagliano, A.C.; Rafele, C. Investigating the Integration of Industry 4.0 and Lean Principles on Supply Chain: A Multi-Perspective Systematic Literature Review. *Appl. Sci.* **2022**, *12*, 586. [CrossRef]
10. Zobolas, G.I.; Tarantilis, C.D.; Ioannou, G. Extending Capacity Planning by Positive Lead Times and Optional Overtime, Earliness and Tardiness for Effective Master Production Scheduling. *Int. J. Prod. Res.* **2008**, *46*, 3359–3386. [CrossRef]
11. *Supply Chain Management and Advanced Planning*; Stadtler, H.; Kilger, C. (Eds.) Springer: Berlin/Heidelberg, Germany, 2008. [CrossRef]
12. Díaz-Madroñero, M.; Mula, J.; Peidro, D. A Review of Discrete-Time Optimization Models for Tactical Production Planning. *Int. J. Prod. Res.* **2014**, *52*, 5171–5205. [CrossRef]
13. Dolgui, A.; Ivanov, D.; Sethi, S.P.; Sokolov, B. Scheduling in Production, Supply Chain and Industry 4.0 Systems by Optimal Control: Fundamentals, State-of-The-Art and Applications. *Int. J. Prod. Res.* **2018**, *57*, 411–432. [CrossRef]
14. Englberger, J.; Herrmann, F.; Manitz, M. Two-Stage Stochastic Master Production Scheduling under Demand Uncertainty in a Rolling Planning Environment. *Int. J. Prod. Res.* **2016**, *54*, 6192–6215. [CrossRef]
15. Caner Taşkın, Z.; Tamer Ünal, A. Tactical Level Planning in Float Glass Manufacturing with Co-Production, Random Yields and Substitutable Products. *Eur. J. Oper. Res.* **2009**, *199*, 252–261. [CrossRef]
16. Inderfurth, K.; Lindner, G.; Rachaniotis, N.P. Lot Sizing in a Production System with Rework and Product Deterioration. *Int. J. Prod. Res.* **2005**, *43*, 1355–1374. [CrossRef]
17. Loukil, T.; Teghem, J.; Tuyttens, D. Solving Multi-Objective Production Scheduling Problems Using Metaheuristics. *Eur. J. Oper. Res.* **2005**, *161*, 42–61. [CrossRef]
18. Tometzki, T.; Engell, S. Systematic Initialization Techniques for Hybrid Evolutionary Algorithms for Solving Two-Stage Stochastic Mixed-Integer Programs. *IEEE Trans. Evol. Comput.* **2011**, *15*, 196–214. [CrossRef]
19. Amorim, P.; Günther, H.-O.; Almada-Lobo, B. Multi-Objective Integrated Production and Distribution Planning of Perishable Products. *Int. J. Prod. Econ.* **2012**, *138*, 89–101. [CrossRef]
20. Karimi-Nasab, M.; Aryanezhad, M.B. A Multi-Objective Production Smoothing Model with Compressible Operating Times. *Appl. Math. Model.* **2011**, *35*, 3596–3610. [CrossRef]
21. Karimi-Nasab, M.; Ghomi, S.M. Multi-Objective Production Scheduling with Controllable Processing Times and Sequence-Dependent Setups for Deteriorating Items. *Int. J. Prod. Res.* **2012**, *50*, 7378–7400. [CrossRef]
22. Mansouri, S.A.; Golmohammadi, D.; Miller, J. The Moderating Role of Master Production Scheduling Method on Throughput in Job Shop Systems. *Int. J. Prod. Econ.* **2019**, *216*, 67–80. [CrossRef]
23. Golmohammadi, D. A Neural Network Decision-Making Model for Job-Shop Scheduling. *Int. J. Prod. Res.* **2013**, *51*, 5142–5157. [CrossRef]

24. Guillaume, R.; Thierry, C.; Zieliński, P. Robust Material Requirement Planning with Cumulative Demand under Uncertainty. *Int. J. Prod. Res.* **2017**, *55*, 6824–6845. [CrossRef]
25. Sahin, F.; Powell Robinson, E.; Gao, L.-L. Master production scheduling policy and rolling schedules in a two-stage make-to-order supply chain. *Int. J. Prod. Econ.* **2008**, *115*, 528–541. [CrossRef]
26. Powell Robinson, E.; Sahin, F.; Gao, L.-L. Master production schedule time interval strategies in make-to-order supply chains. *Int. J. Prod. Res.* **2008**, *46*, 1933–1954. [CrossRef]
27. Sawik, T. Multi-objective master production scheduling in make-to-order manufacturing. *Int. J. Prod. Res.* **2007**, *45*, 2629–2653. [CrossRef]
28. Nedaei, H.; Mahlooji, H. Joint multi-objective master production scheduling and rolling horizon policy analysis in make-to-order supply chains. *Int. J. Prod. Res.* **2014**, *52*, 2767–2787. [CrossRef]
29. Yin, Y.; Stecke, K.E.; Li, D. The evolution of production systems from Industry 2.0 through Industry 4.0. *Int. J. Prod. Res.* **2018**, *56*, 848–861. [CrossRef]
30. Gu, X.; Koren, Y. Mass-Individualisation—The twenty first century manufacturing paradigm. *Int. J. Prod. Res.* **2022**. [CrossRef]
31. Kundu, K.; Rossini, M.; Portioli-Staudacher, A. Analysing the impact of uncertainty reduction on WL methods in MTO flow shops. *Prod. Manuf. Res.* **2018**, *6*, 328–344. [CrossRef]
32. Lee, C.H.; Chen, C.H.; Lin, C.; Li, F.; Zhao, X. Developing a Quick Response Product Configuration System under Industry 4.0 Based on Customer Requirement Modelling and Optimization Method. *Appl. Sci.* **2019**, *9*, 5004.
33. Rahman, H.F.; Janardhanan, M.N.; Nielsen, I.E. Real-Time Order Acceptance and Scheduling Problems in a Flow Shop Environment Using Hybrid GA-PSO Algorithm. *IEEE Access* **2019**, *7*, 112742–112755. [CrossRef]
34. Micieta, B.; Binasova, V.; Lieskovsky, R.; Krajcovic, M.; Dulina, L. Product Segmentation and Sustainability in Customized Assembly with Respect to the Basic Elements of Industry 4.0. *Sustainability* **2019**, *11*, 6057. [CrossRef]
35. Woschank, M.; Dallasega, P.; Kapeller, J.A. The Impact of Planning Granularity on Production Planning and Control Strategies in MTO: A Discrete Event Simulation Study. *Procedia Manuf.* **2020**, *51*, 1502–1507. [CrossRef]
36. Garey, M.R.; Johnson, D.S. *Computers and Intractability: A Guide to the Theory of NP-Completeness*; W.H. Freeman and Co.: New York, NY, USA, 2003.
37. Demartini, M.; Tonelli, F.; Pacella, M.; Papadia, G. A Review of Production Planning Models: Emerging Features and Limitations Compared to Practical Implementation. *Procedia CIRP* **2021**, *104*, 588–593. [CrossRef]
38. Pacella, M.; Papadia, G. Evaluation of Deep Learning with Long Short-Term Memory Networks for Time Series Forecasting in Supply Chain Management. *Procedia CIRP* **2021**, *99*, 604–609. [CrossRef]
39. Bernocco, M.; Schenone, M.; Queirolo, F.; Tonelli, F. A supervised multi-agent approach for aps in multi-site production systems for demand validation and evaluation. *Simul. Ser.* **2003**, *35*, 823–828.
40. Amaro, G.; Hendry, L.; Kingsman, B. Competitive Advantage, Customisation and a New Taxonomy for Non Make-To-Stock Companies. *Int. J. Oper. Prod. Manag.* **1999**, *19*, 349–371. [CrossRef]
41. Rohaninejad, M.; Sahraeian, R.; Nouri, B.V. Minimising the Total Cost of Tardiness and Overtime in a Resumable Capacitated Job Shop Scheduling Problem by Using an Efficient Hybrid Algorithm. *Int. J. Ind. Syst. Eng.* **2017**, *26*, 318. [CrossRef]

Article

Two-Dimensional, Two-Layer Quality Regression Model Based Batch Process Monitoring

Luping Zhao * and Xin Huang

College of Information Science and Engineering, Northeastern University, Shenyang 110819, China; hx1970617@163.com
* Correspondence: zhaolp@ise.neu.edu.cn; Tel.: +86-188-4255-2385

Abstract: In this paper, a two-dimensional, two-layer quality regression model is established to monitor multi-phase, multi-mode batch processes. Firstly, aiming at the multi-phase problem and the multi-mode problem simultaneously, the relations among modes and phases are captured through the analysis between process variables and quality variables by establishing a two-dimensional, two-layer regression partial least squares (PLS) model. The two-dimensional regression traces the intra-batch and inter-batch characteristics, while the two-layer structure establishes the relationship between the target process and historical modes and phases. Consequently, online monitoring is carried out for multi-phase, multi-mode batch processes based on quality prediction. In addition, the online quality prediction and monitoring results based on the proposed method and those based on the traditional phase mean PLS method are compared to prove the effectiveness of the proposed method.

Keywords: batch process; partial least squares; multi-phase; multi-mode; process monitoring

---

Citation: Zhao, L.; Huang, X. Two-Dimensional, Two-Layer Quality Regression Model Based Batch Process Monitoring. *Processes* 2022, *10*, 43. https://doi.org/10.3390/pr10010043

Academic Editor: Sergey Y. Yurish

Received: 4 December 2021
Accepted: 23 December 2021
Published: 27 December 2021

1. Introduction

Batch process is a way of production closely related to people's life in the modern process industry. It is now widely used in the production and preparation of small-batch and high value-added products to meet the rapidly changing market demand, such as in the production of fine chemical industry, food, polymer reaction, metal processing, biopharmaceutical, etc. Batch operation process characteristics are more complex and have richer data statistical characteristics than the continuous industrial process. Firstly, a batch cycle usually consists of several fixed phases, and different phases in each batch may show different potential behaviors, which is called multi-phase characteristic. Multi-phase characteristic is one kind of intra-batch characteristic evolution; it is one characteristic evolution along the time direction within each batch. Secondly, in the process of industrial production, due to the change of operating conditions, there are many different stable working points. Certain number of batches within the same working points belong to the same mode, which is called multi-mode characteristic. Multi-mode characteristic is one kind of inter-batch characteristic evolution; it is one characteristic evolution along the batch direction through the process. Multi-phase characteristic and multi-mode characteristic exist in batch processes simultaneously, which makes the batch processes complex and interesting for researchers.

At the end of the 1980s, principal component analysis (PCA) [1] and partial least squares (PLS) [2], as the focus of multivariate statistical modeling methods, began to be applied to statistical modeling, online monitoring, quality control, and fault diagnosis based on process data [3–5]. In order to maximize their effectiveness, it is necessary to extend these methods to batch processes. In the mid-1990s, Nomikos and MacGregor proposed multilinear principal component analysis (MPCA) and multiway partial least squares regression (MPLS) [6,7]. After, many international research groups invested a lot of manpower and material resources to carry out the research work for batch processes. In 1998, Wold et al. described an approach to multivariate batch process modeling and

---

monitoring, which was focused on following the evolution of the batch and oriented to the monitoring of the individual time points [8]. In 2000, a new methodology for analyzing batch and semi-batch process variable trajectories was proposed for process development and optimization, which was aimed at identifying trajectory features, such as cumulative effects and time-specific effects, of process variables on the final product quality [9]. In 2003, Hyun-Woo et al. proposed a new method for predicting the future observations of the batch that is currently being operated. In their work, the past batch trajectory which is deemed the most similar the new batch was selected from the batch library and used as the basis for predicting the unknown part of the new batch [10]. In 2004, Lu et al. proposed a stage-based sub-PCA modeling method for multistage batch processes, based on the recognition of a batch process may be divided into several operation stages [11]. In 2014, Zhao et al. analyzed the inter-batch evolution and proposed a process monitoring strategy based on inter-batch mode division, where reference windows were used to judge the variance of the process and identify new process modes [12].

In a great deal of the work, multi-phase and multi-mode problems are usually investigated and handled separately as two key characteristics of batch processes.

The multi-phase nature is an important nature of batch processes, and many research studies have been done for process monitoring and quality prediction of batch processes [13–15]. Phases have their own characteristics different from each other, which should be captured by different models. As the basic component of the batch process, phases are the research focus, and even more important than the production modes because phases are indispensable to finish one batch production and obtain the final product. Around the multi-phase characteristic, phase division [11], and uneven-durations [15] are the main problems that the researchers are interested in, and they have been resolved by different strategies. Recently, a multi-phase residual recursive model was established using each quality residual of the phase mean models to connect the contributions of the multiple phases together for quality prediction [16]. An evolutionary PLS method was proposed for process monitoring to deal with the calibration and modeling problems about operation switching [17].

Because of different production requirements, multiple production modes happen and should also be modeled correctly. Different from the continuous process, a mode for batch processes includes several batches, which is a classification along the batch direction. Integrated model [18] and specific model [19] are the traditional strategies to handle the multi-mode problem. Moreover, then, the relationship between modes was analyzed for mode division in process monitoring of time-varying batch processes [12]. In addition, mode relationship analysis has been improved for multi-mode batch process quality prediction [20]. Recently, a multi-mode Fisher discriminant analysis based process monitoring method was proposed to overcome the limitation of the single operation mode assumption [21].

As stated before, the multi-phase characteristic and the multi-mode characteristic are both important natures in batch processes of great significance for process monitoring and quality prediction. Although a great deal of work has been done around these two problems, they are usually investigated and handled separately in each work, or only one is the research focus, and the other is dealt with by some simple methods. That is, the previous analysis of the multi-phase and multi-mode problems tends to focus on only one direction, intra or inter batches, rather than processing simultaneously in both two directions. It has interested the author to deal with these two problems at the same time. Recently, a complete set of process modeling methods attempted to cover the process characteristic evolution problem in both the intra-batch and inter-batch directions [22]. Due to the complexity of the problem, the idea of the proposed method is firstly to handle the problem in two directions separately, and then combine the two strategies. This kind of approach is flexible to select appropriate modeling methods in two directions. However, choosing the method to deal with problems in two directions, respectively, rather than solving the problem from a unified perspective, to some extent, is easy to cause the separation of methods. It is hard to give a comprehensive explanation of the process characteristics under the two irrelevant methods. In addition, when the methods in both directions need to be jointed, the method structure is relatively complex and difficult to comprehend and implement. Therefore, it is necessary to further improve the existing methods by proposing a novel united strategy framework according to the two-dimensional evolution characteristics of the batch processes. That is, it is necessary to study the method of solving the process evolution in both directions from the overall perspective of the two-dimensional evolution of batch processes. Based on the above analysis, the research focus in this paper is to establish the overall framework of tracking the two-dimensional evolution based on a two-dimensional regression model. Moreover, the basic one-dimensional regression model used in the above framework should be selected appropriately to have the ability to establish the relationship between the target process and historical process (modes or phases). The between-mode quality analysis [20], which has a two-layer mode analysis structure, is adopted, and the object is changed to processes (modes or phases), rather than modes. By the two-layer regression analysis strategy, the relationship between the target process and historical processes will be established.

Therefore, in this work, by expanding two-layer regression analysis strategy from dealing with one direction process evolution into two direction process evolutions under the two-dimensional framework, both the intra-batch and inter-batch characteristics will be traced for multi-phase and multi-mode processes, based on which a process monitoring strategy is developed. The two-dimensional regression traces the intra-batch and inter-batch characteristics, while the two-layer structure establishes the relationship between the target process and historical modes and phases. Firstly, aiming at the multi-mode problem and the multi-phase problem simultaneously, the relations among modes and phases are captured through the analysis of the characteristics of each mode and each phase in the batch process by establishing a two-dimensional regression model. This model is expanded from the two-layer between-mode analyzing model which can extract useful quality-related information from historical modes. In the first layer of relation analysis, for all historical processes, different quality regression models are developed by PLS and according regression parameters are obtained. In addition, by applying these regression parameters to the process variables of the target process, a series of assumed quality predictions would be obtained. Then, in the second layer, PLS is conducted between the assumed quality predictions and real qualities of the target process to judge the relationship between the target process and the historical processes. Based on the two-dimensional, two-layer regression model, for the target process, quality prediction can be conducted using all available quality information underling both the intra-batch and inter-batch evolutions. Consequently, online monitoring is carried out for multi-phase, multi-mode batch processes based on quality prediction. In addition, the online quality prediction and monitoring results based on the proposed method and those based on the traditional method are compared to prove the effectiveness of the proposed method. In the traditional strategy [16,23], historical modes are considered together and modeled by PLS method to include more mode information and avoid the overfitting problem. However, for the multi-phase problem, several phase mean models are built for different phases, respectively. To be fair, all concerned modes are available for modeling in both the proposed method and the traditional method.

The remaining work of this paper includes the following aspects: firstly, the Section 2 introduces the establishment of two-dimensional, two-layer regression model and the principle of online monitoring. The Section 3 briefly introduces the characteristics of the injection molding process, and the application of the proposed method is illustrated by the online monitoring of the injection molding process. The Section 4 is conclusions.

2. Methodology

2.1. Batch Process Monitoring Based on Phase Mean PLS

Generally, process data of batch processes are stored in data matrix $\mathbf{X}(I \times J_x \times K)$, where I refers to the number of batches; J_x refers to the number of process variables; and K

refs to the sample times. Based on the phase characteristic, the average variable matrix of phase c can be obtained,

$$\mathbf{X}_c = \sum_{k=1}^{K_c} \mathbf{X}_{c,k} / K_c, \tag{1}$$

where K_c is the data length of phase c, and $\mathbf{X}_{c,k}$ is the time-slice process data matrix.

The PLS algorithm is used to decompose the average process data matrix $\mathbf{X}_c$ and quality data matrix $\mathbf{Y}$ linearly, and the model is as follows:

$$\mathbf{X}_c = \mathbf{T}_c \mathbf{P}_c^{\mathrm{T}} + \mathbf{E}_c = \sum_{a=1}^{A} \mathbf{t}_{c,a} \mathbf{P}_{c,a}^{\mathrm{T}} + \mathbf{E}_c, \tag{2}$$

$$\mathbf{Y} = \mathbf{U}_c \mathbf{Q}_c^{\mathrm{T}} + \mathbf{F}_c = \sum_{a=1}^{A} \mathbf{u}_{c,a} \mathbf{q}_{c,a}^{\mathrm{T}} + \mathbf{F}_c, \tag{3}$$

where $\mathbf{T}_c$ is the score matrix of $\mathbf{X}_c$, $\mathbf{U}_c$ is the score matrix of $\mathbf{Y}$, $\mathbf{P}_c$ is the load matrix of $\mathbf{X}_c$, $\mathbf{Q}_c$ is the load matrix of $\mathbf{Y}$, $\mathbf{E}_c$ is the fitting error matrix of $\mathbf{X}_c$, and $\mathbf{F}_c$ is the fitting error matrix of $\mathbf{Y}$.

The previous model can be expressed by the regression model as:

$$\hat{\mathbf{Y}}_c = \mathbf{X}_c \mathbf{B}_c, \tag{4}$$

where $\mathbf{B}_c$ is the regression parameter matrix, and $\hat{\mathbf{Y}}_c$ is the predicted quality. When considering a single quality variable $\mathbf{y}(I \times 1)$, the regression model can be simply expressed as:

$$\hat{\mathbf{y}}_c = \mathbf{X}_c \boldsymbol{\beta}_c, \tag{5}$$

where $\boldsymbol{\beta}_c$ is the regression parameter.

In this paper, PLS model is used for online monitoring. The commonly used statistics are Hotelling-T^2 statistics and square prediction error SPE statistics. Hotelling-T^2 statistics reflects the deviation degree of latent variables from the established model in amplitude and process data development trend. SPE describes the deviation degree of the measured value of the input variable from the latent variable space in the batch process [13].

In the online monitoring of batch process, the current operating condition can be determined by observing the Hotelling-T^2 and the control limit of the square prediction error SPE. The definitions of T^2 statistics and SPE statistics are as follows:

$$T_c^2 = \mathbf{x}_c^{\mathrm{T}} \mathbf{R}_c \left(\frac{\mathbf{T}_c^{\mathrm{T}} \mathbf{T}_c}{I-1} \right)^{-1} \mathbf{R}_c^{\mathrm{T}} \mathbf{x}_c, \tag{6}$$

$$SPE_c - \|\widetilde{\mathbf{x}}_c\|^2 = \left\| (\mathbf{I}_{J_x} - \mathbf{P}_c \mathbf{R}_c^{\mathrm{T}}) \mathbf{x}_c \right\|^2, \tag{7}$$

where $\widetilde{\mathbf{x}}_c$ is the residual vector, $\mathbf{R}_c = \mathbf{W}_c (\mathbf{P}_c^{\mathrm{T}} \mathbf{W}_c)^{-1}$, and $\mathbf{W}_c$ is the weight matrix. The detailed properties and calculations can be found in reference [8].

The corresponding control limits are:

$$T_{c,\alpha}^2 = \frac{H(I^2-1)}{I(I-H)} F_{c,\alpha}(H, I-H), \tag{8}$$

$$SPE_{c,\alpha} = g_c \chi_{c,h,\alpha}^2, \tag{9}$$

where $F_{c,\alpha}(H, I-H)$ means the F distribution with the confidence level α and the degrees of freedom H and $I-H$, and H refers to the number of retained latent variables; $g_c \chi_{c,h,\alpha}^2$ means the χ^2 distribution with the confidence level α and the proportional coefficient $g_c = s_c / 2\mu_c$; $h_c = 2\mu_c^2 / s_c$; μ_c refers to the mean value of SPE_c; s_c is the variance of SPE_c.

After obtaining the $\mathbf{x}_{c,k}$ vector at the k-th moment, the online T^2 statistics and online *SPE* statistics are calculated, and the calculation formula is as follows:

$$T_{c,k}^2 = \mathbf{x}_{c,k}^{\mathrm{T}}\mathbf{R}_c\left(\frac{\mathbf{T}_c^{\mathrm{T}}\mathbf{T}_c}{I-1}\right)^{-1}\mathbf{R}_c^{\mathrm{T}}\mathbf{x}_{c,k}, \tag{10}$$

$$SPE_{c,k} = \|\widetilde{\mathbf{x}}_{c,k}\|^2 = \|(\mathbf{I}_{J_x} - \mathbf{P}_c\mathbf{R}_c^{\mathrm{T}})\mathbf{x}_{c,k})\|^2, \tag{11}$$

where $\widetilde{\mathbf{x}}_{c,k}$ is the residual vector at the k-th moment.

2.2. Framework of Two-Dimensional, Two-Layer Regression Modeling Strategy

To perform process monitoring based on the three-dimensional data matrix of the batch process, in this paper, a novel two-dimensional, two-layer regression modeling strategy is established, where the process evolution that exists in both directions within and between batches can be tracked, and important quality information in historical processes can be extracted.

In the process of batch process modeling and online application, the most direct embodiment of the process characteristics is the data, and the main characteristic of the batch process is the three-dimensional data matrix. Extraction and analysis of process characteristics from the data is the basis of subsequent process modeling and process analysis. On the basis of understanding the characteristics of the actual three-dimensional data matrix of the batch process, appropriate scientific methods should be adopted to extract the process evolution information intelligently and accurately, so as to deeply understand the characteristics of the process and further establish the process model. The analysis of process evolution characteristics reflects the respective characteristics in both intra-batch direction and inter-batch direction but also the consistency of the two directions from the three-dimensional data matrix.

The intra-batch evolution problem and inter-batch evolution problem are shown in Figure 1. In early work, the researchers proposed different methods to track the intra-batch evolution and inter-batch evolution, respectively. In general, these methods are individual and can be selected simultaneously in evolution analysis and used for modeling in one of the two directions, as shown in Figure 1a. However, it is relatively complex to choose different methods when processing jointly in both directions. In addition, the most important is that the evolution nature in the both directions is the same. Although the evolutions in the two directions may be raised because of different physical significances, they are both process characteristic evolutions. It is a straightforward idea to adopt one method to handle both problems in the two directions. Therefore, a thinking framework is proposed, that is, expanding the method to inter-batch evolution based on intra-batch evolution, and maintaining the unity of the method under a reasonable premise, so as to achieve the method simplification. It should be noted that the model building method should follow the physical significances of both the inter-batch evolution and the intra-batch evolution during application. Following this framework, the proposed methods in both directions are consistent in core ideas and will be more convenient when uniting.

Secondly, from the overall perspective of batch process three-dimensional data matrix, the other kind of method can solve the process evolution in two directions simultaneously. As shown in Figure 1b, this kind of method can be split into two directions, and performs under the thought of handling the whole first, and dealing with the parts later. This kind of strategy adopts a unified core algorithm to extract and analyze the evolution in two directions. The core method here should be as suitable for the tracking of process evolution characteristics as much as possible. Attention is paid to the use of intelligent analysis method to avoid computational redundancy and reflect the superiority of the algorithm starting from the 3-dimensional data matrix.

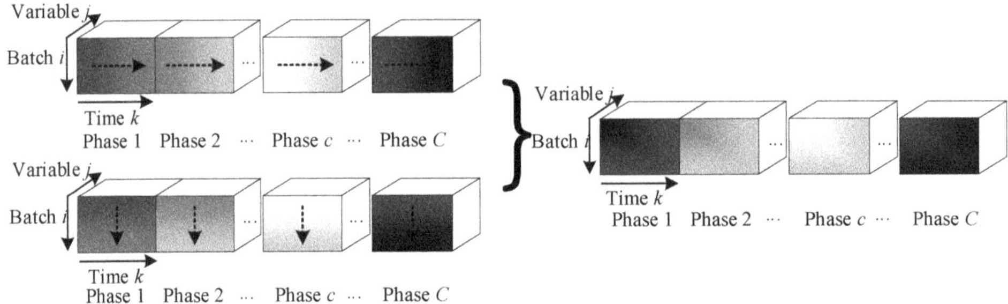

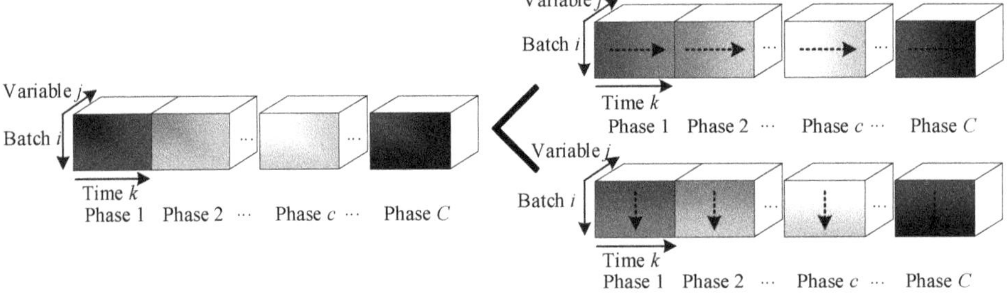

Figure 1. Process modeling based on 3-D data matrix analysis.

Thus, this work is to explore more effective data modeling methods on the basis of preliminary work and trace the intra-batch evolution and inter-batch evolution simultaneously, as shown in Figure 1a, while, for the strategy shown in Figure 1b, the research will be talked about in later works. Next, the modeling method based on two-layer multi-mode regression analysis will be expanded to deal with the multi-mode and multi-phase problem at the same time, and a two-dimensional, two-layer regression model will be proposed.

In the multi-mode batch process, when a new mode occurs, the single mode model will no longer match the target process, resulting in serious overrun of T^2 and SPE statistics when monitoring. Therefore, the statistical modeling and online monitoring of the batch process should not only consider a single mode but also carry out the quality analysis of multiple modes.

The main strategy to deal with the multi-mode problem is to extract the relationship between the historical modes and the target process. By doing this, more between-mode information would be captured by the model, and the multi-mode process can be better predicted and monitored. In this two-layer regression strategy, first, the PLS regression models are built for historical modes. Then, using those regression models and the process variables of the target process, a series of assumed quality predictions can be obtained in the first regression layer. In the second layer, based on these assumed quality predictions and real quality values of the target process, the PLS regression model is established for quality prediction.

To adopt the same strategy to handle the multi-mode problem and the multi-phase problem at the same time, the strategy above will be applied to the intra-batch batch direction and handle the multi-phase problem by analyzing the intra-batch evolution. The framework of the two-dimensional, two-layer regression modeling strategy is shown in Figure 2. In the first regression layer, first, the PLS regression models are built between the process variables and quality variables of the historical processes with phase and mode indexes to get regression parameters. Then, using those regression parameters and the process variables of the target process, a series of assumed quality predictions can be obtained. Here, it is assumed that the target process belongs to each modeled process. This step connects all historical process information with the target process. Whether the information captured is related to the final quality of the target process will be judged in next regression layer. Consequently, in the second regression layer, based on these assumed quality predictions and real quality values of the target process, the PLS regression model is established, which captures the valuable information existing in the historical processes related to the final quality of the target process. Finally, by applying the two-dimensional, two-layer regression model, the final prediction quality can be obtained.

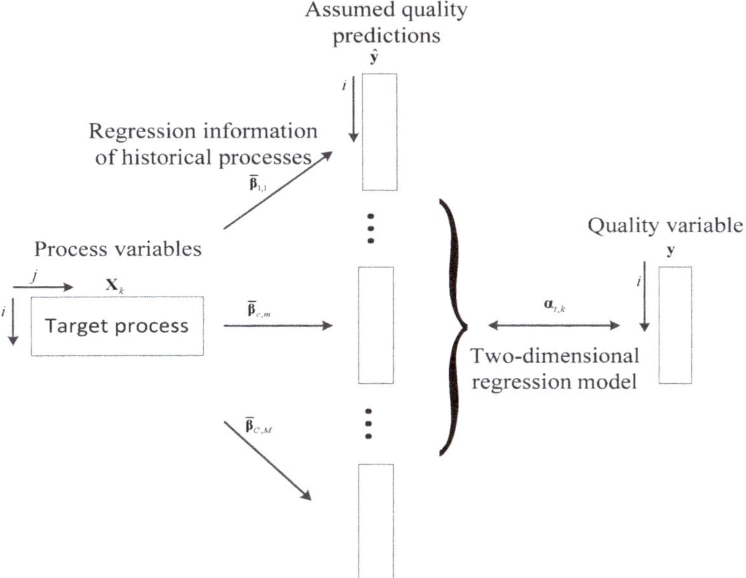

Figure 2. Illustration of two-dimensional, two-layer regression modeling.

In the proposed method, the two-dimensional regression, which is mainly reflected by the regression parameters of the historical processes with both phase and mode indexes, traces the intra-batch and inter-batch characteristics, while the two-layer structure establishes the relationship between the target process and historical processes. Compared with the between-mode model, the two-dimensional, two-layer regression model handles the inter-batch evolution and intra-batch evolution simultaneously, and, at the same time, maintaining the unity of the method. However, the between-mode method deals with only the inter-batch multi-mode problem, which means other methods dealing with the intra-batch multi-phase problem or phase information are necessary. Compared with the traditional strategy [16,23], where historical modes are considered together and modeled by phase mean PLS method, and several phase mean models are built for different phases, respectively, the proposed method mixes all process mode and phase information together in the first regression layer, then using the second regression layer to extract valuable information for the target process, which is more reasonable to use regression method to judge the relationship between the processes, rather than excluding the information by phase division.

The details of the proposed two-dimensional, two-layer quality regression modeling strategy are introduced in the next section.

2.3. Two-Dimensional, Two-Layer Regression Model

In this section, after establishing the models representing the quality related information of each historical processes, the regression relationship between the process variables and the quality of the target process is analyzed through the regression model of the historical phases of historical modes, and, on this basis, online monitoring is carried out.

Firstly, the models representing the quality related information of each historical processes are established. Within phase c of mode m, $X_{c,m,k}(I_{c,m} \times J_x)$ and $\mathbf{Y}_{c,m}$ are the normalized time-slice process variables and quality variables. The PLS algorithm is used to capture the relationship of the process data matrix $X_{c,m,k}$ and quality data matrix $\mathbf{Y}_{c,m}$, and the model is as follows:

$$X_{c,m,k} = \mathbf{T}_{c,m,k}\mathbf{P}_{c,m,k}^{\mathrm{T}} + \mathbf{E}_{c,m,k},\tag{12}$$

$$\mathbf{Y}_{c,m} = \mathbf{U}_{c,m,k}\mathbf{Q}_{c,m,k}^{\mathrm{T}} + F_{c,m,k},\tag{13}$$

where $\mathbf{T}_{c,m,k}$ and $\mathbf{U}_{c,m,k}$ are the score matrices, $\mathbf{P}_{c,m,k}$ and $\mathbf{Q}_{c,m,k}$ are the loading matrices, and $\mathbf{E}_{c,m,k}$ and $\mathbf{F}_{c,m,k}$ are the residual matrices.

The prediction model can be expressed as below:

$$\hat{\mathbf{Y}}_{c,m,k} = X_{c,m,k}\mathbf{B}_{c,m,k}^{\mathrm{T}}\tag{14}$$

where $\mathbf{B}_{c,m,k}$ is the regression parameter matrix.

When only one final quality variable is considered, the prediction model can be written as:

$$\hat{y}_{c,m,k} = X_{c,m,k}\beta_{c,m,k},\tag{15}$$

where $\beta_{c,m,k}$ is the regression parameter vector.

So, the regression coefficient of historical phases $\overline{\beta}_{c,m}$ is obtained:

$$\overline{\beta}_{c,m} = \sum_{k=1}^{K_c} \beta_{c,m,k}/K_c,\tag{16}$$

where K_c stands for the number of time-slices within phase c of mode m, m stands for the number of the historical modes, $m = 1, 2, \ldots, M$, and c stands for the historical phases in each batch, $c = 1, 2, \ldots, C$.

For the target process, $X_{t,k}(I_t \times J_x)$ and $\mathbf{y}_t$ are the normalized time-slice process variables and quality variable.

The regression model and the process variables of the target process are used to obtain the assumed quality prediction of the target process:

$$\hat{y}_{t,c,m,k} = X_{t,k}\overline{\beta}_{c,m},\tag{17}$$

where t stands for the target process, m stands for the number of the historical modes, $m = 1, 2, \ldots, M$, and c stands for the number of historical phases, $c = 1, 2, \ldots, C$. It should be noted that $\hat{y}_{t,c,m,k}$ is called the assumed quality prediction, which means it is obtained by assuming that the process variables belong to the certain phase of the certain mode with the regression parameter $\overline{\beta}_{c,m}$. By obtaining the assumed quality predictions, the quality information of historical phases and modes is shared by the target process. Further, the quality information of historical phases and modes will be judged and extracted by the next regression.

Then, the relationship between the assumed quality predictions and the quality data of the target process will be established. All these assumed predictions of the historical modes can comprise a new matrix $\mathbf{Z}_{t,k}(I_t \times MC)$, $\mathbf{Z}_{t,k} = \left[\hat{y}_{t,1,1,k},\ldots,\hat{y}_{t,m,1,k},\ldots,\hat{y}_{t,M,1,k},\ldots,\right.$ $\hat{y}_{t,1,c,k},\ldots,\hat{y}_{t,m,c,k},\ldots,\hat{y}_{t,M,c,k},\ldots,\hat{y}_{t,1,C,k},\ldots,\hat{y}_{t,m,C,k},\ldots,\hat{y}_{t,M,C,k}\left.\right]$. Then, the kth time-slice PLS regression model is built between $\mathbf{Z}_{t,k}$ and $\mathbf{y}_t$ as follows:

$$\begin{aligned} \mathbf{Z}_{t,k} &= \mathbf{T}_{t,k}\mathbf{P}_{t,k}^{\mathrm{T}} + \mathbf{E}_{t,k} \\ \mathbf{y}_t &= \mathbf{u}_{t,k}\mathbf{q}_{t,k}^{\mathrm{T}} + \mathbf{f}_{t,k} \end{aligned} \tag{18}$$

where $\mathbf{T}_{t,k}$ and $\mathbf{u}_{t,k}$ are the score matrix and vector of the target process, $\mathbf{P}_{t,k}$ and $\mathbf{q}_{t,k}$ are the loading matrix and vector of the target process, and $\mathbf{E}_{t,k}$ and $\mathbf{f}_{t,k}$ are the residual matrix and vector of the target process. Then, novel predictions are obtained:

$$\hat{\mathbf{y}}_{t,k}^* = \mathbf{Z}_{t,k}\boldsymbol{\alpha}_{t,k}, \tag{19}$$

where $\boldsymbol{\alpha}_{t,k}$ is the regression parameter of the kth time-slice model.

The mean regression parameters of the target process can be obtained from the regression parameters of the time-slice models,

$$\overline{\boldsymbol{\alpha}}_t = \frac{1}{K_t}\sum_{k=1}^{K_t}\boldsymbol{\alpha}_{t,k}, \tag{20}$$

where K_t is the number of the time intervals within phase c. Then, the predictions based on the regression parameter of the whole phase, $\overline{\boldsymbol{\alpha}}_t$, are obtained:

$$\overline{\hat{\mathbf{y}}}_{t,k}^* = \mathbf{Z}_{t,k}\overline{\boldsymbol{\alpha}}_t. \tag{21}$$

Corresponding coefficients can be obtained based on the assumed prediction quality obtained from the above analysis:

$$\overline{\mathbf{T}}_t = \frac{1}{K_t}\sum_{k=1}^{K_t}\mathbf{T}_{t,k}, \tag{22}$$

$$\overline{\mathbf{P}}_t = \frac{1}{K_t}\sum_{k=1}^{K_t}\mathbf{P}_{t,k}, \tag{23}$$

$$\overline{\mathbf{W}}_t = \frac{1}{K_t}\sum_{k=1}^{K_t}\mathbf{W}_{t,k}, \tag{24}$$

$$\overline{\mathbf{R}}_t = \overline{\mathbf{W}}_t(\overline{\mathbf{P}}_t\overline{\mathbf{W}}_t)^{-1}. \tag{25}$$

After obtaining the $\mathbf{z}_{t,k}$ vector at the k-th moment, the online T^2 statistics and online SPE statistics are calculated, and the calculation formula is as follows:

$$T_{t,k}^2 = \mathbf{z}_{t,k}^{\mathrm{T}}\overline{\mathbf{R}}_t\left(\frac{\overline{\mathbf{T}}_t^{\mathrm{T}}\overline{\mathbf{T}}_t}{I-1}\right)^{-1}\overline{\mathbf{R}}_t^{\mathrm{T}}\mathbf{z}_{t,k}, \tag{26}$$

$$SPE_{t,k} = \|\tilde{\mathbf{z}}_{t,k}\|^2 = \|(\mathbf{I}_{J_x} - \overline{\mathbf{P}}_t\overline{\mathbf{R}}_t^{\mathrm{T}})\mathbf{z}_{t,k}\|^2, \tag{27}$$

where $T_{t,k}^2$ is the T^2 statistic of the current k-th moment, and $SPE_{t,k}$ is the SPE statistic of the current k-th moment; $\tilde{\mathbf{z}}_{t,k}$ is the residual vector at the k-th moment.

The corresponding control limits are calculated similarly as Equations (8) and (9).

3. Illustration and Discussions

3.1. Introduction of Injection Molding Process

Injection molding a typical batch process, which is one of the important technologies of plastic processing. To obtain plastic products, an injection molding process mainly consists of mold closing, injection, packing-holding, plasticizing, cooling, mold opening, part ejection, and other processes. Among those phases, injection, packing-holding, plasticizing, and cooling are the most important four operation phases to determine the quality of products: in the injection phase, the molten plastic is injected into the mold; then, in the packing-holding phase, a certain pressure is maintained to fulfill the hollows which may occur due to the temperature decrease; consequently, in the plasticizing phase, the raw material, i.e., plastic particles, is transported forward, plasticized, and melted; finally, in the cooling phase, the plastic is cooled in the mold until the product becomes sufficiently rigid for ejection. As the development of computers, all important process variables can be obtained online by high-precision sensors.

The material used in this experiment is high density polyethylene (HDPE). The quality analyzed in this experiment is the weight of injection molded parts. The operating conditions are shown in Table 1. The selected process variables are shown in Table 2, which are used to establish the model. The variable data shown in the table can be collected by sensors. The mode information of this experiment includes the packing pressure (PP) and the barrel temperature (BT). Five different modes are obtained. The experimental conditions are shown in Table 3. The data used in the modeling process are all real experiment data.

Table 1. Operating condition settings for injection molding process.

Operating Parameter	Set Value
Material	High density polyethylene (HDPE)
Packing pressure	25 Bar, 30 Bar, 35 Bar
Packing time	3 s
Mold cooling water temperature	25 °C
Injection velocity	24 mm/s
Barrel temperature	180 °C, 200 °C
Cooling time	15 s

Table 2. Process variables of injection molding process in start-up process.

Number	Variable Description	Unit
1	Screw speed	Mm/s
2	Plasticizing pressure	Bar
3	Nozzle temperature	°C
4	Cylinder pressure	Bar
5	SV2 valve opening	%
6	SV1 valve opening	%

Table 3. Different operation modes caused by different packing pressure (PP) and different barrel temperature (BT).

Modes	PP/Bar	BT/°C
Mode 1	25	180
Mode 2	35	180
Mode 3	25	200
Mode 4	30	200
Mode 5	35	200

3.2. Normal Batch Monitoring

The two-dimensional, two-layer regression model proposed in this paper can track the process evolution that exists in both directions within and between batches and extract important quality information from historical processes. To illustrate the superiority of this method, real injection molding process data with both multi-mode and multi-phase characteristics are used for analyzing.

In this simulation, 18 batches in mode 1, mode 2, and mode 3, 22 batches in mode 4, and 17 batches in mode 5 are selected, respectively, as normal batches of historical modes. Mode 3, with 18 batches, is used as the target mode for modeling. Five batches of mode 3 are selected as the normal test batches. During regression modeling, the four-fold cross-validation method is used to determine the number of reserved latent variables of the traditional method and the proposed method, which is 2 after analyzation. The confidence level of α is set to 0.99.

To illustrate the advantages of the proposed method, it is compared with the traditional multi-mode and multi-phase strategy. In this kind of traditional strategy, historical modes are considered together and modeled by phase mean PLS method, and several models are built for different phases, respectively. To be fair, all concerned modes are available for modeling in both the proposed method and the traditional method.

The mean RMSE of the quality predictions of the test batches from mode 3 by the traditional method is 0.0077, while the mean RMSE obtained by the proposed method is 0.0071, so the proposed method provides a more accurate prediction. The RMSE values of online quality predictions of the test batches using the two methods are shown in Figure 3. The dot dashed line is the RMSE value of the traditional method, the full line is the RMSE value of the proposed method, and vertical dotted lines are the phase boundaries. It can be seen from Figure 3 that the RMSE values of the traditional method have obvious phase variation, which indicates that the precision of the traditional method may be affected by different phase models. Therefore, the proposed method provides better quality prediction results than the proposed method. The RMSE values of the traditional method reflect obvious different phase characteristics because, in the traditional method, four models have been established for the four phases, while, for the proposed method, the RMSE values are not only lower than those of the traditional method within each phase but also keep a continuous low value throughout the four phases. This is because the proposed method mixes all process mode and phase information together in the first regression layer and then uses the second regression layer to extract valuable information for the target process. The results prove that using the two-layer regression method to judge the relationship between the target process and the historical processes is better than the traditional method based on phase division.

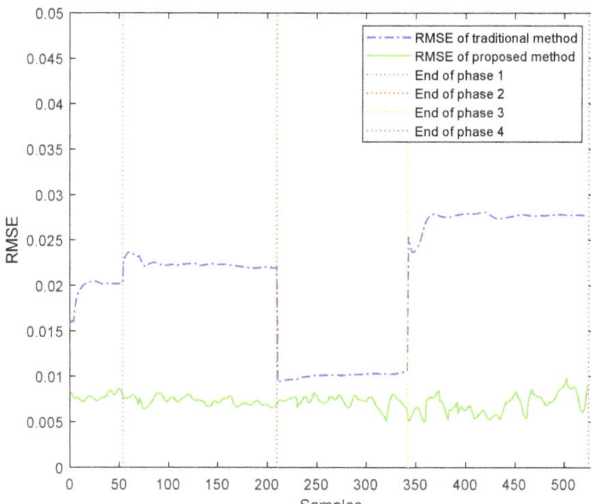

Figure 3. RMSE values of online quality predictions.

Further, one of the five test batches from mode 3 is selected to show the online quality prediction and monitoring results. The online quality prediction results of one test batch are shown in Figure 4. Similar conclusions can be obtained from Figure 4 with those from Figure 3, which proves the correctness of the conclusion from Figure 3.

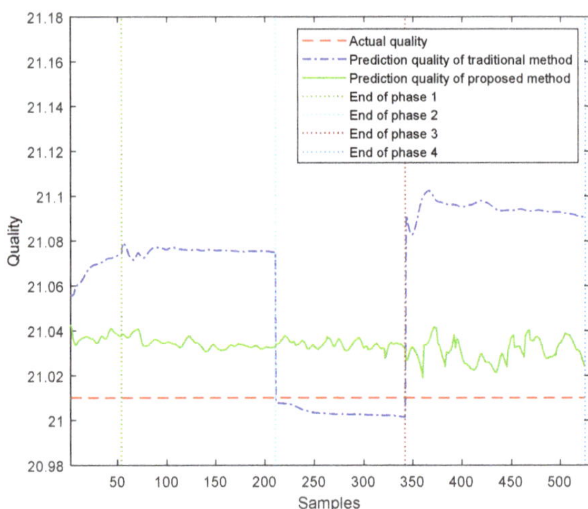

Figure 4. Online quality prediction results of one test batch.

The monitoring results of the four phases of one normal test batch from mode 3 are shown in Figures 5–8, respectively. This batch, which has some fluctuation, may be identified as anormal batch by the traditional method, and false alarm will occur due to the process fluctuation. While, by the propose method, this normal batch can be normally monitored. The details are as follows: In Figures 5 and 6, it can be seen that both T^2 and SPE of the proposed method and the traditional method do not exceed the control limits in the injection phase and the packing-holding phase. In Figure 7, it can be seen that, in the plasticizing phase, T^2 and SPE of the proposed method do not exceed their respective control limits, while, for the traditional method, although T^2 does not exceed its control limit, there is an obvious period at the beginning of the phase during which the SPE value exceeds its control limit. In fact, the fluctuation of this batch is caused by the transition between the packing-holding phase and the plasticizing phase. During the transition, process characteristic change from the packing-holding phase characteristic to the plasticizing phase characteristic. This characteristic fluctuation is natural progress and does not affect the production, so this batch is normal. Because the traditional method which uses the phase mean model to represent the phase characteristic is more easily affected by the characteristic fluctuation away from the phase mean characteristic, this fluctuation is identified as a fault by the traditional method. However, this is not consistent with the conclusion from the real production. Because the proposed method comprehensively deals with the multi-phase and multi-mode problems by involving all quality-related phase and mode information in historical processes in the regression model, it is not easily affected by phase fluctuation, and it can offer the right monitoring result, in which this batch is normal during the plasticizing phase. In Figure 8, there is one point of SPE value of the proposed method exceeds the control limit. Generally, this point will not be considered as a fault. Therefore, it can be concluded from the monitoring results of the normal test batch that the proposed method is better than the traditional method because, by the two-dimensional, two-layer quality regression, all quality-related phase and mode information in historical processes have been extracted and utilized for quality prediction and monitoring.

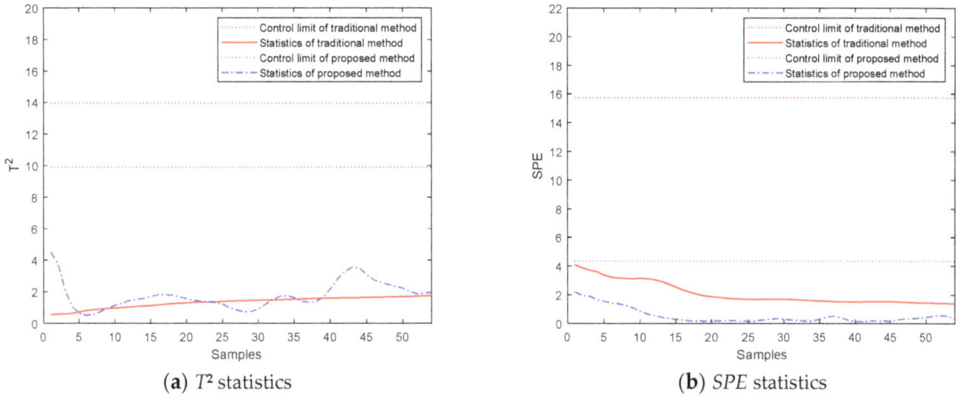

(a) T² statistics   (b) SPE statistics

Figure 5. Online monitoring of the injection phase.

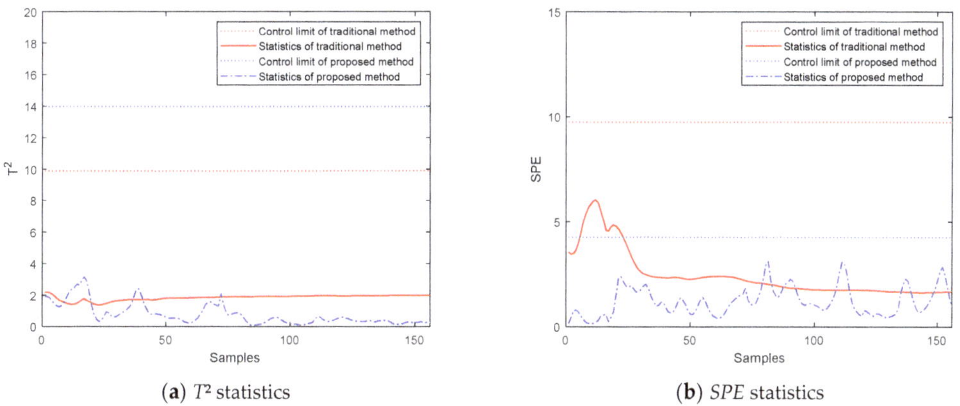

(a) T² statistics   (b) SPE statistics

Figure 6. Online monitoring of the packing-holding phase.

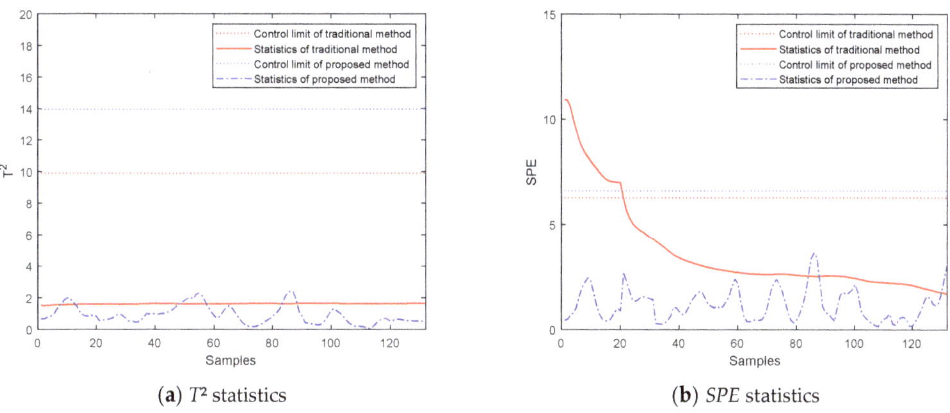

(a) T² statistics   (b) SPE statistics

Figure 7. Online monitoring of the plasticizing phase.

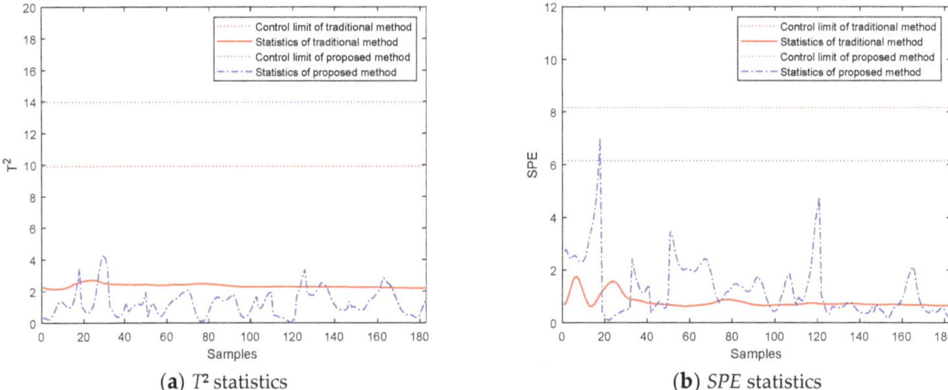

(a) *T²* statistics　　　　　　**(b)** *SPE* statistics

Figure 8. Online monitoring of the cooling phase.

3.3. Abnormal Batch Monitoring

In the injection molding process, the sensor fault is one of the common faults. Because of the sensor fault, some important process variable cannot be detected, the control system cannot know the real progress, and wrong control commands may be delivered, leading to serious dangerousness of production.

To simulate the sensor fault, a test batch with the pressure variable removed during the packing-holding phase is monitored. The T^2 and *SPE* monitoring effects of the traditional method and the proposed method are shown in Figure 9, respectively. Compared with the traditional method, the amplitudes of the statistics of the proposed method are relatively larger. It can be concluded that the sensitivity of the proposed method is not affected, although it involves all quality-related phase and mode information in historical processes in the regression model, and it can identify the fault batch as fast as the traditional method, with an even larger alarm signal. So, the proposed method is better than the traditional method.

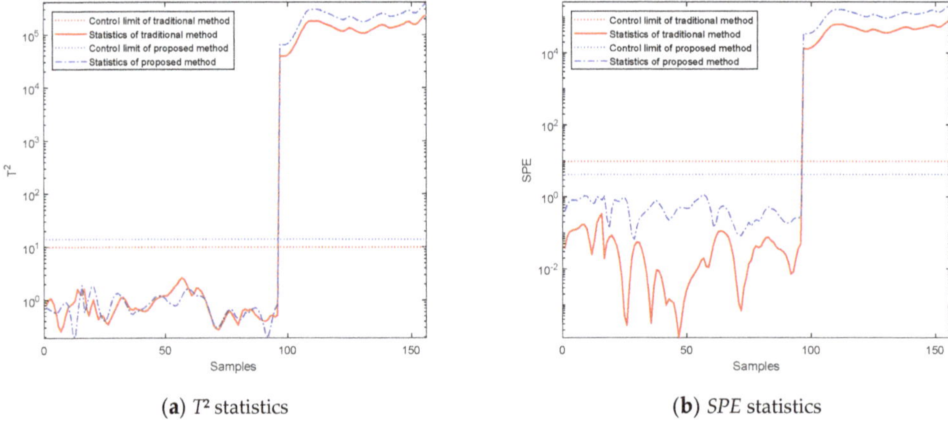

(a) *T²* statistics　　　　　　**(b)** *SPE* statistics

Figure 9. Online monitoring of the sensor fault.

4. Conclusions

In this work, a two-dimensional, two-layer quality regression model is established to monitor multi-phase, multi-mode batch processes. First, traditional batch process monitoring method based on phase mean PLS is introduced. Then, aiming at the multi-mode problem and the multi-phase problem simultaneously, the framework of two-dimensional, two-layer quality regression modeling strategy is proposed by expanding the multi-mode analysis strategy into the multi-phase problem and dealing with the intra-batch evolution and inter-batch evolution at the same time. Accordingly, the two-dimensional, two-layer regression model is established, based on which the quality predictions are obtained, and the process is monitored. Through the application to an injection molding process, it is proved that the proposed methodology can provide better quality prediction and process monitoring results for multi-phase, multi-mode batch processes than the traditional method.

Author Contributions: Conceptualization, L.Z. Data curation, L.Z.; Formal analysis, L.Z.; Funding acquisition, L.Z.; Investigation, L.Z.; Methodology, L.Z. and X.H.; Project administration, L.Z.; Resources, L.Z.; Software, L.Z. and X.H.; Supervision, L.Z.; Validation, L.Z.; Visualization, L.Z.; Writing—original draft, L.Z.; Writing—review & editing, L.Z. All authors have read and agreed to the published version of the manuscript.

Funding: This research was funded by the National Natural Science Foundation of China (No. 61503069) and the Fundamental Research Funds for the Central Universities (N150404020).

Institutional Review Board Statement: Not applicable.

Informed Consent Statement: Not applicable.

Data Availability Statement: The data presented in this study are openly available in https://share.weiyun.com/Jbc9pbWF.

Acknowledgments: This work is supported in part by the National Natural Science Foundation of China (No. 61503069) and the Fundamental Research Funds for the Central Universities (N150404020).

Conflicts of Interest: The authors declare no conflict of interest.

References

1. Dunteman, G.H. *Principal Component Analysis*; SAGE publication Inc.: Los Angeles, CA, USA, 1989.
2. Geladi, P.; Kowalshi, B.R. Partial least squares regression: A tutorial. *Anal. Chim. Acta* **1986**, *185*, 1–17. [CrossRef]
3. Wang, X.Z. *Data Mining and Knowledge Discovery for Process Monitoring and Control*; Springer: Berlin/Heidelberg, Germany, 1999.
4. Kourti, T.; MacGregor, J.F. Process analysis, monitoring and diagnosis, using multivariate projection methods. *Chemom. Intell. Lab. Syst.* **1995**, *28*, 3–21. [CrossRef]
5. Wang, H.W. *Partial Least Squares Regression Method and Its Application*; National Defense Industry Press: Beijing, China, 1999.
6. Nomikos, P.; MacGregor, J.F. Monitoring batch processes using multiway principal component analysis. *AIChE J.* **1994**, *40*, 1361–1375. [CrossRef]
7. Nomikos, P.; MacGregor, J.F. Multi-way partial least squares in monitoring batch processes. *Chemom. Intell. Lab. Syst.* **1995**, *30*, 97–108. [CrossRef]
8. Wold, S.; Kettaneh, N.; Fridén, H.; Holmberg, A. Modelling and diagnostics of batch processes and analogous kinetic experiments. *Chemom. Intell. Lab. Syst.* **1998**, *44*, 331–340. [CrossRef]
9. Duchesne, C.; MacGregor, J.F. Multivariate analysis and optimization of process variable trajectories for batch process. *Chemom. Intell. Lab. Syst.* **2000**, *51*, 125–137. [CrossRef]
10. Cho, H.W.; Kim, K.J. A method for predicting future observations in the monitoring of a batch process. *Qual. Technol.* **2003**, *35*, 59–69. [CrossRef]
11. Lu, N.Y.; Wang, F.L.; Gao, F.R. Sub-PCA modeling and online monitoring strategy for batch processes. *AIChE J.* **2004**, *50*, 255–259. [CrossRef]
12. Zhao, L.P.; Zhao, C.H.; Gao, F.R. Inter-batch-evolution-traced process monitoring based on inter-batch mode division for multi-phase batch processes. *Chemom. Intell. Lab. Syst.* **2014**, *138*, 178–192. [CrossRef]
13. Dong, D.; McAvoy, T.J. Multi-stage batch process monitoring. In Proceedings of the 1995 American Control Conference, Seattle, WA, USA, 21–23 June 1995; pp. 1857–1861.
14. Ündey, C.; Cinar, A. Statistical monitoring of multistage, multi-phase batch processes. *IEEE Control Syst. Mag.* **2002**, *22*, 40–52.
15. Lu, N.Y.; Gao, F.R.; Yang, Y.; Wang, F.L. PCA-based modeling and on-line monitoring strategy for uneven-length batch processes. *Ind. Eng. Chem. Res.* **2004**, *43*, 3343–3352. [CrossRef]
16. Zhao, L.P.; Wang, F.L.; Chang, Y.Q.; Wang, S.; Gao, F.R. Phase-based recursive regression for quality prediction of multi-phase batch processes. In Proceedings of the 13th IEEE International Conference on Control and Automation, Ohrid, Macedonia, 3–6 July 2017; pp. 283–288.
17. Gunther, J.C.; Conner, J.S.; Seborg, D.E. Process monitoring and quality variable prediction utilizing PLS in industrial fed-batch cell culture. *J. Process. Control* **2009**, *19*, 914–921. [CrossRef]

18. Hwang, D.H.; Han, C.H. Real-time monitoring for a process with multiple operating modes. *Control Eng. Pract.* **1999**, *7*, 891–902. [CrossRef]
19. Chen, J.; Liu, J. Mixture principal component analysis models for process monitoring. *Ind. Eng. Chem. Res.* **1999**, *38*, 1478–1488. [CrossRef]
20. Zhao, L.P.; Zhao, C.H.; Gao, F.R. Between-mode quality analysis based multi-mode batch process quality prediction. *Ind. Eng. Chem. Res.* **2014**, *53*, 15629–15638. [CrossRef]
21. Qin, Y. *Research on Data Driven Batch Process Monitoring and Quality Control*; Zhejiang University: Hangzhou, China, 2018.
22. Zhao, L.P.; Huang, X.; Yu, H. Quality-analysis-based process monitoring for multi-phase multi-mode batch processes. *Processes* **2021**, *9*, 1321. [CrossRef]
23. Lu, N.Y.; Gao, F.R. Stage-based process analysis and quality prediction for batch processes. *Ind. Eng. Chem. Res.* **2005**, *44*, 3547–3555. [CrossRef]

Article

Evaluation of Pull Production Control Mechanisms by Simulation

Nataša Tošanović * and Nedeljko Štefanić

Department of Industrial Engineering, Faculty of Mechanical Engineering and Naval Architecture, University of Zagreb, 10000 Zagreb, Croatia; nstefanic@fsb.hr
* Correspondence: natasa.tosanovic@fsb.hr

Abstract: Today, companies need to continuously improve their production processes, which is a complex task. Lean manufacturing is one of the methodologies for production improvement, and one of the basic goals of any lean implementation is to reduce work-in-process (WIP) and shorten the production lead time. One of the basic lean principles for achieving these goals is pull principle. The adoption of this principle is quite challenging, as it requires a long-term commitment in the application and adoption of various lean techniques and tools that are prerequisites for the successful introduction of the pull principle. Kanban is the most well-known pull production control mechanism, and the first one developed within Toyota production system, but later, other pull control mechanisms were developed. Some of them include Conwip, Hybrid Kanban/Conwip, and Drum Buffer Rope (DBR), and those three, together with Kanban, were the research topic of this study. These four mechanisms were not explored and compared all together not for these specific production configurations considered in this research but also with regard to optimal parameters of control mechanisms. The goal was to analyze and compare how these pull control mechanisms affect lead time in different production conditions. For this purpose, simulation experiments were performed. The results showed that for different production conditions, different pull control mechanisms are optimal in terms of shortening lead time. This finding could help companies as a guideline for making a decision in terms of which pull control mechanism to choose.

Keywords: lean manufacturing; lean principles; pull principle; production control mechanisms; production processes; lean implementation

Citation: Tošanović, N.; Štefanić, N. Evaluation of Pull Production Control Mechanisms by Simulation. *Processes* **2022**, *10*, 5. https://doi.org/10.3390/pr10010005

Academic Editor: Sergey Y. Yurish

Received: 30 November 2021
Accepted: 17 December 2021
Published: 21 December 2021

Publisher's Note: MDPI stays neutral with regard to jurisdictional claims in published maps and institutional affiliations.

1. Introduction

As demand for different types of products is advancing exponentially, more efficient production processes, innovative manufacturing models, and methods are becoming more important than ever. It goes without saying that manufacturing companies that want to survive must work continuously to improve their production. Many companies are introducing lean manufacturing to boost their competitiveness; thus, the study conducted by Industry Week and the Manufacturing Performance has found that the largest number (39%) of North American companies considered to be world-class manufacturers use lean manufacturing as the methodology for production improvement [1].

The concept of lean manufacturing originated from the Toyota production system (TPS). The main goal of the concept is to achieve production processes that respond quickly to changes in customer requirements, and this is possible if all waste is eliminated from production processes and with short production lead time [2,3]. John Krafcik coined the term "lean manufacturing" to describe TPS as lean production [4].

However, recently, the most researched topic in the field of manufacturing is Industry 4.0, lean manufacturing is still an important and significant topic both for manufacturing practices and scientific research. Thus, Buer and Strandhagen conducted extensive research concerning the relationship between Industry 4.0 and Lean manufacturing and made a conceptual framework that represents the relationships between Industry 4.0 and lean manufacturing, as well as their implications on performance and environmental factors. One of the relationships described states that Industry 4.0 can support and further develop lean manufacturing practices, which well demonstrates that the lean manufacturing concept is still very important in the context of manufacturing practices [5].

M.P. Ciano, P. Dallasega, and G.O.T. Rosii have also investigated the relationship between Industry 4.0 and lean manufacturing. They state that most studies find that lean manufacturing is a prerequisite for Industry 4.0, which can overcome barriers of lean manufacturing, but most studies miss an in-depth study of this topic on a practice technology level. As such, they conducted multiple case studies investigating these relationships. One of the relationships investigated is the importance of one-piece -flow, the concept from lean manufacturing, and its relationship to a successful implementation of the industry 4.0 [6]. One-piece-flow is one of the ideally achieved goals in any lean implementation. In his book, J. Womack defines lean as a set of five principles, with the two most important and most complex to achieve being the principles of flow and pull. Ideally, flow is a one-piece flow and the way to manage it is by pulling material in the production process. The first production control mechanism for achieving pull was Kanban.

The Kanban system was being developed by Toyota for over ten years; that is, it took Toyota over ten years in the 1950s to implement the idea of pulling material using signal cards. The reason for that was that there are several prerequisites for the successful implementation of such a system, and these are primarily small series or ideally one-piece flow, short set-up times, stable production process, i.e., production process with little variation and no downtime caused by any equipment failure, or some other disturbances [7,8].

In the last few decades, scientists and engineers from the field began to develop other systems similar to the Kanban signaling system for controlling material flow through the manufacturing process. Thus, to this day, several pull control mechanisms have been developed. The purpose of this study to was enable easier decision-making regarding pull control mechanism. The choice of pull strategy is dependent on the configuration of production processes. Pull control mechanisms have been researched extensively. The idea of this study was to evaluate four mechanisms that, according to researched literature, were not evaluated together specifically in production settings described in this study. In addition, most of the reviewed papers do not take into account optimal parameters of control mechanisms themselves which was also considered in this research.

Literature Review

Many studies have been conducted in order to define the optimal parameters of these control mechanisms [9–13], but also there are several studies answering the question of which control mechanism is a better choice in a given production setting. [14–24]. Some studies contradict each other [15,22], and despite the large number of comparisons of individual mechanisms, no research was done that would include a higher number of mechanisms and consider parameters that significantly affect the defined prerequisite for achieving pull. Therefore, it is necessary to conduct research that would consider the parameters that significantly affect the prerequisites for the successful application of production control mechanisms, and to determine which mechanism is a better choice for different values of these parameters.

Since the Kanban was developed in Toyota's production facilities, the application of this pull control mechanism has become more present in different types of production processes. That was the reason behind the need for changing the characteristics of the pull control mechanism in order to adjust to individual needs of production processes. Thus, Monden has concluded that the traditional Kanban system has certain shortcomings, namely that it requires certain preconditions to be successfully applied, which are sometimes not so easy to achieve [7]. Some of these prerequisites are: variability in demand must be low (demand should be almost constant); product variety (production program) must be low; the process must be stable, without variations and machine set-up times should be kept to a minimum.

In 1990, Spearman, Woodruff and Hopp proposed another pull control mechanism, called "Constant Work in Process" (Conwip), which, unlike the classic Kanban system, limits the total amount of work-in-process, and signal cards exist only between the warehouse (where orders first arrive) and the first operation, and the further flow of material takes place according to FIFO (First in-First out) rule [25].

Boonlertvanich introduces another variant of the production management mechanism, which he concludes, has an advantage over both Kanban and Conwip. Boonlertvanich combines some features of Kanban, Conwip and Base Stocks and introduces a new mechanism called the "Extended Conwip-Kanban System" [26].

The fact that thirty-two different variants of the Kanban system have been developed and analyzed to date (of which nine variants have come to life in practice) shows how interesting the field of development of new variants of the Kanban signaling system has been to scientists and engineers [27]. Furthermore, to date, four different variants of the Conwip system have been developed [28].

The choice of the appropriate production management mechanism is very important. In some production conditions, the application of one machine will be more favorable in terms of achieving the shortest possible production lead time and as little work in process, which is the main goal of lean production, while in other conditions another mechanism will work better. It is up to the production manager to decide which mechanism to apply and define the optimal parameters of the selected mechanism. But how to make a good decision? What are the factors to consider when choosing a production control mechanism? If the process is not stable, i.e., the variability of individual parameters is quite high, will some other mechanism be more appropriate than a mechanism that would be appropriate in the same process but with small variations in parameters? According to Chao and Shih [29], there are forty-one parameters that affect the behavior of the production process. This piece of data alone indicates how complex the production process is and that defining the mechanisms which control the flow of materials through the production process is not such a simple task.

As the field of pull production control mechanisms has evolved, more and more scientists are comparing them to answer the question of which mechanism is more appropriate in the specific conditions of the production process. In general, all papers dealing with the comparison of production control mechanisms can be divided into two groups: research dealing with the study of production processes in which only one product is made [14–24], and research dealing with processes in which two or more products are produced [19,22,23,30].

The disadvantage of most comparisons is the lack of a unified framework for comparison. Namely, many studies do not consider the optimal parameters of the mechanisms they compare [15]. This is most likely why the conclusions are contradictory. For example, Lavoie, Gharbi and Kenne [15] conclude that in some situations Kanban is better, which is contrary to Bonvik's conclusion [24]. Cheraghi et al. noted that Gastettner and Kuan analyzed Kanban and Conwip and concluded that choosing the best distribution of Kanban cards achieves less work in the process than with Conwip, which is contrary to Spearman's work [22,25].

In order for the comparison to be correct, Amos et al. believes that each of the mechanisms should have optimal settings with respect to the given criteria and thus propose a methodology for comparing different mechanisms based on simulation experimentation and multicriteria optimization. In that work, Amos et al. compare Kanban, Conwip, and DBR in the production process of one product. They simulated an unbalanced production line consisting of 14 workstations and experimented by changing the location of a bottleneck in the process. The conclusion is that DBR is generally better than the Kanban and Conwip systems [14].

As an experimental factor in the comparison, many authors take the amount of stock in the buffers, which is defined by the size of the signal cards [15,17,21]. Lavoie, Gharbi and Kenne, monitor how the cost function behaves for different buffer sizes and conclude that the hybrid mechanism is better than Kanban and Conwip if the cost of storage is taken into account, but if these costs are not taken into account and only the cost of holding stocks is observed, then the hybrid mechanism and Conwip are equally good. Kanban becomes more financially viable if the cost of holding inventory and the number of operations increase [15].

Pettersen also takes the level of stocks in the buffers as an influential factor, but the frequency of machine downtime is taken as another parameter that influences the choice of mechanism. Pettersen compares Kanban and Conwip, simulating the production of a single product on a four-workstations production line and confirmed the advantage of Conwip over Kanban [16].

Enns and Rogers [21] also dealt with the problem of choosing a production control mechanism in the process of making a single product. They observed processing time and variability in demand by changing the rate of arrival of orders. In the end, they do not give an unambiguous conclusion as to which mechanism would be better under the stated conditions in production.

Bonvik, as well as Enns and Rogers, analyze the production of one product in conditions of different levels of demand, comparing Kanban, Conwip and Hybrid Kanban/Conwip mechanism, concluding that in most cases the Hybrid mechanism gives better results if the goal is to achieve less work in process and better service level [24].

Sharma and Agrawal [18] analyzed the behavior of the production system under different mechanisms (Kanban, Conwip, and Hybrid), also under conditions of variable demand. The production line of four workstations and the production of one product were analyzed. The results obtained by the AHP method were ranked and it was analyzed which alternatives are more suitable according to each criterion (minimum work in process, maximum production rate, etc.). Given the required criteria, the best choice of mechanism for different demand regimes (four different statistical distributions) was determined. The conclusion is that Kanban is better for three distributions, and Conwip for only one distribution [18].

Kabadurms [19] analyzed the production process of five different products by simulation experimentation and concluded that in conditions of variable demand the POLCA system (Paired Cell Overlapping by Card with Authorization) gives better results compared to Conwip, if the criterion is a small amount of work in the process and as short lead time. Kabadurms used the factorial design of the experiment by changing five parameters: coefficient of variation of processing time, orders arrival rate, batch size, number of process delays, and product variations.

Cheraghi [22] also analyzed a production line that produces more products, two different products specifically. He analyzed how batch size, demand frequency, and machine maintenance mode affect the total amount of work-in-process and the rate of production. He concluded that the frequency of demand significantly affects the behavior of production control mechanisms and concluded that pull production is not always the best choice compared to push when it comes to the criterion of reducing the total number of work-in-process.

In his doctoral dissertation, Terrence [23] also analyzes the production of several different products on the production line by comparing Kanban with the Economic Production Series (EPQ). The criteria for comparison were the work in process, the production cycle, the total production costs, and the influence of the machine set-up time. Terrence concludes that the time it takes to set up the machines significantly affects the production costs in both the Kanban and EPQ cases. However, in the case of a longer machine set-up time, EPQ is more cost-effective than Kanban. This is partly due to the larger production series of the EPQ, which reduces the frequency of machine adjustments. Costs are lower in the case of Kanban only when the set-up time is less than 15 min. Terrence concludes that the advantage of Kanban is that it gives flexibility to production, but does not take into account the variability of the level or frequency of demand which is a very important factor, but keeps demand constant and also does not take into account certain factors that can significantly influence behavior of production process, as well as on the choice of production control mechanism (variations in the duration of operations, process delays, etc.).

One of the factors that significantly affect the applicability and effectiveness of the production control mechanism is bottleneck. Among all researched literature, only Amos et al. [14] have investigated its influence on the performance of pull control mechanisms, Kanban, Conwip and DBR. But the influence of other factors, together with bottleneck, as well as their interaction, were not considered. In addition, the greater part of the researched literature considered two or three control mechanisms. The aim of this study was to evaluate only influence of the bottleneck together with another two factors, found in the literature, but also factors important for the successful implementation of pull principle from the authors' experience. Those factors are variability in process and the time of operations. Together with those three factors, the influence of the selection of pull production control mechanism was evaluated. The idea was to analyze how the change in the characteristics of the production process, thus different levels of influence factors and the selection of production control mechanism affects the performance measurement of the production process, the lead time. The research goal was to determine under what conditions a certain production control mechanism is a better choice given the characteristics of the production process; thus the research questions are:

1. Q1: Is there a performance difference between different pull production control mechanisms in terms of production lead time?
2. Q2: Is the one pull production control mechanism that is optimal for one set of production conditions also optimal for another setting of production conditions?

2. Materials and Methods

In order to analyze the effects of factors (characteristics of the production process) on response (lead time), simulation experimentation and design of experiments were performed. The effect of the decision which production control mechanism to use in different production conditions (different levels of factors) on response criterion (lead time) was examined.

The pull mechanism is one of the basic principles of lean manufacturing [2]. The best way to describe it is to compare the pull mechanism with the push mechanism. Push mechanism is related to Material Resource Planning, and as Hopp and Spearman state, there is no other way of controlling work in the process other than work orders [8]. Figure 1 represents the production push system.

The flow of material in the push system is the same as in the pull system, but the rules defining how the information is distributed towards production stations are the key difference between the push and pull system [31]. Figure 2 represents the production pull system.

The fundamental difference between push and pull production systems, according to Hop and Spearman, is that the pull system explicitly restricts the work in process, while the push mechanism does not. They also state that there is a confusion and wrong perception that the difference between those two systems lies in the fact that push represents a make-to-stock system and pull represents a make-to-order system, which according to Hopp and Spearman is wrong. They state that this perception occurred after the releasing the book, Lean Thinking [2], where the authors define pull system as the rule that no one upstream in the production process can produce a product or service until the customer downstream requests one [8].

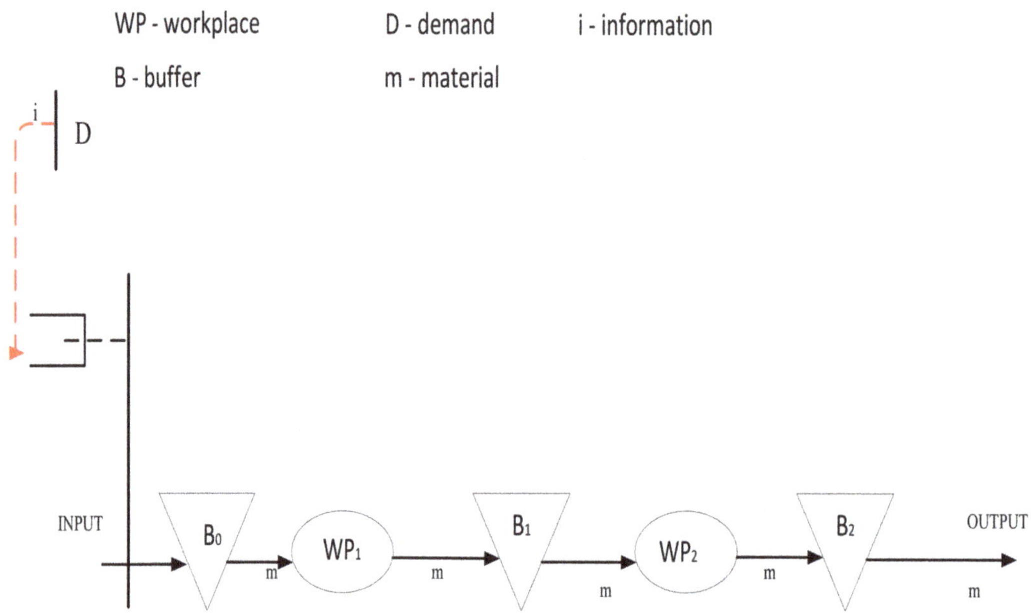

Figure 1. Production push system.

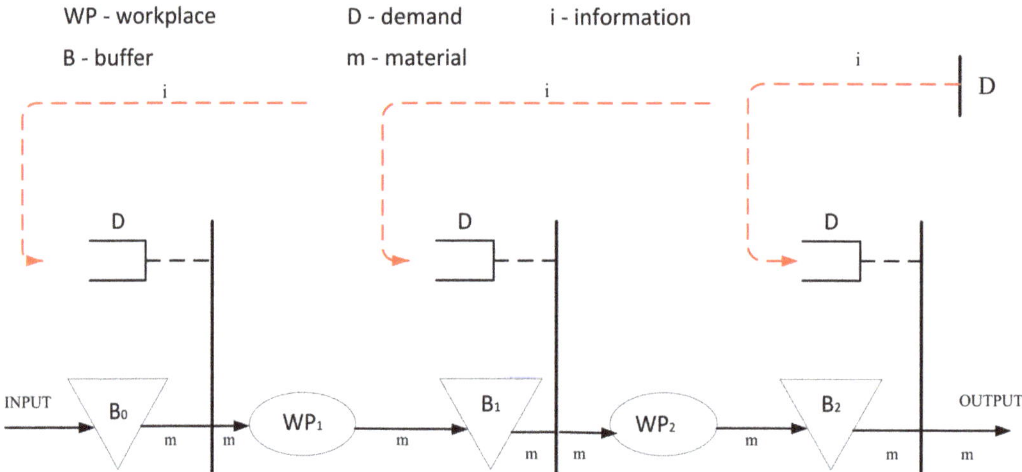

Figure 2. Production pull system.

Hopp and Spearman also state that there is misunderstanding that pull mechanism is the synonym for Kanban system, while the true definition is that the Kanban system is only one of the mechanisms to achieve pull production. Although it indeed was the first production control mechanism developed in Toyota for achieving pull, later, other pull production control mechanisms were developed.

If one looks at the production system as a flow of information and a flow of materials, the flow of information is the one that triggers and control the flow of material. The information flow can be global and local. The global flow represents the information on customer demand, and the local flow is communication between different phases of the production process. [31]. The difference among all production control mechanisms is in the logic of distribution of both local and global information through the production process. Thus, every production control mechanism defines the flow of material by its own unique logic of the flow of information. [8].

Bicheno states that Kanban is one of the three most used production control mechanisms) for achieving a pull system. The other two most widely used mechanisms are CONWIP ("Constant work in process") and DBR (Drum buffer rope) [32]. After reviewing the literature, it was decided to research these three mechanisms as well as Kanban/Conwip hybrid mechanism. Thus, the researched mechanisms are as follows:

1. Kanban;
2. Conwip;
3. Kanban/Conwip hybrid;
4. DBR.

3. Results

3.1. Simulation Model

In order for the pull production control mechanisms to be compared and evaluated, simulation experimentation had to be performed. The first step was to build production process model. In the researched literature, it was found that the production line consisting of five workplaces present different enough aspects of production relationships and problems [23,33–35]. One such model was used in Enns and Rogers research paper [21]. The model consists of five workplaces, and every station presents a different operation, plus there is inventory between workplaces. This model was used for validation of the model in this research.

Figure 3 represents the model of the production process simulated in this study. The production begins at workplace number 1, then continues respectively through workplaces 2, 3, 4, and 5. The processing time is 60 min, while in the case of the existence of bottleneck, the processing time is 80 min. The production process begins at workplace 1 and continues at workplaces 2, 3, 4, and 5, respectively. The processing time on each machine is 60 min, and as one of the parameters is the existence, that is, the non-existence of a bottleneck. In the case of a bottleneck, the duration of the operation is 80 min. A lognormal distribution was used to generate processing time values. A review of the literature, but also conversations with experts dealing with this area, found that the lognormal distribution well the phenomena of processing time rather well. A new random number seed was used for each simulation run. This ensures the randomness of the generated values. For each level, i.e., a combination of input parameters, five simulation runs (five repetitions) was performed.

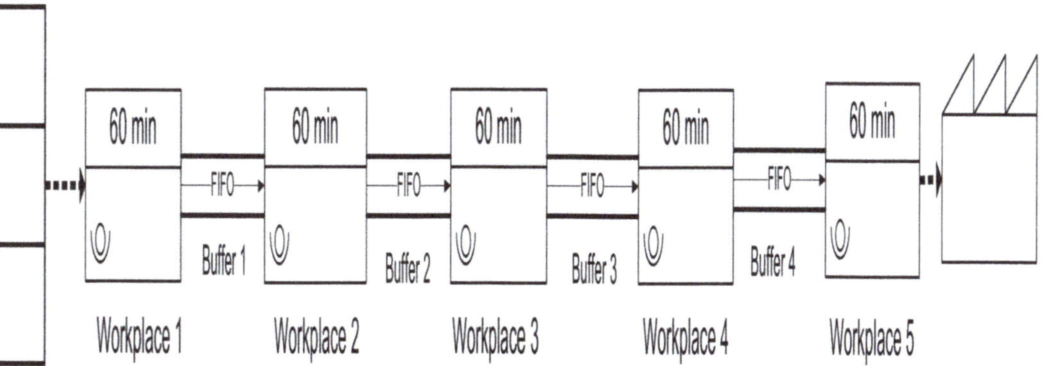

Figure 3. Production process.

The assumptions for the model are consistent with the other studies [8,23,26,36], and are as follows:

1. Each machine can only process one product in a time unit;
2. Transport of the parts from the buffer to the next workplace follows the rule FIFO (first-in-first-out);
3. Transport time between operations is negligible;
4. One-piece flow production;
5. Kanban system is the one-card Kanban system;
6. Set-up time is not the subject of this research, so the production line produces one type of product;
7. There is no storage of finished, good material and the material is always available from the raw material storage;
8. Simulation time is 117,000 h (equals one year);
9. Possible machine downtimes, waiting or any other stoppage of the process that may affect the processing time variability, are "covered" by the randomness of the processing time;
10. Set-up time is not the subject of the research, so it is included in the processing time.

3.2. Validation of the Model

After the production process was modeled, it had to be validated. Two validation techniques were utilized. The first one is the comparison to other models technique and the other is the extreme condition test technique. Furthermore, all models were confirmed by Face validation in the way that three colleagues from the same field of research examined them and confirmed their validity [37]. Since the five-station model is also used in Enns and Rogers paper, their results could be compared with the ones that were gathered in this study. In their study, push and pull models were developed, and the pull model was controlled with the Conwip mechanism. Thus, as a first step, these two models were developed, and the results were compared. Simulation experimentation was performed in Simulink, a special module for the simulation in the MATLAB software, and since the discrete manufacturing production processes is simulated, an add-on for discrete simulation, SimEvents, was used [38]. After modeling and validation of these two models, all other mechanisms were modeled, Kanban, Hybrid Conwip/Kanban and DBR, respectively. The *extreme condition test* and *face validation* were performed.

In order to create the conditions for comparison, as in Enns and Roger's article for comparison, the simulation time for the push and Conwip model was 101,000 min, and five runs were made for both push and Conwip. The processing time was chosen to be stochastic and is defined with operations time and coefficient of variation, c_p, which is defined to be the standard deviation of the processing times divided by the mean processing time, $1/\mu$. The distribution used to generate processing times was the Gamma distribution, is defined by parameters α and β [21]. For the Conwip model, the number of control cards had to be the same; thus, six Conwip cards in the process were modeled. The results of comparison are given in Table 1.

Table 1. Validation of push and Conwip model.

Factors	Results— MATLAB	Results for Comparison	Difference	Difference, %
Lead time-push, min	12.48	12.384	0.096	0.775
WIP-push, pcs	7.6	7.465	0.135	1.8
Lead time-Conwip, min	10.14	9.925	0.215	2.12
Throughput-Conwip, pcs/min	0.618	0.605	0.013	2.1

Since the differences in results are insignificant, it could be concluded that the model for this research has a satisfactory range of accuracy, which is a match to the model presented and validated in the Roger and Enns research [21].

Further on, Kanban, Hybrid Conwip/Kanban and DBR were validated on the existing production model, validated previously by comparison. As already mentioned, the *extreme condition test* was used. In the case of Kanban, two extreme conditions were made:

1.   Extremely long operation time on the second workplace (85,000 min, which is 80% of the whole simulation run);
2.   The number of Kanban on the last workplace was set to be zero.

For the first condition, the result was that only one product came out of the process, which is expected since the second operation took 80% of time of the simulation run. And for the second condition, none of the products came out of the process, which also was expected. Since there were not any Kanban cards in the last workplace, thus there was no signal for processing. These results showed that the Kanban was well modeled.

The hybrid Kanban/Conwip control mechanism utilize both Conwip and Kanban cards in the process. Conwip cards regulate the overall amount of WIP, in a way that the information about demand is sent to the first stage of the production process, while the Kanban card sends the information about demand upstream starting from the last stage of the process up to the first one. Thus, Kanban cards regulate the amount of WIP at every production stage and unlike the Conwip cards that regulate the overall amount of WIP, thus the total amount of WIP in the whole process. The first extreme condition was as follows:

1.   The number of Conwip cards equals zero;
2.   The number of Kanban cards stays the same.

The second extreme condition was as follows:

1.   The number of Conwip cards stays the same;
2.   The number of Kanban cards changes in the fourth workplace and equals zero.

As expected, in both cases none of the products left the production process. In the first scenario, as the number of Conwip cards defines the total number of WIP in the production process, the WIP was zero, so the process could not produce any product. In the second scenario, just like in Kanban validation, as there were no control cards in the workplace, so there was no information for starting the production on that station. Again, none of the product left the process.

The same extreme condition, namely zero control cards, was used for DBR model and the same results were obtained.

3.3. Experiment Set-Up

In order to conduct the experiments, the level of the factors had to be chosen. But first, allow the decision of selected factors to be explained. There is an exact definition, found in literature, on what the prerequisites for successful implementation of pull production are:

1.   Balanced production process;
2.   Short set-up time;
3.   Constant customer demand;
4.   Production without downtimes [7].

Very often, these conditions are hard to achieve. Bottlenecks are quite a challenge in regard to production control. Thus, the question is whether, in these kinds of processes, a different control mechanism other than the Kanban is more suitable in terms of the efficiency of the production process. In addition, a smooth process, without stoppages is desirable, but the real processes are often faced with unexpected stoppages. From the authors' experience, it is also known that the planned operation times are often different, due to many reasons, for example, a worker's skills or frequent change of worker on the machines, requiring time for training, etc. All of this led to experimental factors chosen in this research:

1. Variability of operation;
2. Existence of a bottleneck in the process;
3. Operation time;
4. Number of control cards.

The last factor, the number of control cards, had been chosen because the production control mechanisms that were going to be examined in different production conditions (different level of experimental factors) have their own parameters that influence response function (lead time) that is being observed. Actually, the number of control cards, Kanban cards, Conwip cards, etc., is the parameter that had to be taken into account when considering efficient production flow. This parameter has not been considered much in previous research. Thus, many authors in the reviewed literature take the fixed number of cards, which they define as optimal and vary all other factors, but the optimal number of cards for one setting of factor levels does not have to suit different settings of factors level. This is the reason why the number of control cards has been taken as one of the experimental factors in this research.

The response function in this research is production lead time. The lead time is the total time elapsed from the moment when the material, i.e., the raw material, enters the production process to the moment when the finished product is ready for delivery to the customer [39]. Why has this response function been chosen? The goal of every lean implementation is to shorten the lead time [2], and since the topic of this research is one of the five basic principles of lean thinking [2], namely pull production, so this response function has been selected.

The levels of input parameters were defined after an extensive review of the literature and based on previous experience in various manufacturing companies. The levels are shown in Table 2.

Table 2. Factor levels.

Factors	Level 1	Level 2
Operation time	5	60
Coefficient of variation	0.06	0.18
Existence of a bottleneck in the process	NO	YES
Number of control cards	10	15

As previously mentioned, the simulation model was developed in MATLAB, more precisely by Simulink and SimEvents, which are the features of MATLAB for simulation.

The Response Surface Methodology (RSM) was used to determine the effect of factors on response function; thus, the mathematical model was developed to describe the relationship between factor and response. The general factorial design was chosen to conduct the experiments. This experimental design was chosen because some of the varied factors (input variables) are numerical variables, and some are categorical variables. In such a case, it is convenient to use the general factorial design plan of the experiment. In the software package, Design Expert [40], which was used to analyze the results, this experimental plan is also called "multilevel categoric".

After performing simulations followed by a designed experiment, an analysis of variance (ANOVA) was performed to determine the significance of the factors, and the mathematical model (response function) was developed by regression analysis. The factors of models A, B, C, and D are in order:

1. A—coefficient of variation;
2. B—operation time;
3. C—the existence of a bottleneck;
4. D—number of control cards.

3.4. Experiment Results

In the next part of the chapter, results of the experiments conducted Kanban-controlled processes will be described. This chapter will present the results of data analysis to describe the impact of processing time, coefficient of variation, bottleneck, and number of control cards on the lead time of the production process in the case of Kanban control mechanism. RSM was used to investigate the impacts of factors. Based on this method, a mathematical model was generated to describe the variable response.

Before analyzing the variance, it was necessary to make data transformation. In this data set, the ratio of the maximum and minimum measured value was greater than 10, so the transformation was required [40]. In this way, the homogeneity of variance over the experimental space is satisfied [41]. Data were transformed according to the equation:

$$y' = \ln (y + k), \qquad k = 0, \tag{1}$$

Variable y' in the equation presents transformed value, and variable y presents the real value.

The bull hypothesis H_0 for the experiment design was as follows: variability of processing time, processing time, bottleneck, and number of production control cards do not affect the lead time of the production process. Table 3 presents an analysis of variance (ANOVA) for the design of experiments. ANOVA generates significant factors of the model for the response while its significance was evaluated according to the probability levels. F-value shows that the model is significant. p-value also shows that the model is significant and that the null hypothesis can be rejected. Namely, the null hypothesis is rejected when the p-value is less than 0.05, which is the case here [40]. Furthermore, p-values of the factors A, B, C, and D, as well as their interactions AB, AD, BC, BD, CD, ABD, and BCD show their significance.

Table 3. ANOVA—Kanban.

Source of Variation	Sum of Squares	Df [1]	Mean Square	F-Value	p-Value	Significance
Model	165.31	11	15.03	1 169.75	<0.0001	significant
Factors:						
A	2.48	1	2.48	193.02	<0.0001	significant
B	148.80	1	148.80	11,581.87	<0.0001	significant
C	8.58	1	8.58	667.45	<0.0001	significant
D	3.39	1	3.39	264.03	<0.0001	significant
AB	0.0526	1	0.0526	4.10	0.0469	significant
AD	0.6140	1	0.6140	47.79	<0.0001	significant
BC	0.1743	1	0.1743	13.57	0.0005	significant
BD	0.2573	1	0.2573	20.03	<0.0001	significant
CD	0.6793	1	0.6793	52.87	<0.0001	significant
ABD	0.1425	1	0.1425	11.09	0.0014	significant
BCD	0.1465	1	0.1465	11.41	0.0012	significant
Residual	0.8736	68	0.0128			
Deviation of the model	0.0197	4	0.0049	0.3685	0.8302	not significant

[1] Df—degree of freedom.

The next step was regression analysis, which is a method for estimating the relationships between a dependent variable and one or more independent variables. In this study dependent variable is lead time, while independent variables are: coefficient of variation, processing time, and existence of bottleneck. Based on the results obtained by simulation experimentation, the coefficients of the mathematical (regression) model were estimated.

The value of the coefficient of determination, R^2, which is 0.9947 implied that the model is successful in explaining 99.47% of the experimental variables. In addition, the adjusted value of determination, the value of which is 0.9939 declared high significance of the generated model. These coefficients are in good relation. This is indicated by the necessary condition, which is that the difference between the adjusted R^2 value and determination fitting R^2 value is less than 0.2. When this condition is achieved it means that the obtained regression model (response function) is different from random phenomena (Table 4).

Table 4. Regression analysis—Kanban.

Name	Value
Coefficient of variation, %	2.03
R^2—coefficient of determination	0.9947
R_{adj}^2—adjusted coefficient of determination	0.9939
R_{pre}^2—predicted coefficient of determination	0.9927
Adequate precision	94.4254

As the natural logarithm was used for data transformation, so the obtained mathematical model represents a mathematical function for calculating the lead time in logarithmic form. Thus, the calculated value should be transformed from a natural algorithm into a real number in order to obtain the actual value of the lea time variable.

$$\ln(\text{LT}_{\text{Kanban}}) = 3.66826 - 0.002615\text{Cv} + 0.043041\text{T} + 0.194001\text{Nr} + 0.00809\text{CvT} + 0.008481\text{CvNr} \tag{2}$$

$$\ln(\text{LT}_{\text{Kanban-BN}}) = 3.17249 + 0.244123\text{Cv} + 0.052751\text{T} - 0.327\text{Nr} - 0.001974\text{CvT} + 0.195583\text{CvNr} \tag{3}$$

Variables from Equations (2) and (3) are as follows:

1. $\text{LT}_{\text{Kanban}}$—lead time for the process with without bottleneck, controlled by Kanban, min;
2. $\text{LT}_{\text{Kanban-BN}}$—lead time for the process bottleneck controlled by Kanban, min;
3. C_{V}—coefficient of variation;
4. T—processing time;
5. Nr—number of control cards.

The value obtained by Equation (2) represents the natural logarithm of the lead time, and in order to get the actual value of the lead time, this value needs to be transformed into a real number using the equation:

$$\text{LT}_{\text{Kanban}} = e^{\ln(\text{LT}_{\text{Kanban}})} \tag{4}$$

The same mathematical relation as in Equation (4) has to be used for transforming natural logarithm values obtained by Equation (3) in order to get real number values.

Figure 4a,b presents contour plots of mathematical model which describes the influence of factors and their interactions on the response function, in this case, lead time, for the production process controlled with Kanban. The first figure is for the process in which there are two Kanban cards in each phase of production, thus, for each operation and the second figure represents the process with five Kanban cards in each operation. Lead time, in the case of five Kanban cards, is much longer for the same level of the other factors in regards to the process with two Kanban cards. This was expected since the number of Kanban cards influences the level of work in the process, so the time needed for one piece of material to pass through the whole process is longer. Further experiments with other control mechanisms will show whether there is any difference between mechanisms for these same conditions.

A three-dimensional 3D response surface plot showing the effect of coefficient of variation and operations is presented in Figure 5a,b—both figures present process with bottleneck with a difference in the number of Kanban cards. Number of Kanban cards affects the lead time quite significantly, as demonstrated below.

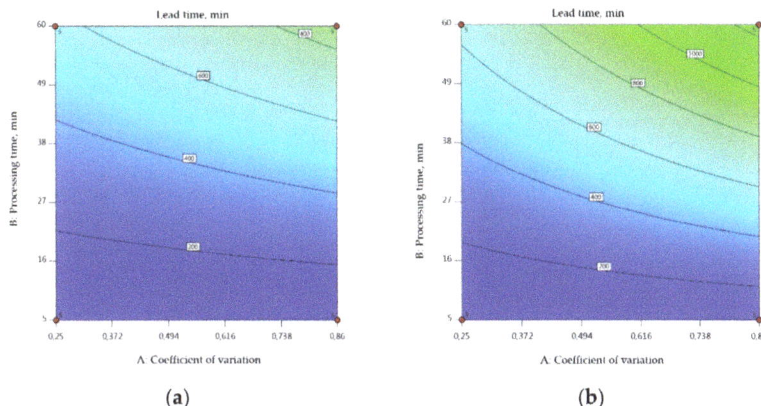

Figure 4. Response surface contour plots: (**a**) Effects of coefficient of variation and operation time on lead time for the process without bottleneck and two Kanban cards in each operation; (**b**) Effects of coefficient of variation and operation time on lead time for the process without a bottleneck and five Kanban cards in each operation.

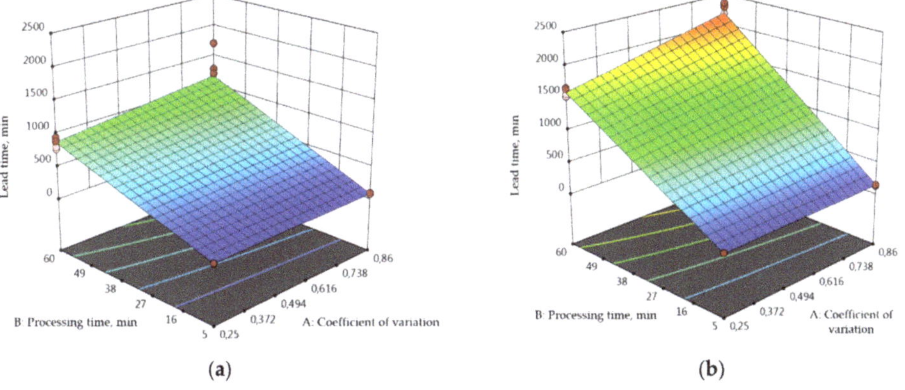

Figure 5. Response surface 3D plots: (**a**) Effects of coefficient of variation and operation time on lead time for the process with bottleneck and two Kanban cards in each operation; (**b**) Effects of coefficient of variation and operation time on lead time for the process with bottleneck and five Kanban cards in each operation.

The same analysis, as presented for the case where Kanban mechanism was controlling simulated production process, was also conducted for Conwip, Hybrid Kanban/Conwip and DBR. The results are going to be presented further in this chapter.

For all three mechanisms, Conwip, Hybrid Kanban/Conwip and DBR, respectively, ANOVA has shown that both model and the factors are significant. Also, *p*-values of the factors A, B, C, and D, as well as their interactions AB, AD, BC, BD, CD, ABD, and BCD show their significance. As for the Kanban, based on the results obtained by simulation experimentation, the coefficients of the mathematical (regression) models were estimated. The coefficient of determination declared the significance of all models.

As for the analysis of variance for Kanban, in the case of Conwip, Hybrid Kanban/Conwip, and DBR, data transformation needed to be made for the same reason. In this way, the homogeneity of variance over the experimental space is satisfied [41]. In case of Conwip, the data were transformed according to the equation:

$$y' = (y + k)^{\lambda}, \qquad \lambda = 0.25 \tag{5}$$

Variable y' in the equation presents transformed value, and variable y presents the real value.

In case of Hybrid and DBR data were transformed according to the equation:

$$y' = \frac{1}{\sqrt{y + k}}, \qquad k = 0 \tag{6}$$

The analysis of variance and regression analysis are given in the Tables 5 to 7.

Table 5. ANOVA—Conwip.

Source of Variation	Sum of Squares	Df	Mean Square	F-Value	*p*-Value	Significance
Model	143.46	12	11.95	1106.90	<0.0001	significant
Factors:						
A	0.5033	1	0.5033	46.60	<0.0001	significant
B	133.04	1	133.04	12317.76	<0.0001	significant
C	6.81	1	6.81	630.09	<0.0001	significant
D	1.79	1	1.79	165.94	<0.0001	significant
AB	0.0058	1	0.0058	0.5400	0.4650	significant
AC	0.0685	1	0.0685	6.34	0.0142	significant
AD	0.0129	1	0.0129	1.19	0.2789	significant
BC	0.5327	1	0.5327	49.33	<0.0001	significant
BD	0.0201	1	0.0201	1.86	0.1774	significant
CD	0.6044	1	0.6044	55.96	<0.0001	significant
ABD	0.0441	1	0.0441	4.09	0.0472	significant
ACD	0.0330	1	0.0330	3.05	0.0851	
Residual	0.7236	67	0.0108			significant
Deviation of the model	0.0644	3	0.0215	2.09	0.1109	not significant

Table 6. ANOVA—Hybrid Kanban/Conwip.

Source of Variation	Sum of Squares	Df	Mean Square	F-Value	*p*-Value	Significance
Model	0.2172	15	0.0145	3266.36	<0.0001	significant
Factors:						
A	0.0016	1	0.0016	359.70	<0.0001	significant
B	0.1908	1	0.1908	43,033.48	<0.0001	significant
C	0.0132	1	0.0132	2970.13	<0.0001	significant
D	0.0014	1	0.0014	318.39	<0.0001	significant
AB	0.0008	1	0.0008	176.88	<0.0001	significant
AC	0.0008	1	0.0008	188.31	<0.0001	significant
AD	0.0001	1	0.0001	12.06	0.0009	significant
BC	0.0063	1	0.0063	1413.99	<0.0001	significant
BD	0.0003	1	0.0003	76.65	<0.0001	significant
CD	0.0000	1	0.0000	3.53	0.0647	significant
ABC	0.0005	1	0.0005	119.90	<0.0001	significant
ABD	0.0000	1	0.0000	7.72	0.0071	
ACD	0.0003	1	0.0003	72.08	<0.0001	
BCD	0.0000	1	0.0000	3.85	0.0541	
ABCD	0.0003	1	0.0003	75.05	<0.0001	
Residual						
Deviation of the model						not significant

Table 7. ANOVA—DBR.

Source of Variation	Sum of Squares	Df	Mean Square	F-Value	*p*-Value	Significance
Model	0.2142	14	0.0153	3 103.46	<0.0001	significant
Factors:						
A	0.0018	1	0.0018	365.87	<0.0001	significant
B	0.1871	1	0.1871	37,947.93	<0.0001	significant
C	0.0142	1	0.0142	2882.65	<0.0001	significant
D	0.0011	1	0.0011	225.04	<0.0001	significant
AB	0.0011	1	0.0011	213.46	<0.0001	significant
AC	0.0007	1	0.0007	141.21	<0.0001	significant
AD	9.415×10^8	1	9.415×10^{-8}	0.0191	0.8905	significant
BC	0.0064	1	0.0064	1289.18	<0.0001	significant
BD	0.0002	1	0.0002	43.31	<0.0001	significant
CD	0.0009	1	0.0009	178.52	<0.0001	significant
ABC	0.0005	1	0.0005	99.98	<0.0001	significant
ABD	0.0000	1	0.0000	7.92	0.0065	
ACD	0.0000	1	0.0000	3.65	0.0606	
BCD	0.0002	1	0.0002	49.68	0.0001	
Residual	0.0003	65	4.930×10^{-6}			
Deviation of the model	0.0000	1	0.0000	3.24	0.0766	not significant

Table 8 represents the regression analysis for Conwip, Hybrid Kanban/Conwip and DBR. The values obtained for the regression functions are given in Table 9.

Table 8. Regression analysis—Conwip, Hybrid Kanban/Conwip and DBR.

Name	Conwip	Hybrid Kanban/Conwip	DBR
Coefficient of variation, %	2.56	2.49	2.64
R^2—coefficient of determination	0.9950	0.9987	0.9985
R_{adj}^2—adjusted coefficient of determination	0.9941	0.9984	0.9982
R_{pre}^2—predicted coefficient of determination	0.9928	0.9980	0.9977
Adequate precision	85.0434	148.9462	147.9360

Table 9. Coefficients of regression functions—Conwip, Hybrid Kanban/Conwip and DBR.

	Cv	T	Nr	CvT	CvNr	TNr	CvTNr
$(Conwip)^{0.25}$	+2.22138	+2.22138	+2.22138	+2.22138	+0.012985	+0.012985	+0.012985	−0.005601
$(Conwip_{BN})^{0.25}$	+1.47958	+1.47958	+1.47958	+1.47958	+1.47958	+1.47958	+1.47958	+1.47958
Hybrid-y′	+0.165776	+0.165776	+0.165776	+0.165776	+0.165776	+0.165776	+0.165776	+0.165776
$Hybrid_{BN}$-y′	+0.186511	+0.186511	+0.186511	+0.186511	+0.186511	+0.186511	+0.186511	+0.186511
DBR-y′	+0.196219	+0.196219	+0.196219	+0.196219	+0.196219	+0.196219	+0.196219	+0.196219
DBR_{BN}-y′	+0.184555	+0.184555	+0.184555	+0.184555	+0.184555	+0.184555	+0.184555	+0.184555

The value obtained by Equation (2) needs to be transformed into real values by using equation:

$$LT_{conwip} = \sqrt[0.25]{\left(LT_{conwip}\right)^{0.25}} \tag{7}$$

In the case of Hybrid and DBR, transformation should be made by the equation below.

$$y = 1/(y')^2 \tag{8}$$

A three-dimensional 3D response surface plot showing the effect of coefficient of variation and processing time on lead time for the process with bottleneck is presented in

Figure 6a,b. Both variations and processing time affects the lead time in terms that it gets longer, which was expected. Furthermore, the number of control cards has the same effect, which can be seen by comparing these two figures.

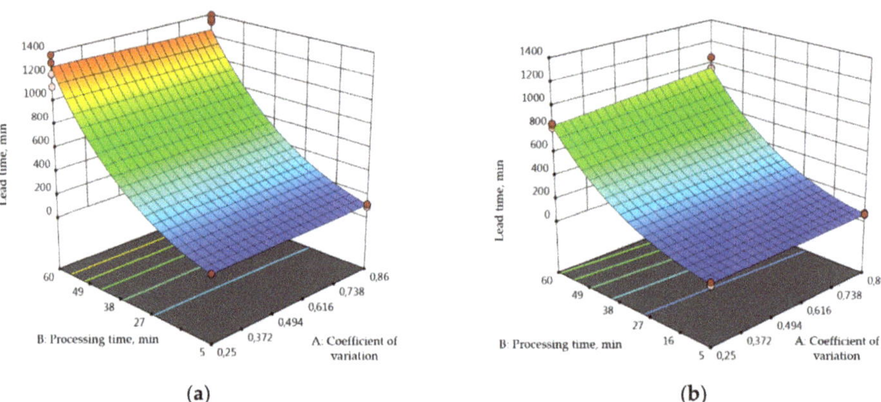

(a)   (b)

Figure 6. Response surface 3D plots: (**a**) Effects of coefficient of variation and operation time on lead time for the process with bottleneck and three Conwip cards in each operation (15 cards for the whole process); (**b**) Effects of coefficient of variation and operation time on lead time for the process with bottleneck and two Conwip cards in each operation (10 cards for the whole process.

In the case of the hybrid mechanism, as presented in Figure 6, the lead time gets longer as the number of control cards is higher (Figure 7a,b). Every card tie one product, and more cards mean more products in the system, thus more work in process. But why would anybody make a decision to utilize more cards, one could ask. The answer is that the higher number of cards, meaning higher WIP, secures unstable processes and acts as a safety buffer, so the optimal solution has to be found.

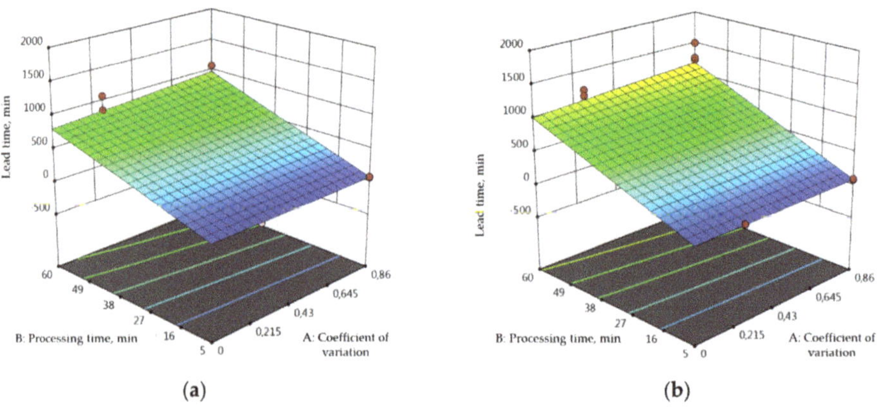

(a)   (b)

Figure 7. Response surface 3D plots: (**a**) Effects of coefficient of variation and operation time on lead time for the process with bottleneck and two cards in each operation; (**b**) Effects of coefficient of variation and operation time on lead time for the process with bottleneck and three cards in each operation.

As well as for Kanban, Conwip, Hybrid Kanban/Conwip, the same relationships between influence factors and response have been found for the DBR, as shown in Figure 8. Now, the question is which mechanism out of these four is optimal in regard to the same production conditions. That was also the research question of this study, and it will be explained further in the text based on the results shown thus far.

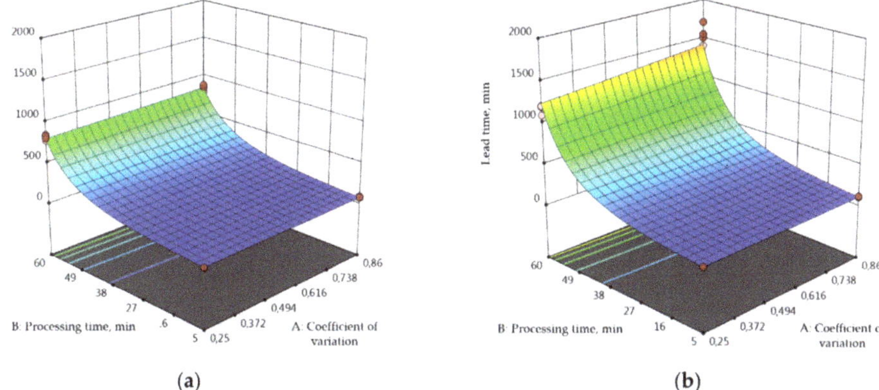

(a)   (b)

Figure 8. Response surface 3D plots: (**a**) Effects of coefficient of variation and operation time on lead time for the process with bottleneck and nine DBR cards in the whole process; (**b**) Effects of coefficient of variation and operation time on lead time for the process with bottleneck and 14 DBR cards in the whole process.

4. Discussion

The pull principle is one of the five basic principles in lean manufacturing. It is the most important condition for achieving goals of lean manufacturing and that is flexible and customer-driven production, meaning ready to quickly respond to changes in demand. Prerequisites for pull are described in Monden's work [7]. Some of these prerequisites, such as balanced production line and reduction of variability are explored in the work of Ertay T. [42]. However, in some cases, certain sources of variability in the process are not manageable at a given moment, but company can still have the positive effects of pull. In addition, balanced line due to existence of a bottleneck is not achievable at the given moment, but still, pull can bring positive effects in the process. As Hopp and Spearman state [8], it is quite common that Kanban is considered as a synonym for pull, but it is just one of the control mechanisms for achieving pull. In this study four different pull control mechanisms were explored, Kanban, Conwip, Hybrid Kanban/Conwip and DBR, respectively. The goal of this study was to evaluate these mechanisms in terms of their effectiveness, that is, how they affect lead time, which is one of the most important performance measures in terms of lean manufacturing. That effectiveness was evaluated by applying those control mechanisms in the same set of different production conditions. The conditions that were explored were existence of bottleneck in the production process, which has not been found a lot in the literature as a factor of research, and from the personal experience of the authors of this study, this factor was found to be one of the obstacles in achieving pull. The other two factors that were explored were variability and processing time as well as the number of signal cards, which is the parameter of control mechanism that ideally has to be optimal for a given condition. The response surfaces as well as regression functions that define relationships between lead time and described parameters, have been generated. Thus, lead time can be calculated for every combination of parameters. Response surfaces in Figures 5–8 show these relations. For example, one possible combination of parameters describing the condition of production process could be as follows: coefficient of variation 0.2; processing time equals 40 min. Figure 9a,b shows the length of lead time for this combination of parameters in case when the overall number of signal cards is 15, where Figure 9a represents a case without bottleneck and Figure 9b with bottleneck. In practice, Kanban is very often the first considered mechanism for achieving pull, but as can be seen in Figures 9 and 10, Kanban is not always the best solution.

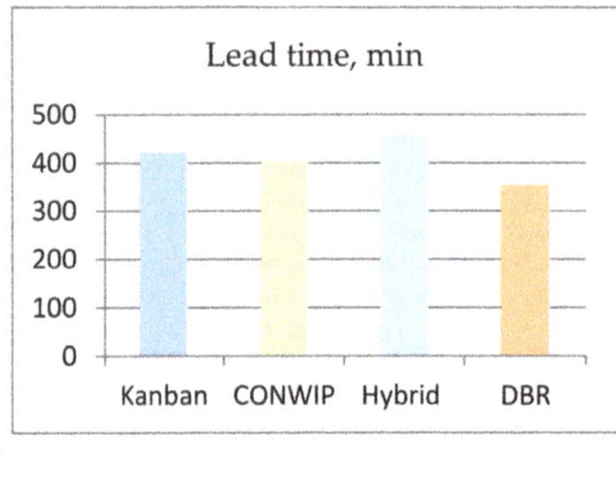
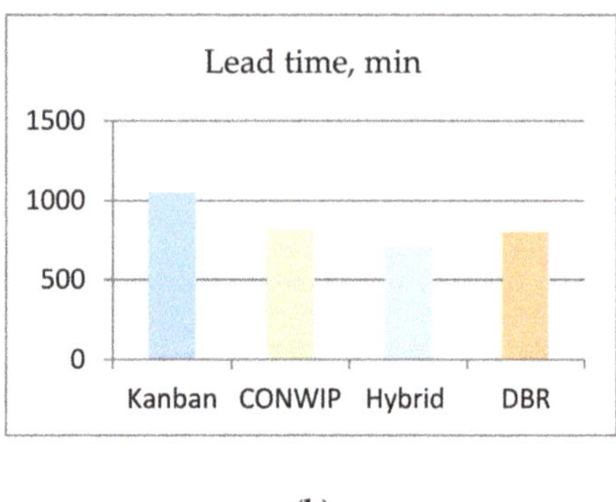

(**a**)   (**b**)

Figure 9. Comparison of the lead time for each control mechanism under the same production condition: (**a**) $C_V = 0.2$ and T = 40 min for the process without bottleneck and WIP = 15; (**b**) $C_V = 0.2$ and T = 40 min, for the process with bottleneck and WIP = 15.

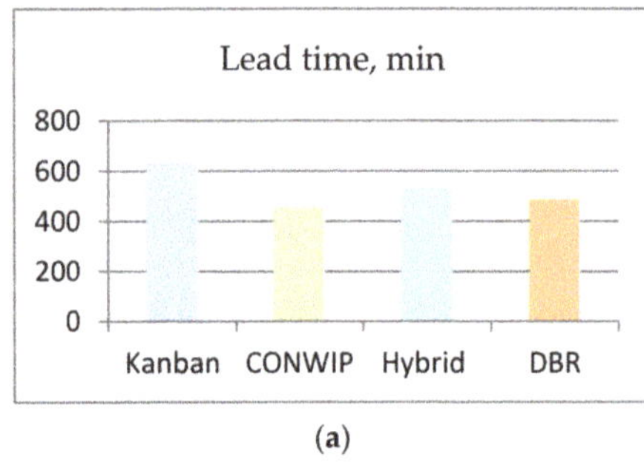
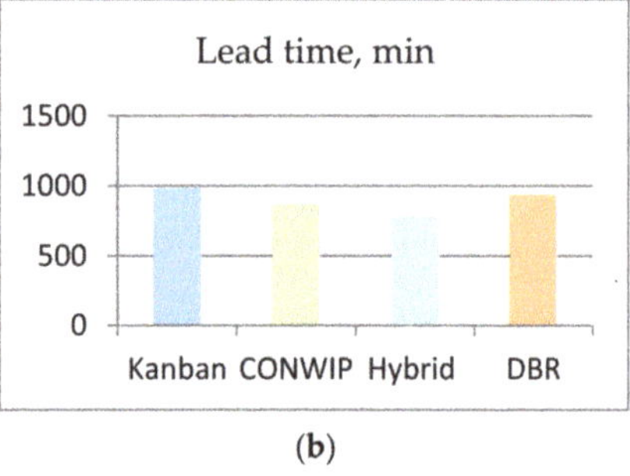

(**a**)   (**b**)

Figure 10. Comparison of the lead time for each control mechanism under the same production condition: (**a**) $C_V = 0.86$, and T = 40 min, for the process without bottleneck and WIP = 15; (**b**) $C_V = 0.86$ and T = 40 min, for the process with bottleneck and WIP = 15.

Figures 9a and 10a represent the relationship of the lead time and all four control mechanisms for the production process with lower and higher variability and with no bottleneck in the process.

Figures 9b and 10b represent the relationship of the lead time and all four control mechanisms for the production process with lower and higher variability, but in this case, with the bottleneck in the process.

Figures 9 and 10 show the length of lead time for all mechanisms for the same level of parameters except the variability. Figure 9a,b present production condition with low variability in the process ($C_v = 0.2$) and Figure 10a,b present production condition with a higher level of variability in the process ($C_v = 0.86$). If one compares the Figures 9a and 10b, it can be concluded that for the same production conditions except one difference and that is variability, different pull control mechanisms contribute to the shorter lead time. Thus, in the process with lower variability the best choice is DBR, and in the process with higher variability the best choice is Conwip (Figures 9a and 10a). The same comparison can be conducted for any combination of researched parameters.

Production managers and lean leaders often do not have time to test which control mechanism to implement in their production process. When lean implementation is in the phase of introducing and implementing pull, the question that arises is which control mechanism to apply. The results presented in this study could be applied in practice, as a guideline for selecting a production control mechanism depending on the condition of the production system. The decision process for the selection of production control mechanism could be as follows:

1. Step 1: Define the current state of the production process in terms of variability, processing time and whether there is a bottleneck in the process;
2. Step 2: Define the goal, that is minimal lead time;
3. Step 3: Select the pull control mechanism using the findings obtained in this study.

Managing production and readiness to continuously improve production processes are challenging and never-ending tasks. As the literature suggests and practical experience supports, lean manufacturing is one of the methods that can help tremendously in achieving that task and is still widely used. Although every production process has its own characteristics, and solutions from one factory cannot be translated to another, some problems and obstacles are similar or the same. Organizing and managing a production process according to pull principles is a huge task for every factory; thus, any finding and guideline for making decisions about pull implementation could be useful. In that context, that was the goal of this research.

5. Conclusions

In this study, an evaluation of four pull production control mechanisms was conducted. The aim was to check if there is any performance difference between these mechanisms in terms of production lead time, and to investigate whether one mechanism which is optimal for certain set of production conditions is also optimal for another different setting. Specifically, for this study Kanban, Conwip, and DBR were chosen, because according to the literature, they are the most widely used pull control mechanisms, but also a hybrid mechanism, which is also commonly used. Parameters that define production conditions and that were in focus in this research were process variability, operation time and existence of bottleneck in the process. Some authors have investigated bottleneck as a parameter, for example, Amos et al., but they studied only Kanban, Conwip and DBR in that context, but not the Hybrid Kanban/Conwip mechanism. They found that DBR outperforms all others.

On the other hand, Bonvik et al. found that the Hybrid Kanban/Conwip outperforms Kanban and Conwip in the case of bottlenecks and high variability. The aim of this study was to include all of these four mentioned mechanisms, but also to consider the variability as a parameter which, for example, in Amos et al.'s study, was not taken into account, but the coefficient of variation was held constant. In addition, the scope of Bonvik et al. was service level and for Amos was a trade-off between throughput and WIP. In this study, regression functions for production lead time are gained, so for the different combinations of different levels of these factors, the optimal mechanism can be chosen.

The results showed that there is the performance difference between different pull production control mechanisms in terms of production lead time. One pull control mechanism that is the best choice for the setting of production conditions is not also the best choice for another set of production conditions. This is the answer to the research questions from the beginning of the study.

The limit of this study is that only the single-product process is explored. For future research, it would be interesting to explore multiple product lines, but also to explore other parameters, such as whether the length of the line together with all of parameters included in this study. Finally, for future research, it would be interesting to investigate these relationships in terms of different performance measures, which could be the goal of the companies in a given moment, such as productivity or service level.

Author Contributions: Conceptualization, N.T. and N.Š.; methodology, N.T.; software, N.T.; validation, N.T. and N.Š.; resources, N.T.; writing—original draft preparation, N.T.; writing—review and editing, N.T. and N.Š. All authors have read and agreed to the published version of the manuscript.

Funding: The Article Processing Charges (APCs) is funded by the European Regional Development Fund, grant number KK.01.1.1.07.0052.

Data Availability Statement: Data available on request.

Conflicts of Interest: The authors declare no conflict of interest.

References

1. North America Manufacturing Benchmarks & Outlook for 2007, Analysis of Data from the IndustryWeek/Manufacturing Performance Institute Census of Manufacturers Conducted in the U.S. and the Canada Manufacturing Study. Available online: http://www-03.ibm.com/marketing/edocument/industrial/lgh_eas_manufacturing_kit/document/print/print.pdf (accessed on 19 April 2021).
2. Womack, J.P.; Jones, D.T. *Lean Thinking: Banish Waste and Create Wealth in Your Corporation*, 2nd ed.; Simon and Schuster: New York, NY, USA, 2003.
3. Murman, E.; Allen, T.; Bozdogan, K.; Cutcher-Gershenfeld, J.; McManus, H.; Nightingale, D.; Rebentisch, E.; Shields, T.; Stahl, F.; Walton, M.; et al. *Lean Enterprise Value: Insights from MIT's Lean Aerospace Initiative*; Palgrave: New York, NY, USA, 2002.
4. Womack, J.P.; Jones, D.; Roos, D. *The Machine That Changed the World*; Simon and Schuster: London, UK, 1990.
5. Buer, S.-V.; Strandhagen, J.O.; Chan, F.T.S. The link between Industry 4.0 and lean manufacturing: Mapping current research and establishing a research agenda. *Int. J. Prod. Res.* **2018**, *56*, 2924–2940. [CrossRef]
6. Ciano, M.P.; Dallasega, P.; Orzes, G.; Rossi, T. One-to-one relationships between Industry 4.0 technologies and Lean Production techniques: A multiple case study. *Int. J. Prod. Res.* **2021**, *59*, 2021. [CrossRef]
7. Monden, Y. *Toyota Production System: An Integrated Approach to Just-In-Time*, 4th ed.; Taylor & Francis Group: Abingdon, UK, 2012.
8. Hopp, W.J.; Spearman, M.C. *Factory Physics*, 3rd ed.; Waveland Press Inc.: Long Grove, IL, USA, 2008.
9. Chan, F.T.S. Effect of Kanban size on just in time manufacturing systems. *J. Mater. Process. Technol.* **2001**, *116*, 146–160. [CrossRef]
10. Jodlbauer, H. Customer driven production planning. *Int. J. Prod. Econ.* **2008**, *111*, 793–801. [CrossRef]
11. Baykoc, O.F.; Erol, S. Simulation modeling and analysis of JIT production system. *Int. J. Prod. Econ.* **1998**, *55*, 203–212. [CrossRef]
12. Enns, S.T. Pull replenishment performance as a function of demand rates and setup times under optimal settings. In Proceedings of the 2007 Winter Simulation Conference, Washington, DC, USA, 9–12 December 2007; pp. 1624–1632.
13. Yang, L.; Zhang, X.P. Design and Application of Kanban Control System in a Multi-Stage, Mixed-Model Assembly Line. *Syst. Eng. Theory Pract.* **2009**, *29*, 64–72. [CrossRef]
14. Amos, H.C.N.; Bernedixen, J.; Syberfeldt, A. A comparative study of production control mechanisms using simulation-based multi-objective optimization. *Int. J. Prod. Res.* **2012**, *50*, 359–377.
15. Lavoie, P.; Gharbi, A.; Kenne, J.P. A comparative study of pull control mechanisms for unreliable homogenous transfer lines. *Int. J. Prod. Econ.* **2010**, *124*, 241–251. [CrossRef]
16. Pettersen, J.A. Pull Based Production Systems—Performance, Modeling and Analysis. Ph.D. Thesis, Lulea University of Technology, Lulea, Sweden, 2010.
17. Pettersen, J.A.; Segerstedt, A. Restricted work-in-process: A study of differences between Kanban and CONWIP. *Int. J. Prod. Econ.* **2009**, *118*, 199–207. [CrossRef]
18. Sharma, S.; Agrawal, N. Selection of a pull production control policy under different demand situations for a manufacturing system by AHP-algorithm. *Comput. Oper. Res.* **2009**, *36*, 1622–1632. [CrossRef]
19. Kabadurmus, O. A comparative study of POLCA and Generic CONWIP production control systems in erratic demand conditions. In Proceedings of the IIE Annual Conference, Miami, FL, USA, 30 May–3 June 2009.
20. Sandanayake, Y.G.; Oduoza, C.F.; Proverbs, D.G. A systematic modeling and simulation approach for JIT performance optimization. *Robot. Comput. Integr. Manuf.* **2008**, *24*, 735–743. [CrossRef]
21. Enns, S.T.; Rogers, P. Clarifying CONWIP versus PUSH system behavior using simulation. In Proceedings of the 2008 Winter Simulation Conference, Miami, FL, USA, 7–10 December 2008; pp. 1867–1872.
22. Cheraghi, H.S.; Dadashzadeh, M.; Soppin, M. Comparative Analysis of Production Control Systems Through Simulation. *J. Bus. Econ. Res.* **2008**, *66*, 88–104. [CrossRef]
23. Terrence, M.J. A Simulation and Evaluation Study of the Economic Production Quantity Lot Size and Kanban for Single Line, Multi—Product Production System Under Various Setup Times. Ph.D. Thesis, Kent State University, Kent, OH, USA, 2008.
24. Bonvik, A.M.; Couch, C.; Gershwin, S.B. A Comparison of Production-line Control Mechanisms. *Int. J. Prod. Res.* **1997**, *35*, 789–804. [CrossRef]
25. Spearman, M.L.; Woodruff, D.L.; Hopp, W.J. CONWIP: A Pull Alternative to Kanban. *Int. J. Prod. Res.* **1990**, *28*, 879–894. [CrossRef]
26. Boonlertvanich, K. Extended CONWIP-Kanban System: Control and performance. Ph.D. Thesis, Georgia Institute of Technology, Atlanta, GA, USA, 2005.
27. Lage, J.M.; Filho, M.G. Variations of the Kanban system: Literature review and classification. *Int. J. Prod. Econ.* **2010**, *125*, 13–21. [CrossRef]
28. Prakash, J.; Chin, J.F. Variation of CONWIP Systems. *World Acad. Sci. Eng. Technol.* **2012**, *6*, 2765–2769.
29. Chao, H.; Shih, W.L. Simulation studies in JIT production. *Int. J. Prod. Res.* **1992**, *30*, 2573–2586. [CrossRef]

30. Torkabadi, A.; Mayorga, R. Evaluation of pull production control strategies under uncertainty: An integrated fuzzy AHP-TOPSIS approach. *J. Ind. Eng. Manag.* **2018**, *11*, 161–184. [CrossRef]
31. Askin, R.G.; Goldberg, J.B. *Design and Analysis of Lean Production Systems*; John Wiley & Sons: New York, NY, USA, 2002.
32. Bicheno, L. *The New Lean Toolbox: Towards Fast Flexible Flow*; Picsie Books: Buckingham, UK, 2004.
33. Darlingtona, J.; Francisb, M.; Foundc, P.; Thomasc, A. Design and implementation of a Drum-Buffer-Rope pull-system. *Prod. Plan. Control.* **2015**, *26*, 489–504. [CrossRef]
34. Marek, R.P.; Elkins, D.A.; Smith, D.R. Understanding the fundamentals of Kanban and CONWIP pull systems using simulation. In Proceedings of the Winter Simulation Conference, Arlington, VA, USA, 9–12 December 2001.
35. Rotaru, A. Performance Analyses of Conwip—Base stock controlled production system using simulation. In *Scientific Bulletin, Automotive Series*; Faculty of Mechanics and Technology, University of Pitesti: Pitești, Romania, 2011; Volume XVI.
36. Sergent, R.G. Verification and validation of simulation models. In Proceedings of the 2010 Winter Simulation Conference, Baltimore, MD, USA, 5–8 December 2010.
37. Matlab, R. Model and Simulate Message Communication and Discrete-Event Systems. 2017. Available online: https://www.mathworks.com/products/simevents/whatsnew.html (accessed on 17 November 2017).
38. Gestettner, S.; Kuhn, H. Analysis of production control systems Kanban and Conwip. *Int. J. Prod. Res.* **1996**, *34*, 3253–3273. [CrossRef]
39. Design-Expert®Software, Version 11. Available online: https://www.statease.com/dx11.html (accessed on 14 November 2017).
40. Pavlić, I. Statistička teorija i primjena, Tehnička knjiga Zagreb. *Color Plates* **2004**, *48*, 239.
41. Cajner, H.; Tonković, Z.; Šakić, N. The significance of data transformation in analysis of experiments. In Proceedings of the 12th International Scientific Conference on Production Engineering (CIM 2009), Biograd, Croatia, 17–20 June 2009.
42. Ertay, T. Simulation approach in comparison of a pull system in a cell production system with a push system in a conventional production system according to flexible cost: A case study. *Int. J. Prod. Econ.* **1988**, *56–57*, 145–155. [CrossRef]

processes

Article

The Analysis of the Spatial Production Mechanism and the Coupling Coordination Degree of the Danwei Compound Based on the Spatial Ternary Dialectics

Zihan Yang, Jianqiang Yang * and Kai Ren

Department of Urban and Rural Planning, School of Architecture, Southeast University, Nanjing 210096, China; 230208012@seu.edu.cn (Z.Y.); 230189012@seu.edu.cn (K.R.)
* Correspondence: 101004529@seu.edu.cn; Tel.: +86-13851513802

Abstract: With the gradual deepening of the development of high-quality urban transformation, the "Danwei Compound" urban space production method constitutes the basis of Chinese current urban spatial transformation. The transformation plan of the original danwei compound "stock" to promote the healthy development of urban society has become the focus of research. First, with the help of Lefebvre's space production theory, combined with the spatial transformation characteristics of its own structural form experienced by the Chinese urban danwei compound, the space production is divided into three stages, namely, the diversity-orderly type average space of the danwei compound system period, dispersed type abstract space of the commercial enclosed community period, and the integrated differential space of a livable community undergoing regeneration and transformation. At each stage, the government, market, and residents have different influences on time-space production. Secondly, using Hefei's typical danwei compound as the research carrier, according to the space ternary dialectics, a multi-level analysis of "representations of space-representational space-spatial practice" is carried out on the production mechanism, and the logic of different types of spaces in different periods are described. Among them, the representations of space of the change of the danwei compound are the interrelationship of multiple governance subjects in different periods, such as changes in the implementation degree of governance strategies, the degree of residents' community governance participation, residents' satisfaction with community governance, etc. The representational space is the residents' community perception and interpersonal relationship at different transition stages, Interpersonal trust, and other social relations' changes. Spatial practice is manifested in changes in the support of public service facilities, public space, per capita living area, building quality, architectural style, and illegal building area. Finally, the three-dimensional space dialectical coupling coordination degree model is used to analyze and compare the representations of space of typical settlements in the three stages and the coupling characteristics of the representational space and the practice of space. On this basis, we provide innovative ideas and put forward relevant measures and suggestions for the regeneration, transformation, and development of livable areas.

Keywords: community transformation; community innovation governance; ternary space; coupling and coordination analysis

Citation: Yang, Z.; Yang, J.; Ren, K. The Analysis of the Spatial Production Mechanism and the Coupling Coordination Degree of the Danwei Compound Based on the Spatial Ternary Dialectics. *Processes* **2021**, *9*, 2281. https://doi.org/10.3390/pr9122281

Academic Editor: Sergey Y. Yurish

Received: 25 October 2021
Accepted: 15 December 2021
Published: 20 December 2021

1. Introduction

The community is the basic unit of urban social life and an important carrier of urban development and change. In the process of social transformation in China, urban communities have undergone profound historical changes: From the danwei compound model that dominated during the planned economy period [1] to the large-scale development of commercialized communities in the market economy period, and then the rise of the construction of livable residential areas during the period of high-quality development has promoted the great tide of urban community construction. In terms of function, the danwei compound centers on the "work" function of the danwei, organizing the living facilities and various welfare facilities within the danwei. In terms of space, "walls" are used to realize the enclosure, closure, and integrity, which form a spatial model of the integration of work and residence. Among them, the danwei refers to the enterprises, government institutions, and public agencies that provide urban residents with various employment opportunities [2], such as factories, schools, hospitals, government agencies, etc. In the period of a socialist planned economy, the danwei compound was the most basic form of social management and organization in Chinese cities [3]. However, due to historical and practical reasons, the traditional danwei compound has not received enough attention and has gradually fallen into a stagnant and declining development dilemma, facing the risk of being demolished, marginalized, or lowered. This not only troubles the daily life of residents in the danwei compound, but also becomes a major problem for the government's community governance. This type of community has attracted wide attention from multiple disciplines such as urban planning, sociology, and management. During the political, economic, and social evolution of community structure, how can multiple forces (governments, danwei, markets, and residents) representing different interests participate in influencing space production at different stages? How can we build a harmonious and livable differentiated space and realize space justice? Based on Lefebvre's theory of space production, this article combines a typical case of a danwei compound that has undergone a series of spatial evolution processes, conducts a qualitative analysis of the space production mechanism, performs quantitative analysis of coupling and coordination of space production in the different development stages of the danwei compound, and proposes possible paths for the livable and sustainable reconstruction of the danwei compound in the new stage.

2. Materials and Methods

2.1. Research Perspective: The Production of Space

With the accelerating development of urbanization in the West, various urban problems have come along. Based on this, the conformity of urban problems' productivity and production relations, the theory of spatial production focuses on what has triggered the "spatialization shift" of social problems since the 1970s and at the end of the 20th century ushered in the formation and culmination of the "space fever" thought [4]. Among them, Lefebvre's theory of space production has played an irreplaceable role in leading and promoting the development of space research. He constructed the theory of space production by leading and promoting space ontology: (1) the history of social space. According to the "production of space in which history participates", three types of space are divided into "absolute space", "abstract space", and "differential space", used to describe the changes in space and corresponding social relations caused by the transformation in production methods [5]. (2) Typology of social space. Changes in the mode of production will cause spatial changes and corresponding changes in social relations. In the "absolute space" in the first historical stage, the space represents the space of ancient society and pre-capitalist society, which is different from an independent natural space and is produced by the interaction of common language, geography, and blood relationship. The "abstract space" in the second historical stage is that the space represents the capitalist society, which is a dominant tool of power and the object of Lefebvre's criticism because this space emphasizes exchange value over use value. Through homogenization, separation and repetition invade daily life practice. The "differential space" in the third historical stage is a space representing socialism, aimed at realizing space justice and emphasizing the "use value" of space users [6]. The history of spatial changes in these three stages is a process of "dematerialization" and full of class conflicts and struggles. (3) Space ternary dialectics. In each of the above stages, space undergoes its own reproduction through specific logic: representation of spaces, that is, the discipline space of the ruling class for daily life, which aims to construct the order of the strong in space. Spaces of representation are complex, disciplined spaces, including the perception and experience of the disciplined class. Spatial practices are spaces for social action [7]. Representations of space and representational space are opposed to each other, and spatial practice is a transformation zone between the two.

Urbanization space is the main practical background for the empirical research of Lefebvre's space production theory. Research on the macro and meso scales mainly focuses on the dynamic mechanism of urbanization from the aspects of capital flow, population flow, and institutional system construction. For example, Paris [8], Chicago [9], Baltimore [10], Los Angeles [11], and Shanghai [12] are used to illustrate the production mechanism of urbanized space. The research on the meso and micro scales mainly starts from the perspectives of sociology and behavior and adopts actor network or social network analysis to analyze how resistance groups or social collective actions shape space under the influence of power and how space affects the shaping of a sense of place and the formation of interest alliances. For example, micro-scale spaces with unique social structures such as "urban-rural fringe" [13], "urban village" [14], "village-transferred community" [15], "tourist community" [16], and "Taobao village" [17] have become popular research objects. The above-mentioned research items fully demonstrate that the theory of space production is effective and practical for explaining the nature of space phenomena and the process of space evolution on a global scale. However, the danwei compound with Chinese characteristics has experienced a major political, economic, and social transformation in China's development at the macro level, and the physical space and social space have been reshaped at the micro level. Therefore, this theory will essentially explain the spatial production process and generation mechanism of the danwei compound.

2.2. Transformation of Danwei Compound and Spatial Production

The theory of space production is an important tool applied to the analysis of social space problems. Lefebvre's analysis of the evolution of absolute space, abstract space, and difference space deduces the context of the socialist revolution, aims to resist the homogenization of social space and pursue space justice, and also interprets the changes in social relations in social space [18]. However, the changes in social relations are not limited to the great changes in human society. For example, the "danwei compound" in China is based on the socialist space but its development and changes also show "discipline" and "anti-discipline" and the conflict and opposition between "use value" and "exchange value" [19], which outline the evolution process of the dominant force in space production and the evolution of community social relations. In this process of space production, the transition of community residents from being disciplined to being active participants in space production is also a process of pursuing space justice. Therefore, if not limited to the specific type of space in the historical stage of social space, focusing the theoretical transformation on "use value, exchange value, and social relations" can provide an analytical framework for the study of space production in closed settlements. According to the development stage of the danwei compound and the change of the dominant space production force, it will be divided into three stages.

2.2.1. Homogeneous Absolute Space

The residential space during the danwei compound period was similar to the absolute space. In the planned economy period, housing was standardized by the danwei in accordance with the country's "quota index system", unified distribution, and unified management. In the danwei compound, the families are linked by the "danwei", and the long-term, stable neighborhood structure has spawned an "acquaintance society" based on occupation, life, emotion, and social relationships [20]. Since urban land is owned by the state, the government prohibits the transfer of land use rights, denying the exchange value of space [3]. On the other hand, danwei takes over the government's decentralized rights, which have maximized the value of land use, and the needs of danwei employees can basically be met through negotiation and discussion [21]. Although there is a certain degree of social gap and spatial isolation outside the danwei compound, within the compound, the identity, status, production work, daily life, and leisure entertainment of the people are in a homogeneous, single, and stable state. However, it is different from Lefebvre's absolute space and the author defines it as "homogeneous absolute space".

2.2.2. Atomic Abstract Space

In the context of the transformation of the economic system, the danwei compound is facing disintegration. With the privatization of housing property rights, the original danwei's funds and management support for the community are withdrawn. Under the combined effect of the government and the market, it has gradually evolved into ordinary commercial communities. In this stage, space planning takes the exchange value as the core and ignores the use value of space, just like the colonization of absolute space by the bourgeoisie in abstract space [22]. At this time, the danwei compound has become a low-level gathering area in terms of income, consumption, social status, taste, etc. due to the departure of people with market consumption power and the influx and settlement of low-level urban residents [23]. At the same time, the mutual understanding and close communication relationship of the former danwei compound have been diluted by the market economy, and the sense of community and place has gradually disappeared [24]. Due to the mobility of residents enhancing greatly, along with the struggle between usage valve of living space and the exchange value of physical space, the social relations of settlements are discrete, so it is defined as an abstract discrete space.

2.2.3. Integrated Differential Space

In the context of the transformation of the value orientation of urban development with a high-quality development, as the people's livelihood project, "reconstruction of old communities and construction of livable residential area" is carried out. In the new round of production of space in residential areas, the "government–market" interest alliance has transformed into a "government–market–public" multi-participation and co-dominant model, showing the characteristics of social space from contradictions and conflicts to inclusive and co-governance [25]. The social spatial relationship has shifted from the confrontational "conquer-be-conquered" to the co-governance "governance-be-governed". Spatial conflicts are gradually dissolving, and spatial integration is gradually realized through multi-party negotiation. Residents unite in order to strive for common spatial interests, so that a new social network of community is gradually established, forming a sense of community belonging. The government–resident relationship has gradually shifted to mutual respect, consultation, and co-governance, and space use value has become the core of community space production. Therefore, it is defined as an "integrated differential space".

From atomic abstract space to integrated differential space, as Lefebvre said, space was an anti-discipline tool and had the ability to resist abstract homogeneity [26]. Lefebvre found that class struggle was a means to change homogeneity, and in the evolution of closed settlements, the game of different interest groups also showed a path to resist homogeneity hegemony and produced differential anti-discipline [27]. At the same time, the three space types divided in the historical sequence of space production carry out self-production in accordance with the ternary dialectics of space.

2.3. Analysis of Coordination Degree of Three-Dimensional Space Coupling in the Evolution of Danwei Compound

2.3.1. Index Selection

Based on the previous theoretical framework and case evolution analysis, the evaluation system of representations of space, representational space, and spatial practice in the theory of ternary space dialectics is established. Among them, representations of space is a place where space rulers who have the ability to control and create space produce discipline and order [28]. Therefore, the number of subjects in the governance of the compound space at different stages, the degree of participation of residents in community governance, and the degree of residents' satisfaction with community management and services are used to quantify and evaluate the dominance and control situation of the social space of the compound by the dominant power order in the compound in different periods. Among them, the participation of multiple entities in the space governance of the compound is quantified based on the entities participating in governance in different governance models such as danwei–family committee–building owner, danwei–market, danwei–property–ownership–residents, etc. The degree of participation of the resident community is mainly quantified and calculated by the annual average number of residents' substantial participation, symbolic participation, and non-participatory participation. The questionnaire survey is used to evaluate the residents' satisfaction with community management and services to reflect the acceptance of space production controlled by the dominant power order by space users (Table 1).

The representational space is the space of the "dweller" and the space of the "user", presenting a living and empirical part of the social space. However, because its spatial status is completely opposite to the "representation of space", it is dominated and can only be passively experienced by us [29]. Therefore, community social networks, community social support, community cohesion, and community sense of belonging are used to quantify and evaluate the passive shaping of the representational space. Among them, the "number of neighbors you know in the compound", "the number of neighbors who will greet each other in the compound", "the number of community residents who have a good relationship to visit", and other indicators are used to quantitatively evaluate the community social network construction situation in different periods. "Do you often seek advice from your neighbors?", "Can you borrow what you need from your neighbor's house smoothly?", "Did the residents of the community ever provide you with help in the past three months?" are the quantitative evaluation indicators of community social support. "Most people in the compound are willing to help each other", "Most of the residents of the compound have a high spirit of participation", and "Generally speaking, the relationship between the residents of the community is harmonious" are in the quantitative evaluation index of community cohesion. "I feel at home in the compound", "Like my residential area", and "Tell others where I am proud to live" are the quantitative evaluation indicators of the community's sense of belonging.

Lefebvre emphasized: Every social lifestyle produces its own space [6]. Spatial practice includes specific locations in a space, as well as spatial settlements unique to each social form. Therefore, the basic living conditions, living conditions and human settlement environment, and other physical space shaping conditions reflect the production of space under different lifestyles in different periods. Among them, "number of households", "proportion of non-supply and marketing cooperatives and their family members", "proportion of household separation in the compound" are the quantitative evaluation indicators for the basic living conditions of the compound. Take "building quality score", "building style score", "infrastructure allocation score", "number of parking spaces for non-motor vehicles", and "number of compound life service facilities" as quantitative evaluation indicators of the compound's living conditions. The "greening rate", "proportion of public activity space area", "number of public activities and leisure facilities", and "proportion of non-illegal construction area" are the quantitative evaluation indicators of the compound's living environment.

Table 1. Ternary space evaluation system.

First Level Indicator	Second Level Indicators	Third Level Indicators	Index Number
Representations of space	Diversified governance of space governance	Number of main interest groups of space governance	x1
	Resident community governance participation	Annual average number of substantive participations	x2
		Annual average number of symbolic participations	x3
		Annual average number of non-participation participation	x4
	Residents' satisfaction with community management and services	Residents' satisfaction with community management	x5
		Residents' satisfaction with community services	x6

Table 1. *Cont.*

First Level Indicator	Second Level Indicators	Third Level Indicators	Index Number
Representational space	Community social network	The number of neighbors you know in the compound	y1
		The number of neighbors in the compound who will greet each other when meeting	y2
		The number of community residents who are familiar with each other and can visit	y3
	Community social support	Do you often seek advice from neighbors? 1 (almost never)–5 (often)	y4
		Can you borrow the needed Dong from the neighbor's house smoothly? 1 (not possible), 3 (sometimes possible), 5 (yes)	y5
		In the past three months, how often did you receive help from community residents? 1 (almost never)–5 (often)	y6
	Community cohesion	Most people in the compound are willing to help each other 1 (disagree)–5 (strongly agree)	y7
		Most of the residents of the compound have a high spirit of participation 1 (disagree)–5 (strongly agree)	y8
		In general, the relationship between the residents of the community is harmonious 1 (disagree)–5 (strongly agree)	y9
	The sense of belonging in the community	Have a sense of belonging in the compound 1 (disagree)–5 (strongly agree)	y10
		Like our compound 1 (disagree)–5 (strongly agree)	y11
		Tell others that I am very satisfied with my housing 1 (disagree)–5 (strongly agree)	y12
Spatial practice	Basic living conditions	Number of households	z1
		Percentage of non-supply and marketing cooperatives and their family members	z2
		Percentage of residence and separation from household registration in the compound	z3
	Living conditions	Construction quality 1 (poor)–5 (good)	z4
		Architectural style 1 (poor)–5 (good)	z5
		Infrastructure configuration 1 (poor)–5 (good)	z6
		Number of parking spaces for non-motor vehicles	z7
		Number of life service facilities	z8
	Living Environment	Percentage of green space	z9
		Proportion of public activity space	z10
		Number of public activities and leisure facilities	z11
		Proportion of illegal construction area	z12

2.3.2. Research Methods

- Coordination degree of spatial production coupling.

The range standardization method is used to standardize the original indicators. If the index data are larger, it is more conducive to the development of the system to adopt positive index processing. Otherwise, the reverse index will be used to deal with. The standardized formula is as follows [30]:

Positive indicators:

$$X_{ij} = \left(x_{ij} - x_{min}\right) / \left(x_{max} - x_{min}\right), \ X_{ij} \in [0, 1] \tag{1}$$

Negative indicators:

$$X_{ij} = \left(x_{max} - x_{ij}\right) / \left(x_{max} - x_{min}\right), \ X_{ij} \in [0, 1] \tag{2}$$

After standardizing the indicators, calculate the proportion of the standard value of the No. i evaluation object under the No. j index [31]. The entropy and weight of the No. j index are:

$$P_{ij} = X_{ij} / \sum_{i=1}^{n} X_{ij} \tag{3}$$

$$H_{ij} = -\frac{1}{\ln n} \sum_{i=1}^{n} P_{ij}(\ln n) P_{ij} (j = 1, 2, \ldots, n) \tag{4}$$

$$x_j = a_j / \sum_{i=1}^{m} a_j \tag{5}$$

In the formula, $0 \le P_{ij} \le 1$, and, assuming $P_{ij} = 0$, $P_{ij} \ln P_{ij} = 0$, the difference coefficient of the No. j index is $a_j = 1 - H_j$.

Further construct a three-dimensional space dialectical coupling coordination degree model [32].

$$D(P_T, P_E, P_G) = \sqrt{C \times T} \tag{6}$$

$$C = \sqrt[3]{\frac{P_T \times P_E \times P_G}{P_T + P_E + P_G}} \tag{7}$$

$$T = \alpha P_T + \beta P_E + \gamma P_G \tag{8}$$

In the formula, D is the degree of coupling coordination, C is the coupling degree of the three systems, T is the comprehensive coordination index of the three systems, P_T, P_E, and P_G are the comprehensive evaluation indexes of the three systems of representations of space, spatial practice, and representational space, respectively, and $\alpha, \beta,$ and γ are undetermined coefficients. Considering that the three systems are of equal importance, we do equal weight processing here, that is, α, β, γ are all $1/3$. Using Liao Chongbin's "ten-cent method" classification standard for coupling coordination degree [33], 0–0.1 means extreme dissonance, 0.1–0.2 means severe dissonance, 0.2–0.3 means moderate dissonance, 0.3–0.4 means mild dissonance, 0.4–0.5 is on the verge of imbalance, 0.5–0.6 is basic coordination, 0.6–0.7 is primary coordination, 0.7–0.8 is intermediate coordination, 0.8–0.9 is good coordination, and above 0.9 is high-quality coordination.

- Geographic detector.

Use differentiation and factor detection to identify the extent to which the influencing factors (X) of the space production quality of the danwei compound explain the level of space production (Y), measured by the q value, and the expression is [34]:

$$q = \frac{\sum_{h=1}^{L} N_h \sigma_h^2}{N \sigma^2} = 1 - \frac{SSW}{SST} \tag{9}$$

$$SSW = \sum_{h=1}^{L} N_h \sigma_h^2, SST = N \sigma^2 \tag{10}$$

In the formula, $h = 1, 2, \cdots, L$, L is the stratification of factor X; N_h and N are the number of units in layer h and study area, respectively; σ_h^2 and σ^2 are the variances of the spatial production effectiveness (Y) of layer h and the study area, respectively; SSW is the sum of variance within the layer; and SST is the total variance of the study area. The value range of q is $[0, 1]$. The larger the value of q, the stronger the explanatory ability of the influencing factor (X) is on the effect of space production (Y), and vice versa. When $q = 1$, it means that the influencing factor (X) completely controls the space production effect (Y). When $q = 0$, it means that the influencing factor (X) has no relationship with the space production effect (Y). The calculation of the q value and a significance test can be realized by GeoDetector software. By comparing the numerical value of each factor q, the dominant factor of the coupling level of space production in the danwei compound can be detected.

2.4. Analysis Framework

As the sub-central city of the Yangtze River Delta metropolitan cluster, Hefei has been driven by the internal drive of the structural transformation of the economic system and the external influence of the continuous deepening of the urbanization process since 1978 (the reform and opening), and the urban danwei compound has quickly realized the spatial transformation of its own structural form. In this context, the Supply and Marketing Cooperative Compound is in the central area of the city (Luyang District). It was built in the 1950s and has a total of 16 buildings. There were 556 households at the initial stage of completion, and there are 469 households at present [35]. The current residents are mainly provincial supply and marketing employees or retired employees and some foreign tenants. The authors conducted research and surveyed the community from June 2018 to October 2020 (30 in-depth interviews, 190 documents, and 156 were effectively recovered). Combining the transformed space production theory, it is divided into three development stages and space types according to the development stage of the community and the corresponding changes in the leading force of space production (Figure 1).

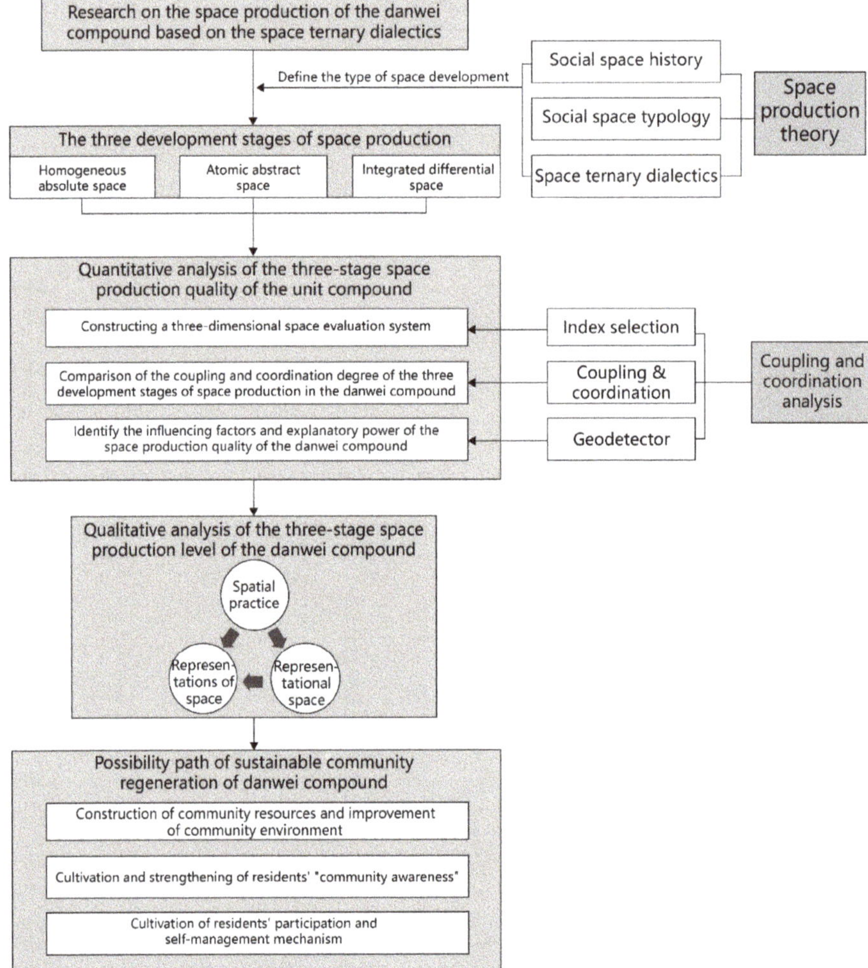

Figure 1. Framework diagram of research (source: self-drawn by one of the authors).

3. Results

3.1. Comparison of the Coupling and Coordination Degree of the Three Development Stages of Space Production in the Danwei Compound

According to Lefebvre's ternary space dialectical theory [36], we calculated the coordinated relationship between representations of space, spatial practice, and representation in the different spatial history of the danwei compound. We used Geographic detector to analyze the dominant factors affecting the ternary spatial coupling of the danwei compound's space production. Calculating the spatial production coupling degree of the three stages of spatial development of the supply and marketing cooperative compound, we obtained 0.28, 0.16, and 0.51, respectively (Table 2). From the perspective of coupling coordination types, the period of homogeneous absolute space was moderately dissonant, the atomic abstract space was severely dissonant, and the integrated differential space was basic coordination. In the period of homogeneous absolute space production, the danwei was in the role of "parent", and the danwei's overall management of the life-long needs of the individual made danwei members strongly dependent on the danwei in order to obtain various resources. Therefore, even if danwei members feel a strong sense of relative deprivation, they have not considered leaving the danwei organization, forming a dependence on the danwei organization [37]. This deprivation–deprivation, dependence–dependence relationship between the danwei and danwei members makes the power of the compound in the process of space power have great disparity, and individual needs for spatial differentiation cannot be expressed. However, relative to the various resources and life opportunities that the danwei can provide to individuals, danwei members voluntarily give up their voices for differentiated needs. Therefore, although the space production of the danwei compound in the first stage can operate normally and efficiently, this does not mean that the spatial conflicts and crises are not hidden in it. Therefore, this stage presents a type of coupling and coordination with moderate dysregulation in the ternary space.

Table 2. Coupling and coordination degree of space production in the danwei compound.

Space Development Stage	T Coordination	C Coupling	D Coupling and Coordination	Coupling Coordination Type
Differential absolute space	0.284460044	0.275603466	0.279996739	Moderate disorder
Atomic abstract space	0.151667004	0.172727011	0.161854837	Severe imbalance
Integrated differential space	0.563872952	0.455314666	0.506694805	Basic coordination

In the stage of atomic abstract space production, with the advancement of reform and opening up, the market economy has gradually replaced the planned economy, and the danwei system has tended to disintegrate. The Danwei has already withdrawn from the vacancies such as housing management and community maintenance. Due to various historical problems, no new social organization has come forward to take over the danwei compound at this stage, which has brought about a vacuum in management and the decline of the physical space environment. Driven by the market economy, capable compound residents have rented and sold compound housing and moved out of the compound. The composition of compound residents is facing the trend of heterogeneity, aging, and marginalization. With changes in the structure of residents, on the one hand, the close community social network during the planned economy period was separated, and a strong sense of community identity and belonging was lost, making the relationship between the residents of the compound present an atomized state of separation. On the other hand, the differentiated composition of residents has brought a variety of spatial differentiation needs, and spatial contradictions have become more prominent. In addition, the combination of local governments, the danwei, and developers has severely squeezed the living space of residents in the danwei compound and intensified spatial conflicts. Therefore, this stage is a period of decline in the development of the danwei compound, presenting a state of coupling and coordination with a serious imbalance in the ternary space.

In the period of integrated differential space production, take the opportunity of the transformation of the old community to improve the physical space environment and facilities of the compound and introduce the property company to market-based management of the compound to provide professional community services for the residents of the compound. The use value of residents' living space and social space has received more attention and improvement. On this basis, the residents of the compound gradually form a sense of community and self-management and take the initiative to participate in community affairs. The compound gradually shows a virtuous cycle of development. Although more and more demands continue to be fulfilled, new demands will continue to arise. This means that although the spatial conflict has been relieved to a certain extent, it has never completely disappeared. Therefore, this stage presents a state of coupling and coordination in which the ternary space is basically coordinated.

3.2. Identification of the Influencing Factors and Explanatory Power of the Space Production Quality of the Danwei Compound

Through survey methods such as satisfaction questionnaire surveys and in-depth interviews of community management departments, residents of the supply and marketing cooperative compound, and danwei management departments, the data indicators obtained were used for geographic detection. We Obtained five leading factors that affect the quality of space production in the unit compound, including basic living conditions (0.89), living conditions (0.89), community social support (0.85), sense of community belonging (0.71), and residents' satisfaction with community management and services (0.69) (Table 3).

Table 3. Three-dimensional spatial coupling influencing factor q of spatial production determines the value of force.

Geodetector Indicators	q Determining Power Value	Geodetector Indicators	q Determining Power Value	Geodetector Indicators	q Determining Power Value
Diversified governance of compound space	0.33	Community social network	0.67	Basic living conditions	0.89
Resident community governance participation	0.21	Community social support	0.85	living conditions	0.89
Residents' satisfaction with community management and services	0.69	Community cohesion	0.67	Living Environment	0.58
		Sense of community	0.71		

According to the analysis of the specific situation of the case community, in the evolution of the spatial production of the supply and marketing cooperative compound, the basic living conditions of the "proportion of non-supply and marketing cooperative employees and their family members" and the "proportion of households in the compound" and other indicators were equal. This Reflects the trend of the compound population from homogenization to heterogeneity. The heterogeneity has led to the alienation of the relationship between the residents of the compound and weakened the residents' sense of community identity and belonging and is the main factor that caused the compound to lose the original governance advantage of social capital enrichment. In the current stage of the integrated differential space production of the case compound, the old community transformation was used as an opportunity to improve the living conditions and human settlement environment of the compound. On the premise of meeting the needs of residents' space use, infrastructure improvement projects such as water and electricity entering the home are carried out. At the same time, living service facilities such as parking spaces, centralized garbage collection points, streetlights, and community publicity boards have been increased. Therefore, the living conditions have been greatly improved, and, therefore, the positive influence on space production is greater. However, due to the limited area of the case compound and the shortage of land, it is unable to provide sufficient public venues for residents; so, the degree of improvement in this old community reconstruction was not significant. At the same time, at this stage, the compound introduced a market-oriented property company to manage the compound uniformly; so, the residents' satisfaction with community management and services increased significantly, and the impact was greater. Based on the improvement of the living conditions of the compound and the improvement of management and public services, the residents' sense of community and self-management was cultivated, and the community's social support and sense of belonging to the community increased significantly.

4. Discussion

4.1. Production of Homogeneous Absolute Space (Led by Traditional Danwei Compound "Government–Danwei–Employee")

In the period of the danwei compound, the internal flow of personnel in the Supply and Marketing Cooperative Compound was almost static. It was based on the danwei system and the common working and living environment formed an acquaintance society [38]. Although there is a certain degree of social gap and spatial isolation outside the danwei compound, within the compound, the identity, status, production work, daily life, and leisure entertainment of the people are in a homogeneous, single, and stable state. However, it is different from Lefebvre's absolute space; therefore, it is defined as "homogeneous absolute space" (Figure 2).

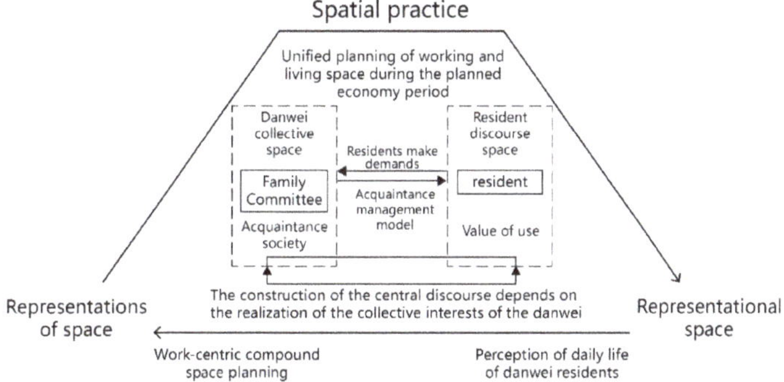

Figure 2. Homogeneous absolute space production path (source: self-drawn by one of the authors).

4.1.1. Representations of Space

Representations of space are the places where the discipline order of space rulers takes place [39]. However, after investigation, it was found that there is no obvious discipline–disciplined mutually antagonistic relationship between the danwei and employees in the Supply and Marketing Cooperative Compound. Danwei compounds often take collective work and life as the center for spatial planning. The work–life discourse space formed in the danwei compound, and residents can express their demands within the scope of the available conditions. The family committee in charge of logistics is an autonomous organization of residents in the compound that is autonomous and connected to the management department of the danwei and, in principle, reflects the wishes and needs of the residents. This kind of work–living space not only represents the physical space and the boundary of the walls inside the Supply and Marketing Cooperative Compound, but is also a kind of identity, belonging to the same danwei, which has long integrated similar work backgrounds, welfare benefits, and living environment, forming a social relationship of acquaintances with certain tribal and rural characteristics [40].

In such a spatial field, the construction of the central discourse depends on the realization of the collective interests of the danwei [20]. This "danwei–family committee–masses" hierarchical model of the power structure in which residents participate in the management of the life affairs of the compound, under the premise of ensuring collective interests, balances the expression of residents' rights demands and the realization of personal interests. Therefore, the planning of the danwei compound space presents a high degree of homogeneity and equalization and efficiently reflects the use value of the working and living space, rather than the heterogeneous economic exchange value. Even when the power discourse of the family committee and the residents of the compound is in conflict, the resilience of the management model based on acquaintances and the constraints of the ideology of the supremacy of collective interests at that time enable the contradiction to be resolved to the greatest extent. Meanwhile, families in the compound with additional difficulties will also receive certain care and support. Therefore, conflicts of interest in space are always easy to resolve, and cracks in social relations are not easily accumulated and deepened. This is also very different from the alienation development of atomic abstract space after reform.

4.1.2. Representational Space

The living field of the residents of the danwei compound is the representational space [41]. Affected by the planned economic system at the time, housing was built, deployed, and managed uniformly by the danwei, and residents were all employees of the danwei. Employee housing is allocated with certain rules and regulations, which also take the need of employees into account. As the family population increases, housing will be adjusted. Therefore, in the homogeneous absolute space at this stage, the living life inside the compound is relatively fair and harmonious. At the same time, due to the high level of consistency between the mode of work and lifestyles of the residents in the danwei compound, it is easy to form similar values, which can then evolve into a solid collective consciousness, forming mutual help, mutual understanding, friendship, and a good self-dedication atmosphere, and conflicts between people are easier to resolve [42]. In addition, the "dialogue" nature of the acquaintances and social characteristics between the family committee and the residents of the compound makes it play a role in reflecting demands and wishes of the residents to the danwei and reconciling the conflicts of the masses. Therefore, the representational space at this stage is more like a community with the meaning of "home" based on a work relationship [43].

Take the living experience of this period dictated by the resident Uncle Li during the interview as an example. Uncle Li was just in his early 30s when he was transferred to the Supply and Marketing Cooperative Compound. He and his wife and daughter were assigned to a 60-square meter house on the first floor. Aunt Zhang was staying in the same danwei with them at the time, with her son and elderly mother. According to the number of people in the family, housing of the same size on the fifth floor was allocated. However, Aunt Zhang's elderly mother was very old and found it inconvenient to go up and down. Therefore, Aunt Zhang relayed her request to the family committee. After negotiation and mediation, Uncle Li agreed to exchange houses with Aunt Zhang and moved to the fifth floor. In Uncle Li's view, we are all neighbors and colleagues in the same danwei, so we should negotiate and help others if others have difficulties. Additionally, if he moved to the fifth floor, his little girl's room would have better lighting. Therefore, why not do it?

4.1.3. Spatial Practice

In the process of the danwei compound spatial practice in this period, the management structure of danwei–family committee–residents based on the society of acquaintances was not only limited by the rules and regulations of danwei management but also relied more on the close interpersonal relationship network and similar values formed in long-term accompanied work and daily life experience [44]. During the investigation, Uncle Zhang, a former employee of the Supply and Marketing Cooperative, recounted such personal experiences. During the danwei–system compound period, because the danwei followed strict rules and regulations on the area of individual housing, according to regulations, ordinary employees calculated the total area for one family according to the standard of 25 square meters for each family member, based on weddings and funerals. For the newly increased population and relocation of households, the housing area can be increased or recovered. Secondly, the employees of the supply and marketing cooperative compound had a unified orientation to enjoy educational service facilities such as kindergartens and Primary School as well as medical facilities such as the supply and marketing cooperative health stations. At the same time, there were public spaces and living service facilities, such as a large auditorium, playground, canteen, etc., in the compound [35]. Thirdly, guided by the spiritual consciousness of collective honor, there will be basically no private construction and illegal building of public space in the compound. The residents of the compound mostly abide by the regulations of the family committee, actively maintain the overall style and cleanliness of public space, and maintain the accessibility and integrity of street space in the compound. Finally, in terms of building quality and architectural style, in the planned economy period, the country's overall economic development was in its infancy. Therefore, the economic levels had not yet solved the food and clothing problems of the vast majority of people [45]. Therefore, during the construction of the danwei compound during this period, the goal of economy, practicality, efficiency, and beauty was achieved in the order of hierarchy in order to pursue efficient construction and use and solve the lack of housing for a large number of poor employees as soon as possible. Therefore, the building structures are mostly brick-concrete multi-story houses, with poor building quality, and the architectural style fails to reflect more humanistic care and local characteristics.

4.2. The Production of Atomic Abstract Space (Led by "Government–Market" in Commercial Settlements)

In the 1990s, under the influence of the reform of the economic system, from the privatization reform of the danwei compound to the establishment of the housing marketization system, housing commercialization policies took root in the hearts of the people. The housing price separated the social composition relationship of the communities [46]. During this period, the Supply and Marketing Cooperative Compound responded to the national housing reform policy and began to gradually implement the privatization process of the public housing. Individuals could buy out the right to use the houses, which they lived in at that time, based on the assessed housing price. At the same time, because the Anhui Provincial Supply and Marketing Cooperative was located in the center of Hefei, the transportation was convenient, and it was close to Hefei's best primary schools, middle schools, hospitals, and other educational and medical public service facilities, with great investment-development potential [35]. Therefore, during this period, housing prices in the Supply and Marketing Cooperatives compound gradually increased. The original residents began to sell their housing in the compound in order to obtain a large amount of capital income or to improve their living conditions.

As a large number of original residents moved out of the compound, the composition of residents during this period showed serious differentiation in terms of educational level, occupational background, income level, and lifestyle, and the characteristics of differentiation were obvious. The needs of community residents showed a diversified trend, and the diversified needs could not be sustained by the government alone. With the intervention of capital, the Supply and Marketing Cooperative Compound blindly pursued the maximization of economic benefits in the control of government power and commercial interests and showed the dual attributes of consumption and daily life. During the "acquaintance society" period, the social network relationship of community residents was completely broken and there was a trend of discrete atomization among residents. Since space is the embodiment of social relations, the transformation of social relations also brings about the transformation of space production in this stage. From the "danwei–family committee–masses" model, coordinating with each other and participating in co-governance for the common realization of the collective interests of the danwei transformed into a "government–market" model, which showed a confrontational subject–object relationship of "discipline–disciplined". The exchange value of the community space crushed the use value and became the dominant value of the space evolution, and an atomic abstract community space was produced (Figure 3).

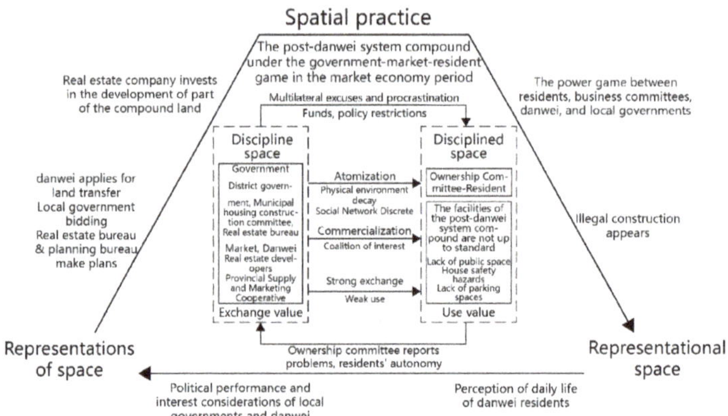

Figure 3. Atomic abstract space production path (source: self-drawn by one of the authors).

4.2.1. Representations of Space

People who have the ability to command space, through various means to create tangible space, representations of space and representational space, recreate and strengthen their own power [47]. The Supply and Marketing Cooperative Compound has united market forces (real estate developers, chain stores, etc.) and the local government forces (including district government, housing construction bureau, real estate bureau, etc.) to gradually dismember the living space of the original compound and jointly produce a closed commercial community with gentrification characteristics. The original danwei compound was incorporated into the commercial urban space, while the residents of the compound at this time were disciplined by a discourse system dominated by the exchange value.

Since 1990, Supply and Marketing Cooperatives Compound has redeveloped the houses along the street on the side of Lujiang Road in the compound and introduced commercial businesses such as chain hotels, restaurants, and supermarkets in order to obtain rental profits, after obtaining approval from the government departments in their jurisdiction (including the district government, the Housing Construction Bureau, and the Real Estate Bureau). In the process of renovation, in order to adapt to the usage of the commercial and service industry, the depth of the original residential buildings was expanded to the inside and outside of the courtyard, respectively, occupying the public activities and passage spaces of the compound. Driven by market-oriented reforms, in 2001, public service facilities such as kindergartens, canteens, restaurants, auditoriums, activity rooms, and guest houses in the original Supply and Marketing Cooperative Compound were replaced by real estate development. A total of 10 hectares of land occupied by the original factory and compound public service facilities along the side of Tongcheng South Road was auctioned at the highest price of 241.95 million yuan/hectare in the entire district, which was later developed into commercial, office, and high-rise residential space, which implemented closed management as an independent modern community [35]. The underground parking lot, public activity venues, and sports facilities in the small area are not shared with the residents of the original compound. Additionally, the opening price of this modern commercial community was 11,000 yuan/m^2, which was equivalent to twice the average price of a new house at 5500 yuan/m^2 in Luyang District at that time and was also much higher than the price of 4000 yuan/m^2 for second-hand houses in the Supply and Marketing Cooperative Compound at that time. The superior geographical location, convenient transportation, high-quality education and medical resources, high-quality residential development, and high real estate prices enabled the newly developed modern community to show a significant tendency towards gentrification [36], which transformed the public areas of the Supply and Marketing Cooperative Compound, such as kindergartens, auditoriums, canteens, and factories.

At the same time, after the redevelopment and reduction in land area, the Supply and Marketing Cooperative Compound, under the market-oriented reform policy and the original housing property usage rights was bought out by individuals. Meanwhile, the danwei management investment, strength, and voice in the compound were not as much as before. Therefore, at this time, the Supply and Marketing Cooperative Compound showed a decline in the situation with no one cleaning up garbage, mottled and cracked roads, aging houses without repairs, and decayed greenery without care. In addition, due to the commercialization of the compound's land and the development of real estate, local governments, danwei, and the market placed too much emphasis on seeking the exchange value of the space, ignoring the living needs of the residents and the use value of the compound and compressing almost all of the public activity area in the compound. Those made the danwei compound only have a single, homogenous, residential function, which was presented as the production of an abstract space with discrete characteristics.

4.2.2. Representational Space

During the transition period of the commodity economy, the residents of the Provincial Supply and Marketing Cooperatives compound were in a situation where the exchange value was greater than the use value. Daily life was full of the symbolic meaning of discipliners. The society of acquaintances disappeared, with the discrete atomization state appearing [48]. The residents of the compound personally experienced the multi-faceted disconnection of daily life needs, behaviors, and spatial patterns.

In terms of physical space, first, there was a lack of public activity space. The original public area of 10 hectares of the compound and the land were auctioned for real estate development without fully considering the interests and living needs of the residents of the original compound, resulting in a significant reduction in the public activity space of the residents of the original compound. The only remaining public activity space can no longer be used due to a long-term lack of management and maintenance resulting in cracked pavement, overgrown weeds, and accumulation of debris. At this stage, the public activity space in the compound is almost zero, which cannot provide the residents with space and opportunities for communication and interaction. Residents represented a danwei–home two-point-and-one-line isolated living state, which intensified the social network, ruptured the community relationship, and reduced the possibility of further repair and regeneration [49]. Second, was the lack of parking space. There was a non-motorized vehicle parking shed in front of the office building in the original Supply and Marketing Cooperative Compound. However, with the development of the market economy, people's living standards were improving day by day. Motor vehicles and non-motorized vehicles have gradually become popular in ordinary households; so, the spatial conflicts have intensified year by year. The root cause is the serious compression of the compound space caused by real estate development. The danwei and local government have not made long-term plans and considerations for the compound's future development and residents' daily needs, which makes the development of parking space in the compound limited, seriously lagging behind the growing parking demand of residents. As a result, cars and bicycles compete for parking space, and the fire channels and the corridors in front of buildings are parked in disorder, which seriously affects the daily commute of residents and poses safety risks. Secondly, there is a lack of daily life service facilities in the compound. The daily activity facilities such as the kindergarten, canteen, and auditorium in the original compound were demolished and real estate development was carried out. At present, children in the compound need to go to a kindergarten that is 1 km away from the compound, which greatly increases the commuting time and distance for parents to pick up their children to and from school, reducing the convenience of life and reducing the contact and interaction of the residents of the compound. The disappearance of the canteen brought inconvenience to the lives of the residents of the compound, especially the elderly residents who account for a large proportion. The cancellation of the auditorium meant the disappearance of the place where collective activities were held, which led to the almost disappearance of collective activities, the collapse of the collective ideology of the danwei, and the disappearance of common emotional connections, which accelerated the loss of the original residents in the compound [50]. Finally, the management and governance of the compound were in a state of suspension. Under the tide of marketization, with the privatization of housing use property rights, the danwei's investment and voice in the management and control of compound residential areas gradually weakened. As the management power of the compound was not transferred to the community, the compound gradually evolved into a "three-regardless" residential area. The lack of long-term management and maintenance of the compound led to confusion in the community's garbage disposal sites. There is no fixed garbage collection point, meaning garbage cannot be treated regularly. The building is seriously aging, the facade is mottled and cracked, and the corridors are covered with small advertisements and graffiti. The public activity space is overgrown with weeds, lacks corresponding activity facilities, piles up debris for a long time, and cannot be used normally.

At the social space level, firstly, due to the long-term detached status of the Supply and Marketing Cooperative Compound, the material environment such as the aging of the building and the decline of the public activity space is rapidly deteriorating. At the same time, because the compound housing was transformed from the danwei public property right to the private-use property right housing, the living area per capita was limited and gradually could not satisfy the residents' increasing pursuit of large-area housing. Therefore, a large number of original danwei residents chose to rent and sell compound housing and go to areas relatively far away from the city center to purchase newly developed residential quarters and carry out improvement-type replacement of houses. The loss of nearly 50% of the original residents, as well as the migration of a large number of migrants such as migrant workers and students, led to the complete disintegration of the original social network of the compound. The original danwei collective community cohesion disappeared, and residents no longer interacted with each other (which they formerly did even if they did not know each other). This represents an atomized and discrete daily life space with individuals [51]. The dilapidated physical environment of the community limits the possibility of communication and contact between residents and further prevents the re-establishment of the social network of the compound.

Secondly, in the context of housing market reforms, since the first land auction in Hefei in 1998, along with the simultaneous advancement of external expansion and internal potential urbanization, the housing value and the "rental difference/rental difference in high-quality areas" the "rent gap" was "activated", which was quickly reflected in the transaction price and spatial difference of commercial housing [52]. In 2008, with the support of the local government, the price of the re-developed real estate was twice the average selling price of a new house in Luyang District at that time, which was much higher than the price of second-hand houses in the original compound. For the residents of the compound involved in the demolition and redevelopment, the Provincial Supply and Marketing Cooperative proposed three types of resettlement compensation methods: cash compensation, in situ resettlement, and participation in group buying. Most of the residents whose homes were demolished did not want to abandon the superior location in the city center occupied by the original houses. At the same time, they did not have enough funds to provide more residential area in the new residential quarters and chose to relocate in situ. Although this kind of resettlement compensation method guaranteed the interests of the residents of the demolished compound to the greatest extent, the newly developed residential districts were already on sale due to their superior location, scarce education and medical resources, and high-quality apartments. Those led to a rapid rise in second-hand housing prices, as high as 66,000 yuan/m^2, which was much higher than the average second-hand housing price in the same area in Luyang District. Additionally, the phenomenon of gentrification was prominent [53]. The rapidly rising housing prices have caused psychological imbalances among the residents still living in the original danwei compound and laid a hidden danger of instability for the orderly development of the social network in the compound.

In short, the community compound at this stage cannot create a sense of community belonging and gradually moves towards the decline and rupture of material space and social networks, which had become the fundamental reason for the discrete evolution of the community. When the identity space in the representational space and the institutional space of the representations of space do not form a synergy, the livable development of the community will be a utopian fantasy [15].

4.2.3. Spatial Practice

The behavior subject constantly interacts with the environment and, while the behavior is disciplined by the space, it also shapes the space [54]. At the level of spatial practice, some employee representatives who still live in the compound play games through the danwei with the local government. Residents occupy and transform space through their own subjective initiative, express their own demands, and highlight the contradiction between public and private space.

The game between residents, danwei, and government is the renovation of dilapidated houses. When the danwei compound was built in the 1950s, all of which are brick-concrete structures, the construction standards were low, some housing facilities did not meet the standards, and the per capita living area was small, which is not suitable for long-term living. With the decline of the material environment in the of the Supply and Marketing Cooperatives Compound, the houses in the compound were seriously damaged. Building 18 was identified as a severely damaged house in the 1990s. Similar to Building 18, there were Buildings 17, 15, 3, and 2. Buildings have very poor earthquake-proof effects and have many hidden safety hazards. These current conditions have long caused concerns for the residents. Since 2010, some employees of the danwei have repeatedly requested in writing to the Supply and Marketing Cooperative to reform their houses for salvage, while the Supply and Marketing Cooperative did not proceed with further acceptance on the grounds of financing difficulties. In July 2014, after several situations and negotiations, the Provincial Supply and Marketing Cooperative asked the Service Center of the danwei to conduct a household survey on the renovation of dilapidated houses in the compound. According to survey data, nearly 93% of the residents of the compound agreed to the renovation, and the masses had a strong will. However, at this time, the Supply and Marketing Cooperative Compound wanted the Hefei Municipal Government to apply for the renovation of dilapidated houses. Due to the demolition of the former Hefei Dilapidated House Reconstruction Office, there were few corresponding support policies, which hindered the market-oriented operation of the overall renovation of the compound. The renovation of the compound once again ran aground. It can be seen that the danwei was the link between the residents of the compound and the local government. In the danwei compound, it represented the end of the administrative management of the government department. However, it usually requires the support and encouragement of the government in terms of policy and finance. The danwei responded to problems with a relatively strong posture. When the community had increasingly complex and diverse needs and difficulties, it was seen as a "dragging" strategy. The local government adopted the attitude of non-active participation in the settlement without the policy instructions of the superior.

Illegal construction/occupation of production by residents of the compound represented that "The poor and the working class have little power over space, but they have the ability to establish a certain situation" [18], although the current main management danwei of the compound (Supply and Marketing Cooperative) has also tried to negotiate with the local government on solving the existing complex contradictions and needs of the compound through applications and written reports. However, the actual problems have not been fundamentally resolved in the delay. Therefore, the contradiction between supply and demand of residential space in the compound has become more and more intensified, causing residents to have to solve the problem in such a way that the weak, occupying space, use the an art of resistance. Unlike space rulers, the weak have no independent place and cannot confront the strong. Therefore, they must use time and opportunities to occupy places in a roundabout, indirect, and scattered manner [41]. Residents of the compound occupy the public space on the first floor by building a sunroom or building a backyard to increase the living area, occupying fire aisles and self-built parking spaces, occupying the inter-building space to build a storage room privately to express anti-discipline behavior and attitudes, and strategically occupy the public space as one's daily life space.

4.3. Production of Integrated Differential Space (Co-Governance of "Government–Market–Residents" in Livable Areas) (Co-Governance of "Government–Market–Residents" in Block-Based Residential Areas)

In the face of the constant complaints from the residents of the compound and the informal resistance of the residents, the material space of the danwei residential area is facing further deterioration and decline. Additionally, the danwei and the local government are facing an increasingly prominent "discipline–anti-discipline" spatial contradiction, which even intensifies the new hidden dangers of social problems [55]. In 2010, in response to the country's policy call to "visit urban settlements built before 2000" and "vigorously promote the transformation of old urban communities", Hefei officially launched the transformation of old communities. The decayed Supply and Marketing Cooperative Compound, which was built in the 1950s, became a key target for the Luyang District Government to show the results of the renovation of the old community. As the main funding provider, the local government has conducted public tenders for the design and construction company for the renovation of the old community. In the process of transformation, in response to the central work requirements of "filling up the shortcomings of the old urban community infrastructure and satisfying the people's yearning for a better life", the local government and danwei put aside the posture of space powerhouses and the danwei agency service centers, with the design and construction company, conduct door-to-door visits and investigations for the immediate transformation needs of the residents of the compound. Therefore, in the final transformation plan of the old community, the use value of residents' living space and social space has received more attention and improvement. In the transition from discrete to space to differentiated space, although more and more demands are constantly being realized, new demands will continue to arise, which means that although the spatial contradiction has been relieved to a certain extent, it has never been completely relieved. Therefore, this period is an integrated differential space: The contradiction is resolved in the game and negotiation, but the ideal terminal state has not been reached (Figure 4).

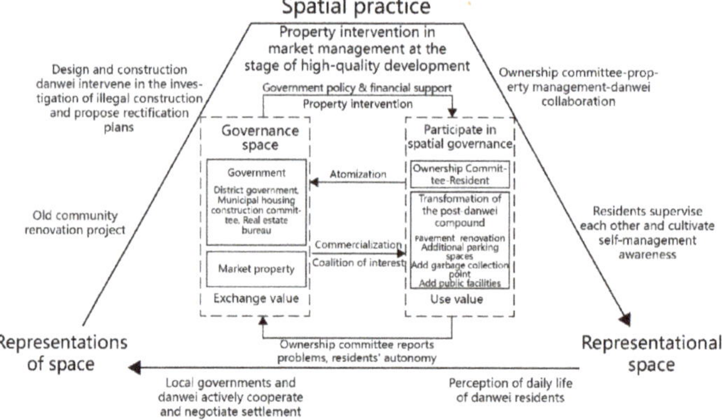

Figure 4. Integrated differential space production path (source: self-drawn by one of the authors).

4.3.1. Representations of Space

The disciplined discrete space turns to a diversified and differentiated space, which is mainly reflected in the introduction of market-oriented properties: the formation of danwei–property management agency–residents multi-participation in the governance model of the compound and the differentiated negotiation of illegal construction/occupation of public opinion.

Firstly, the relationship between the government, danwei, the market, and the residents of the compound has changed. The root cause of space conflict and atomization is that the neatly colluded space exchange value hides the space use value of residents' daily life. Therefore, on the one hand, the residents of the compound require the danwei to declare the renovation of the community in the name of the environmental improvement of the old community and strive for the reform index from the city. On the other hand, they require the introduction of market management. Danwei Agency Service Center, the property management agency, and the residents jointly participate in the governance of the compound. They Introduce the property management agency, and carry out market-oriented management of parking, greening and public space, and regulate the garbage classification and public security management at the centralized garbage disposal point. Residents and danwei jointly participate in the decision making and evaluation of the selection and employment of the property management agency.

Secondly, regarding the demolition and renovation of the original illegal buildings, Although the illegal construction/occupation expressed the resistance of some residents to space discipline and the expression of differentiated space needs, it also encroached on public space and harmed the interests of other residents. For example, occupying the backyard on the first floor, occupying pedestrian passages, adding storage rooms between the buildings, occupying fire-fighting passages, rebuilding a sunroom on the top floor, etc. will bury hidden dangers for the living safety of the compound and, at the same time, conflicts and quarrels among residents intensified. Meanwhile, in the long-term game between the government, danwei, and residents, they gradually realized that forced demolition will only intensify the contradictions among the people, the government, and danwei. It is necessary to respect diverse needs, respect residents' living habits, and negotiate and handle the issue of encirclement construction. Taking advantage of the opportunity of the renovation of the old community in 2018, the design department entered the site to conduct a comprehensive investigation and safety assessment of the enclosure/occupation situation of the housing. Then, the service center of the provincial supply and marketing cooperative agency took the lead to invite the district housing construction committee, district urban management office, district environmental protection office, and other relevant departments, as well as design companies, construction companies, and residents' representatives to organize consultation meetings, based on the safety technical assessment report issued by the design company, combined with the technical safety regulations of the government on community fire protection, house structure, pedestrian passage width, etc., considering the interests of residents and diversified needs, according to the actual situation of multiple illegal constructions The situation was subject to differentiated dismantling, transformation, and retention treatment. The demolition of illegal construction was completed smoothly, eliminating potential safety hazards, beautifying the public space environment, alleviating contradictions and disputes, and curbing the trend of illegal construction/occupying.

4.3.2. Representational Space

Residents' daily life space should not be dominated by commoditized exchanges. Therefore, in the process of the transformation of the old community, through on-site investigations and interviews on the residents' own demands and wishes, a list of transformation projects was formed to realize the "popular opinion" and "anti-discipline" of the space production. The danwei compound was transformed from being disciplined to an active participant in space production, and the use value of space was re-excavated. Specific representational space transformations included (1) exterior wall renovation and corridor cleaning; the painted exterior wall or cement mortar exterior wall was completely removed and then sprayed with real stone paint, covering an area of 17,808 square meters, and the red brick exterior wall was cleaned and varnished with an area of 752 square meters. For the existing residential corridors with dirty walls and a serious "psoriasis" phenomenon, the general walls needed to be polished with sandpaper during the renovation and cleaning, and the severely damaged walls needed to be renovated to the structural layer. The total area of cleaning was about 18,270 square meters. The eradicated area of the old wall was 6395 square meters. (2) Additional fitness venues and fitness equipment, with four additional fitness venues with an area of 186 square meters and 12 additional pieces of fitness equipment. Pruning and transplanting a total of 19 trees in the community, replanting Manila turf, totaling 210 square meters, and repairing the flowerbeds with brick cement mortar, totaling 45 square meters. (3) Mainly laying in front and back of the house and aisles, etc., was a total of 4175 square meters. Existing trash cans were damaged. The rectification adopted anti-aging, hard plastic, wheeled garbage cans. There are 51 trash cans in total and one set per unit. Special storage spaces with a size of 80 cm × 80 cm are reserved along the road. The top surface of the motor vehicle shed was replaced with a color steel plate, an area of 86 square meters. The iron fence was painted with rust, an area of 12 square meters. Based on the current conditions, the roads in the community will be repaired on the original basis. (4) Upgrade the community to add electronic monitoring equipment. Renovate the surrounding walls of the community. The current quality of the surrounding walls is still good, but it is dirty and damaged. The design is sprayed with real stone paint, and the wall sprayed was 766 m with a height of 3 m [56].

4.3.3. Spatial Practice

Lefebvre's space is not only an empirical setting of things in a certain field but also an attitude and customary practice. The spatial practice of creating a differentiated space is mainly reflected in the stimulating of the residents' awareness of active participation in community governance and the development of the habit of actively maintaining the environment and order of the community [53]. In 2018, with the help of the transformation of the old community, the environment of the supply and marketing cooperative compound was rectified by public opinion and the living environment of the residents was greatly improved. However, to transform a new compound environment requires long-term professional maintenance to maintain the sustainable development of the results. Therefore, the supply and marketing cooperative tried to introduce a property company to be responsible for the professional management of the compound. The danwei conducted household surveys and solicited opinions from residents on the introduction of market-oriented property management work. As a result, more than half of the residents disagreed with the establishment of the introduction of properties. Since most of the people living in the compound are elderly people and foreign tenants, a household used to pay only 60 yuan a year for the sanitation management fee, but now it has to pay the property fee on a monthly basis. Therefore, many people cannot accept it ideologically. Under the ideological mobilization of street organizations, danwei, and building managers went house to house, an owner committee was established, and a recommended property company was introduced for test management. Gradually, it has gained more and more recognition from the residents of the compound in terms of maintaining public security in the community, guiding residents' garbage classification, maintenance of public facilities, greening maintenance, and parking order guidance. At the same time, under the positive development trend of continuous improvement of the compound environment, the residents of the compound will take the initiative to report to the property that the sanitary conditions in the corridor need to be dealt with in a timely manner, spontaneously supervise each other's garbage entering the bin, monitor the fire channel parking situation, and consciously maintain the environment and order of the compound [57], Gradually forming a space of difference reflecting diversity.

It can be seen that the maintenance of space rules requires the intervention of multiple forces, but residents feel that it is more important. If the new space habit cannot be brought into play with the help of spatial practice, then the formation of the space of difference is also a fantasy [58]. The cultivation of new habits relies on the support of multiple forces: The policy and financial support of the street and the danwei, the communication between the business committee and the building owner in daily life with the residents, the supervision of urban management, the guarantee of property management, etc., all promote the integration of space Indispensable conditions and means.

The Authors should discuss the results and how they can be interpreted from the perspective of previous studies and of the working hypotheses. The findings and their implications should be discussed in the broadest context possible. Future research directions may also be highlighted.

5. Conclusions

Through the qualitative analysis of the space production stage of the case compound, combined with the quantitative analysis of the coupling and coordination of the space production, and through the geographic detector, the influence intensity difference analysis of the factors that affect the quality of the danwei compound's space production was carried out. It was concluded that the case compound is currently in a period of integrated differential space production. According to the space ternary dialectical theory, the representations of space, representational space, and spatial practice in this period are basically in a state of coupling and coordination, of which the spatial production coupling coordination score is 0.51. In the transition from the atomic abstract space to the fusion difference space, the use value of residents' living space and social space has received more attention and improvement. Although more and more demands continue to be realized, new demands will continue to arise, which means that although the spatial contradiction has been relieved to a certain extent, it has never completely disappeared. Through the analysis of the geo-detector, it can be seen that "basic living conditions", "living conditions", "community social support", "community belonging", and "residents' satisfaction with community management and services" are the five leading factors that affect the quality of space production, and their q determining power values are 0.89, 0.89, 0.85, 0.71, and 0.69, respectively. Therefore, starting from these aspects, we propose the following possible paths for the sustainable community regeneration of the three danwei compounds under the background of the "post-danwei era".

5.1. Construction of Community Resources and Improvement of Community Environment

Community resources include not only tangible capital such as natural conditions, physical environment, and supporting facilities that support the sustainable development of the community but also intangible capital such as cultural connotations, social networks, and spiritual belonging that represent the characteristics of the community [42]. The quantity and quality of community capital are important factors to measure the level of community construction and development, as well as an important dimension of influence on the quality of space production [59]. Therefore, in the "post-danwei era", the construction of livable danwei settlements should have sufficient tangible hardware assets, such as community residents' service facilities, and should also create a distinctive spiritual and cultural core. Community resources include not only tangible capital such as natural conditions, physical environment, and supporting facilities that support the sustainable development of the community but also intangible capital such as cultural connotations, social networks, and spiritual belonging that represent the characteristics of the community [60]. The quantity and quality of community capital are important factors to measure the level of community construction and development, as well as an important dimension of influence on the quality of space production [59]. Therefore, in the "post-danwei era", the construction of livable danwei settlements should have sufficient tangible hardware assets such as community residents' service facilities and should also create a distinctive spiritual and cultural core. With the decline of the traditional danwei system community, the danwei no longer plays the role of the main provider of community resources and lacks effective alternative resource providers to join, resulting in the decline of the compound's material environment, the separation of social networks, and the loss of cultural connotation. In response to this situation, urban local governments should mobilize relevant departments and service agencies from the perspective of overall urban planning and development, conscious supplementing and constructing the community resources of the danwei system community after the change, and acting as the provider and manager of community resources. This move must be maintained for a period. Only in this way can it bring about effective improvement of this type of community environment and improvement of residents' living standards. It is also possible to produce a new cultural spiritual connotation that maintains the integration of the community on this basis.

5.2. Cultivation and Strengthening of Residents' "Community Awareness"

In the danwei system-dominated period, the danwei was the main body of the state for social resource management and distribution, so everyone had to depend on the danwei to meet their various needs in life. Therefore, everyone in the city was a "danwei person" [3]. After the start of the economic system reform, major changes took place in the danwei itself and its interior during the introduction of the market economic system. The most significant change was that the danwei split the social functions and returned it to the society, which was previously undertaken by danwei [61]. This means that danwei employees were pushed out of the scope of protection under the traditional redistribution system. Their reliance on the state and danwei for obtaining various resource of living was cut off, and they changed from "danwei people" to "social people." As a result of this transformation, there was an urgent need for the "community", a new entity that satisfied the urban social functions, and the community was responsible for many social functions that were stripped away from the original danwei. The above-mentioned changes brought about changes in real-life conditions and opportunities for urban residents. At the same time as these objective changes occurring, a subjective change in concept and consciousness was urgently needed. This was to change the high level of dependence on the work of danwei in the past into a strong sense of belonging to the residential community. The cultivation and strengthening of this awareness is a big boost to the construction of urban communities, and it is also a necessary condition for the reconstruction of the traditional danwei system communities [62].

5.3. Cultivation of Residents' Participation and Self-Management Mechanism

After the system reform, the danwei system community was thrown out with the danwei's blessing during the disintegration of the danwei system, and there was a great management crisis [63]. In this traditional danwei-based community, after the system transformation and enterprise restructuring, on the one hand, the danwei was withdrawn from the field of community governance. On the other hand, there was no effective management entity to settle in, which created a vacuum in community governance and also led to the decline of the community on the edge and the bottom [64]. Therefore, on the one hand, we should speed up the introduction of the five market-oriented professional real estate companies to manage and provide services to the compound. On the other hand, instead of waiting for a new community management agency to come in, it is better to rely on residents' self-management. Allowing community residents to better participate in the process of community reconstruction and management may be more worthy of consideration. The basis of this mechanism lies in the meaning contained in the concept of "community", that is, "community is essentially a community of social life. The value of the community and the social basis for its existence and development lies in the common interests and common concerns of public affairs among the residents of the community" [65]. Therefore, the cultivation and functioning of a mechanism for residents' participation and self-management should be an effective condition for promoting community construction and community governance. This is not only a choice in the reconstruction of the traditional danwei system community, but also a long-term plan for the construction and development of the entire urban community.

Author Contributions: Conceptualization, Z.Y.; formal analysis, Z.Y. and K.R.; writing—original draft preparation, Z.Y. and K.R.; writing—review and editing, Z.Y. and J.Y.; supervision, J.Y.; funding acquisition, J.Y. All authors have read and agreed to the published version of the manuscript.

Funding: This paper and the related research were financially supported by the National Natural Science Foundation of China (51778126).

Institutional Review Board Statement: Not applicable.

Informed Consent Statement: Not applicable.

Data Availability Statement: The data used to support the findings of this study are included within the article.

Conflicts of Interest: The authors declare no conflict of interest.

References

1. Liu, T.; Chai, Y. The Meaning, Content and Prospect of Danwei System Research in the Perspective of Geography. *Prog. Geogr.* **2012**, *31*, 527–534.
2. Chai, Y. Danwei-Based Chinese Cities' Internal Life-Space Structure—A Case Study of Lanzhou City. *Geogr. Res.* **1996**, *15*, 30–38.
3. Bjorklund, E.M. The Danwei: Socio-Spatial Characteristics of Work Units in China's Urban Society. *Econ. Geogr.* **1986**, *62*, 19–29. [CrossRef]
4. Brogger, D. Urban diaspora space: Rural-urban migration and the production of unequal urban spaces. *Geoforum* **2019**, *102*, 97–105. [CrossRef]
5. Zylstra, G.D. Productions of Space, Productions of Power: Studying Space, Urban Design, and Social Relations. *J. Urban Hist.* **2017**, *43*, 562–569. [CrossRef]
6. Lefebvre, H. *The Production of Space*; Blackwell Publishing: Oxford, UK, 1991.
7. Henson, Z. Sustainable Community Systems: Commoning and Spatial Production. *Theory Action* **2016**, *9*, 32–51. [CrossRef]
8. Fricke, C.; Gualini, E. Metropolitan regions as contested spaces: The discursive construction of metropolitan space in comparative perspective. *Territ. Polit. Gov.* **2018**, *6*, 199–221. [CrossRef]
9. Dangschat, J.S. Space Matters—Marginalization and Its Places. *Int. J. Urban Reg. Res.* **2009**, *33*, 835–840. [CrossRef]
10. Merrifield, A. The Struggle over Place—Redeveloping American Can in Southeast Baltimore. *Trans. Inst. Br. Geogr.* **1993**, *18*, 102–121. [CrossRef]
11. Sims, J.R. More than gentrification: Geographies of capitalist displacement in Los Angeles 1994–1999. *Urban Geogr.* **2016**, *37*, 26–56. [CrossRef]
12. Wu, F.L. State Dominance in Urban Redevelopment: Beyond Gentrification in Urban China. *Urban Aff. Rev.* **2016**, *52*, 631–658. [CrossRef]
13. Ye, C.; Chen, M.X.; Chen, R.S.; Guo, Z.W. Multi-scalar separations: Land use and production of space in Xianlin, a university town in Nanjing, China. *Habitat Int.* **2014**, *42*, 264–272. [CrossRef]
14. Shenjing, H.E.; Junxi, Q.; Minhua, W.U. Studentification in urban village: A case study of Xiadu Village, Guangzhou. *Geogr. Res.* **2011**, *30*, 1508–1519.
15. Du, P. Space Production of "Village-Turned Community": From Differentiation and Discretization to Integration. *City Plan. Rev.* **2019**, *43*, 64–70.
16. Sun, J.; Zhou, Y. Study on the reproduction of space of tourism community from the perspective of everyday life: Based on theories of Lefebvre and De Certeau. *Acta Geogr. Sin.* **2014**, *69*, 1575–1589.
17. Lin, Y.L. E-urbanism: E-commerce, migration, and the transformation of Taobao villages in urban China. *Cities* **2019**, *91*, 202–212. [CrossRef]
18. Liang, Z.; Chao, Y.E.; Senlin, H.U. Spatial production and spatial dialectic: Evidence from the New Urban Districts in China. *J. Geogr. Sci.* **2019**, *29*, 1981–1998.
19. Zhou, P. A socio-economic-cultural exploration on open space form and everyday activities in Danwei: A case study of Jingmian compound, Beijing. *Urban Des. Int.* **2014**, *19*, 22–37.
20. Chai, Y. From socialist danwei to new danwei: A daily-life-based framework for sustainable development in urban China. *Asian Geogr.* **2014**, *31*, 183–190. [CrossRef]
21. Xiao, Z.; Liu, T.; Chai, Y.; Zhang, M. Corporate-Run Society: The Practice of the Danwei System in Beijing during the Planned Economy Period. *Sustainability* **2020**, *12*, 1338. [CrossRef]
22. Zhang, C.; Chai, Y. The Residential Residual in Danwei Community and Its Impact Factor. *Urban Plan. Int.* **2009**, *24*, 15–19.

23. Zhang, Y.; Chai, Y.; Zhou, Q. The spatiality and spatial changes of danwei compound in Chinese cities: Case study of Beijing No. 2 textile. *Urban Plan. Int.* **2009**, *24*, 20–27.
24. Wu, H. The Danwei Tradition in Chinese Organizational Identification and Its Modern Changes. *J. Soc. Sci. Hunan Norm. Univ.* **2016**, *45*, 49–56.
25. Liu, J.; Deng, X.; Li, C. Community Planning Based on Socio-Spatial Production: Explorations in "New Qinghe Experiment". *China City Plan. Rev.* **2016**, *40*, 9–14.
26. Kolodii, V.V.; Kolodii, N.A.; Chayka, Y.A. Activism and participation: Social technologies of cooperation with urban population in the process of "production" of urban space. *Vestn. Tomsk. Gos. Univ.-Filos.-Sotsiologiya-Politol.-Tomsk State Univ. J. Philos. Sociol. Political Sci.* **2017**, *38*, 175–185. [CrossRef]
27. Galvis, J.P. Remaking Equality: Community Governance and the Politics of Exclusion in Bogota's Public Spaces. *Int. J. Urban Reg. Res.* **2014**, *38*, 1458–1475. [CrossRef]
28. Zhu, X.; Qiao, J. Sustainable Development of Rural Tourism Community:Based on the Analysis from the Perspective of Ternary Dialectics on Production of Space. *Econ. Geogr.* **2020**, *40*, 153–164.
29. Moreno, L. Henri Lefebvre on Space: Architecture, Urban Research and the Production of Theory. *Ann. Assoc. Am. Geogr.* **2012**, *102*, 885–890. [CrossRef]
30. Wang, J. *Spatial Sampling and Statistical Inference*; Science Press: Beijing, China, 2009.
31. Wen, L.; Cheng, B.; Chai, Y.; Wang, J. Application of SAR remote sensing data on land use change monitoring. *Sci. Surv. Mapp.* **2014**, *39*, 65–69.
32. Xiong, J.; Wang, W.; He, S.; Yin, Y.; Tang, C. Spatio-temporal Pattern and Influencing Factor of Coupling Coordination of Tourism Urbanization System in the Dongting Lake Region. *Sci. Geogr. Sin.* **2020**, *40*, 1532–1542.
33. Lu, C.; Li, W.; Pang, M.; Xue, B.; Miao, H. Quantifying the Economy-Environment Interactions in Tourism: Case of Gansu Province, China. *Sustainability* **2018**, *10*, 711. [CrossRef]
34. Wang, J.; Xu, C. Geodetector: Principle and prospective. *Acta Geogr. Sin.* **2017**, *72*, 116–134.
35. Bureau, H.P. *Hefei Urban Planning Record*; HuangShan Books: Huangshan, China, 2013.
36. Phillips, M. The production, symbolization and socialization of gentrification: Impressions from two Berkshire villages. *Trans. Inst. Br. Geogr.* **2002**, *27*, 282–308. [CrossRef]
37. Lu, F. The Origins and Formation of the Unit (DANWEI) System—Preface. *Chin. Sociol. Anthropol.* **1993**, *25*, 7.
38. Wang, D.; Chai, Y. The jobs–housing relationship and commuting in Beijing, China: The legacy of Danwei. *J. Transp. Geogr.* **2009**, *17*, 30–38. [CrossRef]
39. Unwin, T. A Waste of Space? Towards a Critique of the Social Production of Space. *Trans. Inst. Br. Geogr.* **2000**, *25*, 11–29. [CrossRef]
40. Ye, N.; Kita, M.; Matsubara, S.; Okyere, S.A.; Shimoda, M. Socio-Spatial Changes in Danwei Neighbourhoods: A Case Study of the AMS Danwei Compound in Hefei, China. *Urban Sci.* **2021**, *5*, 35. [CrossRef]
41. Liu, H.; Lu, B. A Brief Discussion on the Dialectic and Triple Senses Intrinsic to the Production of Space. *Urban Plan. Int.* **2021**, *36*, 14–22.
42. Li, X.; Kleinhans, R.; van Ham, M. Ambivalence in place attachment: The lived experiences of residents in danwei communities facing demolition in Shenyang, China. *Hous. Stud.* **2019**, *34*, 997–1020. [CrossRef]
43. Bray, D. Building "community": New strategies of governance in urban China. *Econ. Soc.* **2006**, *35*, 530–549. [CrossRef]
44. Ye, N.; Kita, M.; Matsubara, S.; Okyere, S.A.; Shimoda, M. A Study of the Spatial Distribution of Danwei Compounds in the Old Town of Hefei, China. *Urban Sci.* **2021**, *5*, 7. [CrossRef]
45. Hefei Local Gazetteers Compilation Committee. *Local Gazetteers of Hefei; Anhui Sheng Di Fang Zhi Cong Shu*; Anhui Peoples Publishing House: Hefei, China, 1999.
46. Feng, G.; Chen, F. Old wine in new skins? China's neighbourhood transformation from danwei to shequ. *ERDE* **2019**, *150*, 72–85.
47. Harvey, D. *The Condition of Postmodernity*; Blackwell: Cambridge, UK, 1990; pp. 227–240.
48. Hallowell, G.; Baran, P. Neighborhood Dynamics and Longterm Change. *Geogr. Anal.* **2020**, *53*, 213–236. [CrossRef]
49. Wu, F. Neighborhood Attachment, Social Participation, and Willingness to Stay in China's Low-Income Communities. *Urban Aff. Rev.* **2012**, *48*, 547–570. [CrossRef]
50. Aptekar, S. Looking Forward, Looking Back: Collective Memory and Neighborhood Identity in Two Urban Parks. *Symb. Interact.* **2017**, *40*, 101–121. [CrossRef]
51. He, S.; Wu, F. Socio-spatial impacts of property-led redevelopment on China's urban neighbourhoods. *Cities* **2007**, *24*, 194–208. [CrossRef]
52. Glaeser, E.L.; Huang, W.; Ma, Y.; Shleifer, A. *A Real Estate Boom with Chinese Characteristics*; Social Science Electronic Publishing: Rochester, NY, USA, 2017.
53. Chen, K.; Wen, Y. The Great Housing Boom of China. *Am. Econ. J. Macroecon.* **2017**, *9*, 73–114. [CrossRef]
54. Harvey, D. *Social Justice and the City*; Edward Arnold: London, UK, 1973; p. 116.
55. Bray, D. *Social Space and Governance in Urban China*; Stanford University Press: Stanford, CA, USA, 2005.
56. District, H.; Urban-rural, D. *Environmental Comprehensive Improvement Project of Lujiang Road Living Area of Provincial Supply and Marketing Cooperative*; Housing Construction Bureau of Luyang District: Hefei, China, 2018; pp. 2–11.

57. Francis, J.; Giles-Corti, B.; Wood, L.; Knuiman, M. Creating Sense of Community: The role of public space. *J. Environ. Psychol.* **2012**, *32*, 401–409. [CrossRef]
58. Zeng, W. The Social Governance Community in Transforming Neighborhoods: A Spatial Reconstruction Perspective. *Soc. Sci. China* **2020**, *41*, 173–198.
59. Natarajan, L. Socio-spatial learning: A case study of community knowledge in participatory spatial planning. *Prog. Plan.* **2015**, *111*, 1–23. [CrossRef]
60. Thomas, J.M. Neighborhood Planning: Uses of Oral History. *J. Plan. Hist.* **2004**, *3*, 50–70. [CrossRef]
61. Chai, Y.; Xiao, Z.; Zhang, Y. Urban Organization and Planning Transformation in China: A Danwei Perspective. *Urban Plan. Forum* **2011**, *6*, 28–35.
62. Flaherty, J.; Brown, R.B. A Multilevel Systemic Model of Community Attachment: Assessing the Relative Importance of the Community and Individual Levels. *Am. J. Sociol.* **2010**, *2*, 503–542. [CrossRef]
63. Guo, H. Literature Review on Danwei in China. *Mod. Manag. Sci.* **2013**, *41*, 41–45.
64. Yu, F.; Kang, W. Study on the Renewal and Protection Strategy of the Enterprises Danwei Compound in Harbin. *Urban Stud.* **2017**, *24*, 41–47.
65. Demirli, M.E.; Ultav, Z.T.; Demirtas-Milz, N. A socio-spatial analysis of urban transformation at a neighborhood scale: The case of the relocation of Kadifekale inhabitants to TOK Uzundere in zmir. *Cities* **2015**, *48*, 140–159. [CrossRef]

Article

Road Scene Recognition of Forklift AGV Equipment Based on Deep Learning

Gang Liu [1,2,3], Rongxu Zhang [4], Yanyan Wang [1,*] and Rongjun Man [4]

[1] Shenzhen Research Institute, Shandong University, Shenzhen 518000, China; sdulg@sdu.edu.cn
[2] School of Mechanical Engineering, Shandong University, Jinan 250061, China
[3] Key Laboratory of High Efficiency and Clean Mechanical Manufacture, Ministry of Education, Shandong University, Jinan 250061, China
[4] School of Control Science and Engineering, Shandong University, Jinan 250061, China; 201914400@mail.sdu.edu.cn (R.M.); zhangrongxu2014@163.com (R.Z.)
* Correspondence: yanyan.wang@sdu.edu.cn; Tel.: +86-531-8839-2269

Abstract: The application of scene recognition in intelligent robots to forklift AGV equipment is of great significance in order to improve the automation and intelligence level of distribution centers. At present, using the camera to collect image information to obtain environmental information can break through the limitation of traditional guideway and positioning equipment, and is beneficial to the path planning and system expansion in the later stage of warehouse construction. Taking the forklift AGV equipment in the distribution center as the research object, this paper explores the scene recognition and path planning of forklift AGV equipment based on a deep convolution neural network. On the basis of the characteristics of the warehouse environment, a semantic segmentation network applied to the scene recognition of the warehouse environment is established, and a scene recognition method suitable for the warehouse environment is proposed, so that the equipment can use the deep learning method to learn the environment features and achieve accurate recognition in the large-scale environment, without adding environmental landmarks, which provides an effective convolution neural network model for the scene recognition of forklift AGV equipment in the warehouse environment. The activation function layer of the model is studied by using the activation function with better gradient performance. The results show that the performance of the H-Swish activation function is better than that of the ReLU function in recognition accuracy and computational complexity, and it can save costs as a calculation form of the mobile terminal.

Keywords: storage system; forklift AGV; deep learning; semantic segmentation; H-Swish

Citation: Liu, G.; Zhang, R.; Wang, Y.; Man, R. Road Scene Recognition of Forklift AGV Equipment Based on Deep Learning. *Processes* **2021**, *9*, 1955. https://doi.org/10.3390/pr9111955

Academic Editors: Sergey Y. Yurish and Jong-Ho Shin

Received: 16 August 2021
Accepted: 29 October 2021
Published: 31 October 2021

Publisher's Note: MDPI stays neutral with regard to jurisdictional claims in published maps and institutional affiliations.

1. Introduction

In the application of forklift AGV equipment path planning in the distribution center, the main tasks include road scene recognition, path planning, obstacle identification, and local obstacle avoidance, etc. The road scene recognition of forklift AGV equipment is a very important task, especially the use of pure visual methods, as these problems have high complexity and are more challenging when applied to the distribution center warehouse environment. In the aspect of scene recognition and path planning, a large number of scholars have studied and made great achievements in SLAM (simultaneous localization and mapping) [1,2], deep learning [3–7], and other aspects. These methods are often aimed at the general indoor and outdoor life scenes, but the adaptability of AGV equipment for forklift trucks in logistics and distribution centers is low. In order to ensure the smooth operation of the forklift AGV equipment in the warehouse, the equipment needs to obtain the warehouse scene of the distribution center, including the identification information of racks, aisles, people, and other targets in the warehouse environment. This paper mainly uses the semantic segmentation model, based on the deep learning method, to study the warehouse environment of forklift AGV equipment, and proposes a deep convolution neural network for the image segmentation of the warehouse environment in the distribution center. Depthwise separable convolution was used to reduce the parameters, computation, and training time of the model, and the H-Swish activation function was used to improve the accuracy of the network model. The deep convolution neural network can accurately identify equipment in the distribution center warehouse and provide reliable environmental information for the smooth operation of forklift AGV equipment.

Forklift AGV equipment is needed to complete the access operation of goods in the logistics distribution center, as shown in Figure 1. Research on forklift AGV equipment has made great progress over the years. However, the current application of forklift AGV equipment is still limited by the cost and the computing power. Forklift AGV equipment needs to be aware of the environment in almost all cases. The visual system of forklift AGV equipment can be constructed by a deep learning method to identify the shelves, passages, and other environmental information in the working environment. In the aspect of working-environment map construction and vehicle location, forklift AGV equipment obtains environmental information through the camera, lidar, and other environmental sensing equipment, analyzes environmental information data through deep learning and image processing, identifies passable areas and obstacles in the environment, and uses a path planning algorithm to plan the path of forklift AGV equipment, so as to complete the task of forklift AGV equipment. The environmental information obtained by camera or lidar has rich environmental information, which can deeply restore the real environment in the construction of the environmental map. Research on the environmental map construction of forklift AGV equipment has received more and more attention, but each method will have different performances in different environments.

Figure 1. Forklift AGV.

Vision-based scene recognition mainly collects the image of the workspace through the camera, processes the environmental information in the image by means of machine vision and deep learning, transforms the pixel information into understandable feature information, and provides useful feature information for mobile robot scene recognition. In the aspect of mobile robot scene recognition, the robot mainly perceives the surrounding environment through ultrasonic, laser, and vision sensors, and obtains the environmental information of the current position and its own posture through simultaneous localization and mapping (SLAM), or deep learning.

SLAM collects the surrounding environment information with the help of the sensor carried by the mobile robot itself and calculates its current location according to the collected environmental information, which is the basis of the application of the mobile robot. The sensors carried by mobile robots usually include lidar, depth camera, and so on, in which the scene recognition method using the camera as the sensor is called "visual SLAM" [8]. The collected environment pictures have rich texture information, and the information in the pictures can be obtained by using the method of deep learning, such as ORB-SLAM2 [9]. The feature-based method used in visual SLAM cannot express the semantic information of the environment, which is solved by the semantic SLAM [10], based on deep learning, but it uses optical flow to calculate the image, which has a large amount of computation [1]. WANG [2] proposes the use of a binocular camera to obtain the image information of the path of a mobile robot, and then uses SURF to extract feature points from the binocular image as undetected obstacles for matching. Finally, combined with the binocular vision calibration model, the target position is determined by using the parallax between the matching points to realize the real-time navigation of the mobile robot.

The use of the deep learning method for scene recognition, which is different from the SLAM method and target detection, can obtain more abundant image feature information, including image semantic information, texture information, and local features. In the aspect of semantic segmentation using the deep learning method, firstly, Long [3] proposed a deep learning full convolution network FCN, which applies the deep learning method to image segmentation and can produce accurate and detailed image segmentation. Noh [4] proposed a deconvolution network, which is a mirror image relationship between the network structure and the convolution network structure of VGG16 [11], and the inverse operation of the convolution operation. Badrinarayanan [5] proposed that the SegNet network model of the image semantic segmentation network includes two parts: the encoder and the decoder. The encoder is the VGG16 [11] convolutional neural network, which removes the full connection layer and samples the low-input feature map in the first half of the convolutional neural network to produce dense feature maps, which is beneficial to the training of the network. The DeepLab image segmentation model [12] uses the convolution of the upgraded sampling filter to effectively expand the filtering range of the filter, without increasing the number of parameters and the amount of computation. Lin [6] proposed a multipath thinning network, RefineNet, which can make use of multiple levels of shallow features in subsamples to get a better effect of image segmentation. Liu [7] proposed a visual localization algorithm in a dynamic environment, which uses the image segmentation ability of deep learning to screen and predict potential moving objects in the environment. Training a supervised learning model is expensive, time-consuming, and laborious. Chen [13] proposed a weak supervised semantic segmentation method based on dynamic mask generation. Firstly, image features are extracted by CNN, and then multilayer features are iteratively integrated to get the edge of the foreground object to generate a mask. Finally, the mask is modified by CNN. However, the performance of this algorithm is not as good as that of fully supervised learning, and its accuracy is not high. Souly [14] modified the GAN network and created a large number of unreal images for discriminator learning by generating a confrontation network to achieve image semantic segmentation. The adversarial erasing method [15] and the antagonistic complementary learning method [16] are also applied to semantic segmentation, but the generalization ability is poor and only performs well on specific datasets.

Training time is an important factor in the training of convolution neural networks. The deep convolution neural network has a large number of parameters, and the calculation of pixel-by-pixel convolution consumption in image semantic segmentation is also huge, so the convolution neural network based on GPU computing can be trained faster. However, with the increase of the depth of the network, the progress in hardware can no longer support the needs of model training. Courbariaux [17] proposed the method of BinaryConnect to effectively improve the speed of convolution neural networks in the training stage. By forcing the weight binarization used in the forwarding and backpropagation of the convolution neural network, it replaces the traditional method of real weight multiplying real value activation, or gradient, in the process of propagation.

To sum up, although researchers have made a lot of basic research results in image semantic segmentation, they still cannot meet the needs of practical engineering applications. The application environment in the actual logistics system often has the characteristics of high-dynamic and multi-device interactions. In addition, because the forklift AGV equipment needs to run in the unstructured environment scene, there are many complex factors and these external disturbances are often difficult to model. More in-depth and comprehensive systematic research on vision-based road scene recognition is needed.

This paper uses the image semantic segmentation method to segment the pixels in the environment picture of the distribution center according to the semantic label in the image, as shown in Figure 2.

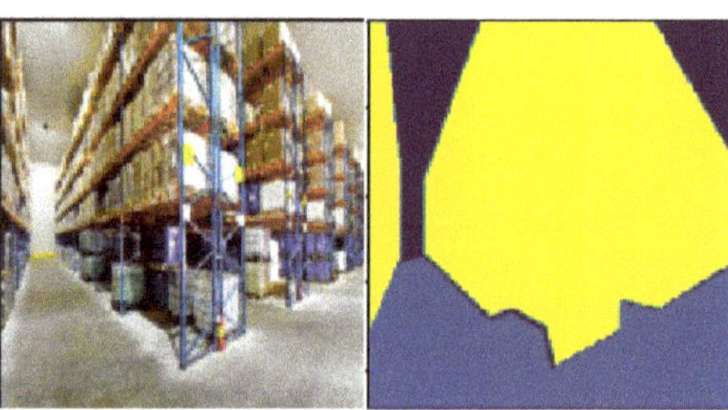

Figure 2. Image semantic segmentation.

The image segmentation methods in mobile robot scene recognition include the pixel "threshold" method [18], the pixel clustering method [19], the pixel image semantic segmentation method [20], and so on. These methods of image segmentation based on pixels have low computational complexity, but the segmentation effect is poor, far below the application level. After the development of the deep learning theory, the image semantic segmentation model, trained by a convolution neural network, has a deeper network structure and can better identify the high-order information of pixels, including the FCN network [3], the SegNet network [5], the DeepLab network [12,21], and so on. However, because of the complexity of the warehouse environment, these methods cannot be universal, so it is impossible to choose a universally applicable method. In this paper, the deep learning method was used to segment the image semantic of the forklift AGV equipment operating environment in the distribution center warehouse in order to achieve the identification of the forklift AGV equipment operating environment.

2. Deep Convolution Semantic Segmentation Network

At present, there are some network architectures for image segmentation, such as SegNet, VGG, and so on. SegNet achieves image segmentation by classifying each pixel in the image. Its idea is very similar to that of FCN, except that the technology of the encoder (Encoder) and decoder (Decoder) is different. Taking the SegNet network as an example, the encoder network uses the network structure of VGG-16. By removing the full connection layer of the network, the encoder network is used to obtain the low-resolution feature map from the high-resolution image. The decoder network converts the low-resolution image feature mapping into the label segmentation of the original image pixels through the training and learning of the dataset.

The structure of the SegNet network is shown in Figure 3. It can be seen from Figure 3 that the structure of the convolution neural network is approximately symmetrical. In the input layer of the network, the light blue layer is the convolution layer, batch normalization, and activation function; the purple layer is the subsampled layer; the light yellow layer is the upsampling layer; the dark blue layer is the deconvolution layer, batch normalization, and activation function; and the subsequent dark yellow layer is the objective function layer. The first half of the whole network is called the "coding" network, which uses the first 13-layer convolution network of VGG-16, and the second half is the "decoding" network. The coefficient feature graph obtained by the coding network is restored to the dense feature graph by deconvolution and upsampling. The last layer of the network is the soft connection layer, and the maximum value of different classifications is the output to obtain the final segmentation graph.

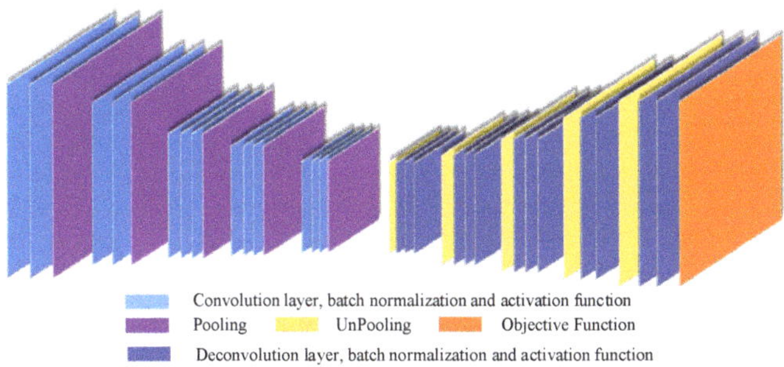

Figure 3. The structure of the SegNet network.

2.1. Encoder

In the coding network, the features of the input image are extracted by the convolution layer, using a convolution kernel of a 3×3 size, and the step size of the convolution kernel is 1. A pixel is added to the edge of the input image so that the input and output pixels of the convolution layer can be kept the same, and the same principle is also used in the decoding network to keep the image pixel size unchanged.

The coding network part uses the first 13 layers of VGG-16, which is different from VGG-16 in that the position information of upsampling is saved in the subsampled operation so that the decoder can use it to do nonlinear upsampling.

2.2. Decoder

The pooling operation can reduce the image in CNN, including in two ways: max-pooling and average-pooling. What is used in this paper is the maximum pool method, in which the value of the largest weight is taken out in a 2×2 filter, and the position of the maximum weight in the 2×2 filter is saved at the same time. From Figure 3 of the network framework, we can see that the purple max-pooling layer and the yellow max-unpooling layer are connected by the pooling index. In fact, the index after the max-pooling operation is output to the corresponding max-unpooling. Because the network is symmetrical, the first max-pooling operation corresponds to the last max-unpooling operation, and so on.

In the max-pooling operation, after the feature map of the input max-pooling layer is maximized, the largest value will be selected in each max-pooling filter, and the remaining three smaller values will be ignored. The max-pooling index is saved in the max-pooling operation so that the position where the maximum value is found in the max-unpooling operation in the network can be decoded for the deconvolution operation.

3. Deep Convolution Network Model Based on Depthwise Separable Convolution

As the basis of the scene understanding of forklift AGV equipment, image semantic segmentation determines the quality of the equipment target recognition and path planning. The convolution neural network has made great achievements in visual tasks so that the network can achieve pixel-by-pixel semantic label classification. In the semantic segmentation of images, in order to achieve a higher effect of semantic segmentation, a deeper convolution neural network is needed, and the deep convolution neural network can improve the abstract expression ability of the network. It can be more suitable for the working environment of large-scale distribution center warehouses. However, the deep convolution neural network also has some shortcomings. With the increase in network depth, the number of network parameters generated by the training convolution neural network increases sharply, which leads to the need for increased storage in the network model, which limits the application of semantic segmentation networks in mobile platforms, such as forklift AGV equipment. At the same time, the deep convolution neural network needs relatively large and complex calculations in the model training stage and model testing stage, which leads to the limitation of the running speed of the semantic segmentation algorithm. There are problems in the path planning algorithm based on real-time semantic segmentation. Therefore, the research reduces the number of parameters and model complexity of the convolution neural network and satisfies a certain semantic segmentation accuracy, which has great practical application value in forklift AGV equipment.

In this chapter, a scene semantic segmentation network is constructed, which is used to identify the path of forklift AGV equipment in the distribution center warehouse and to provide robust and accurate scene recognition for the equipment. The network is a 26-layer convolutional neural network, which is composed of a 13-layer convolutional coding network and a 13-layer deconvolution network. On the basis of the SegNet network model, the number of parameters of the coding network is reduced by using depth separable convolution. In the training stage of the model, the training time is less than that of the original model. The structure and parameters of the network model are shown in Table 1.

Table 1. Neural network model parameters based on depthwise separable convolution.

Name	Type	Input Size	Parameters	Name	Type	Input Size	Parameters
Conv_00	Convolution	224*224	64,3*3,1,1	Convtr_42d	Deconvolution	14*14	512,3*3,1,1
X_0	Max pooling	224*224	2*2,2	Convtr_41d	Deconvolution	14*14	512,3*3,1,1
Conv_10	Convolution	112*112	64,3*3,1,1	Convtr_40d	Deconvolution	14*14	512,3*3,1,1
Conv_11	Convolution	112*112	128,1*1,0,1	X_3d	Max Unpooling	14*14	2*2,2
X_1	Max Pooling	112*112	2*2,2	Convtr_32d	Deconvolution	28*28	512,3*3,1,1
Conv_20	Convolution	56*56	128,3*3,1,1	Convtr_31d	Deconvolution	28*28	512,3*3,1,1
Conv_21	Convolution	56*56	256,1*1,0,1	Convtr_30d	Deconvolution	28*28	256,3*3,1,1
Conv_22	Convolution	56*56	256,3*3,1,1	X_2d	Max Unpooling	28*28	2*2,2
X_2	Max Pooling	56*56	2*2,2	Convtr_22d	Deconvolution	56*56	256,3*3,1,1
Conv_30	Convolution	28*28	256,3*3,1,1	Convtr_21d	Deconvolution	56*56	256,3*3,1,1
Conv_31	Convolution	28*28	512,1*1,0,1	Convtr_20d	Deconvolution	56*56	128,3*3,1,1
Conv_32	Convolution	28*28	512,3*3,1,1	X_1d	Max Unpooling	56*56	2*2,2
Conv_33	Convolution	28*28	512,3*3,1,1	Convtr_11d	Deconvolution	112*112	128,3*3,1,1
X_3	Max Pooling	28*28	2*2,2	Convtr_10d	Deconvolution	112*112	64,3*3,1,1
Conv_40	Convolution	14*14	512,3*3,1,1	X_0d	Max Unpooling	112*112	2*2,2
Conv_41	Convolution	14*14	512,3*3,1,1	Convtr_01d	Deconvolution	224*224	64,3*3,1,1
Conv_42	Convolution	14*14	512,3*3,1,1	Convtr_01d	Deconvolution	224*224	8,3*3,1,1
X_4	Max Pooling	14*14	2*2,2		Softmax		
X_4d	Max Unpooling	7*7	2*2,2				

Note: The input size represents the input feature pixel. The parameters of convolution (Conv_*) and deconvolution (Convtr_*) are separated by commas, which, in turn, represent the number of output channels, the convolution kernel size, and the padding and stride. The parameters of max-pooling and max-unpooling are the pooling window size and step size, respectively.

The parameters of the batch normalization layer and the activation function layer are not listed in Table 1, in which the input parameters of each batch normalization layer are the number of output channels of the previous convolution layer. The formula for calculating the output of the batch normalization operation is as follows:

$$y = \frac{x - E(x)}{\sqrt{Var(x) + \epsilon}} \times \gamma + \beta \tag{1}$$

In the formula, x is the output data of the previous convolution layer, that is, the input data, $E(x)$ and $Var(x)$, of this layer are the mean and variance of the batch data, respectively. ϵ is a variable that prevents zero increase in the denominator, and γ and β are the linear transformations of the input data. The default values for γ and β are 1 and 0, respectively, so that the batch normalization operation does not reduce the model.

The activation function layer uses a modified linear unit (ReLU), whose input is the output of the previous batch normalization layer. In the coding network, the encoder uses depth separable convolution, and the encoder is composed of a channel-by-channel convolution layer, using a 3×3 convolution core, and the number of convolution kernels is the same as the number of input channels: Stride = 1 and Padding = 1. Through depthwise convolution to generate a set of feature images with the same number of channels, and then through a convolution kernel of 1×1, a group of feature graphs are convoluted point by point with the same number of output channels. The feature map is again subjected to BN operation and nonlinear mapping, and, finally, max-pooling is carried out. After the pooling operation, the size of the feature graph becomes 1/2.

In the decoding network, the feature graph is generated by the max-unpooling operation to produce a feature graph with sparse features. The sparse feature images of the upsampled output are convoluted by fractionally-strided convolutions to get dense feature maps. After the BN operation and nonlinear mapping, high-dimensional feature maps are output at the last layer. At the last layer of the network is the Softmax classifier. By classifying each pixel, the probability of each pixel corresponding to each label in the output image is output. The label category with the highest probability is the classification tag of the pixel. The objective function of the classifier is as follows:

$$\Gamma_{softmax\ loss} = -\frac{1}{N} \sum_{i=1}^{N} log \left(\frac{e^{h_{y_i}}}{\sum_{j=1}^{C} e^{h_j}} \right) \tag{2}$$

3.1. Depthwise Separable Convolution

Depthwise separable convolution [22] includes a depthwise convolution and a pointwise convolution. When inputting a 6×6 pixel, three-channel picture data (shape is $6 \times 6 \times 3$), in the convolution operation, use a 3×3 convolution kernel; Padding = 1, Stride = 1. After the convolution operation was completed, there were three feature images, with a feature size of 6×6, as shown in Figure 4. The convolution kernel shape in this case is:

$$Shape = W \times H \times C \tag{3}$$

W is the width of the convolution kernel, H is the height of the convolution kernel, and C is the number of input channels.

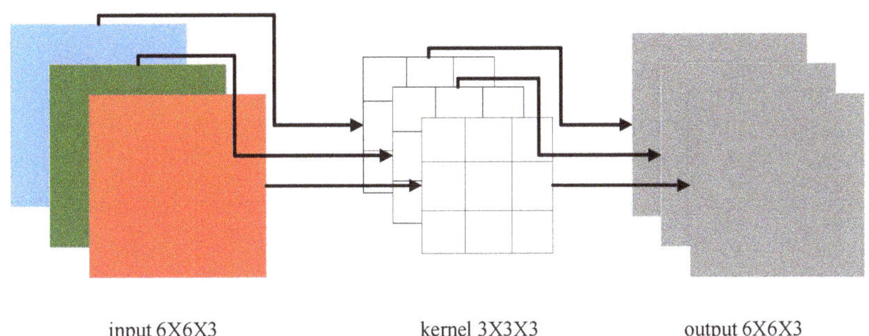

Figure 4. Depthwise convolution.

One of the filters contains only a convolution kernel with a size of 3×3, and the number of parameters in the convolution part is calculated as shown in Equation (4):

$$N_{depthwise} = W \times H \times C \tag{4}$$

The amount of calculation is as follows:

$$C_{depthwise} = W \times H \times (W_p - W + 1) \times (H_p - H + 1) \times C \tag{5}$$

W_p is the width of the input picture, and H_p is the height of the input picture.

After depthwise convolution, the feature graph with the same number of input channels can be output, and each feature graph only represents the corresponding characteristics of one input channel, without merging other channel features. It cannot express the characteristic information of navigation in the same spatial position on different channels. In order to make the convolution network have more abstract expression ability, we need more channel feature graph representation, so we need pointwise convolution to combine these feature graphs into a new feature graph.

Pointwise convolution is a weighted combination of the characteristics of each input channel. The size of the convolution kernel used is $1 \times 1 \times M$. M is the number of input channels in the point-by-point convolution layer. After pointwise convolution, a new feature graph is generated with characteristic information on the channel depth, as shown in Figure 5. The size of the convolution core of this layer is:

$$Shape = 1 \times 1 \times C_{in} \times C_{out} \tag{6}$$

C_{in} is the number of input channels, and C_{out} is the number of output channels.

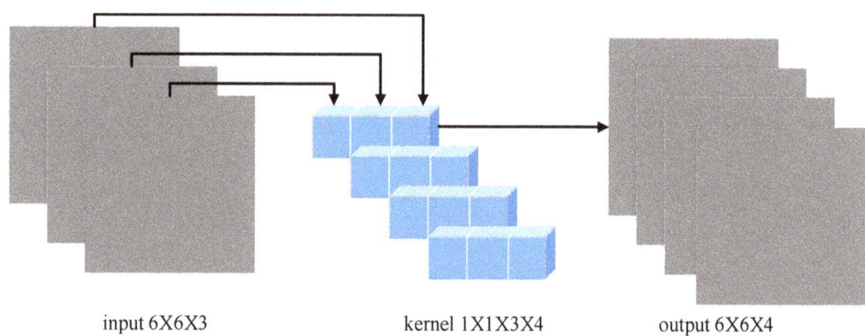

Figure 5. Pointwise convolution.

Because of the convolution operation of 1×1, the number of parameters involved in the point-by-point convolution is expressed as Equation (7):

$$N_{pointwise} = 1 \times 1 \times C_{in} \times C_{out} \tag{7}$$

The amount of calculation is as follows:

$$C_{pointwise} = 1 \times 1 \times W_f \times H_f \times C_{in} \times C_{out} \tag{8}$$

W_f is the width of the input feature layer, and H_f is the height of the input feature layer.

After pointwise convolution, four feature graphs are also output, which is the same as the output dimension of conventional convolution.

Compared with conventional convolution, the number of parameters is expressed:

$$N_{std} = W \times H \times C_{in} \times C_{out} \tag{9}$$

The amount of calculation is as follows:

$$C_{std} = W \times H \times \left(W_p - W + 1\right) \times \left(H_p - H + 1\right) \times C_{in} \times C_{out} \tag{10}$$

The comparison of the number of parameters and the amount of calculation with the conventional convolution algorithm is shown in Table 2:

Table 2. Comparison between depthwise separable convolution and conventional convolution.

Amount of Calculation	Number of Parameters
$C_{std} > C_{depthwise} + C_{pointwise}$	$N_{std} > N_{depthwise} + N_{pointwise}$

As can be seen from Table 2 above, the number of parameters and the amount of computation of depthwise separable convolution are much smaller than that of the standard convolution operation. Therefore, the use of depthwise separable convolution can reduce the number of parameters generated by the convolution neural network in the process of model training while ensuring a certain accuracy, thus reducing the memory consumption of mobile devices and, on the other hand, reducing the amount of computation in the training phase of the model. A smaller amount of calculation correspondingly reduces the time required for model training.

3.2. Batch Normalization

In the process of convolution neural network training, batch normalization (BN) [23] makes the mean value of the output signal of this network layer 0, and the variance is 1. This operation can improve the decreasing speed of the loss value of the model and, at the same time, alleviate the "gradient dispersion" to a certain extent, and make the convolution network model converge more easily in the training stage. BN operations are generally used before nonlinear mapping functions.

3.3. Unpooling

Unpooling is the operation of increasing the image resolution and is the inverse operation of pooling, in which the position index information of the maximum value retained by the red operation is recorded during the max-pooling operation in the coding network part of the convolutional neural network, and the information is used to expand the feature graph when the max-unpooling operation is carried out in the decoding network. The specific operation is the maximum position of the maximum value. Except for the position of the maximum value, the rest is 0, as shown in Figure 6.

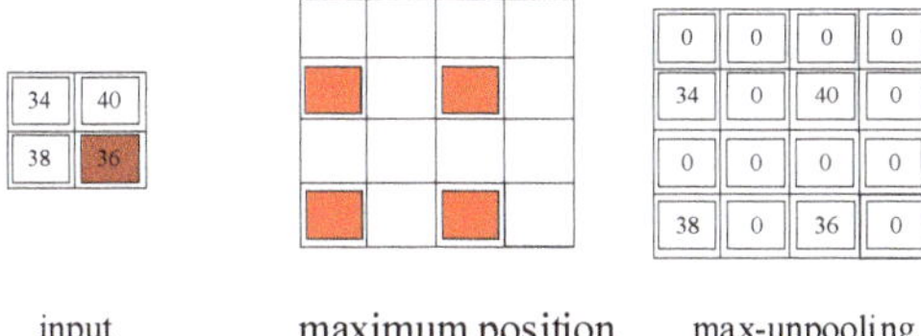

Figure 6. Unpooling.

4. Convolution Neural Network Based on H-Swish Activation Function

The activation function layer uses the activation function to simulate the characteristics of human brain neurons, controls whether the neurons are in an excited state, increases the nonlinear characteristics of the convolution neural network, and increases the ability of the convolution neural network to express the abstract features in the image. The rectified liner unit (ReLU) function [24] is a piecewise function, which is defined as:

$$\text{ReLU}(x) = max\{0, x\} = \begin{cases} x & x \geq 0 \\ 0 & x < 0 \end{cases} \tag{11}$$

As shown in Figure 7, the gradient of the ReLU function is 1 when $x \geq 0$, and vice versa.

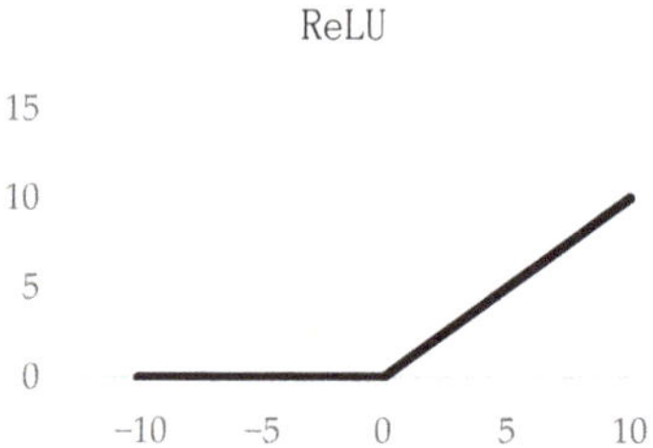

(**a**) ReLU function.

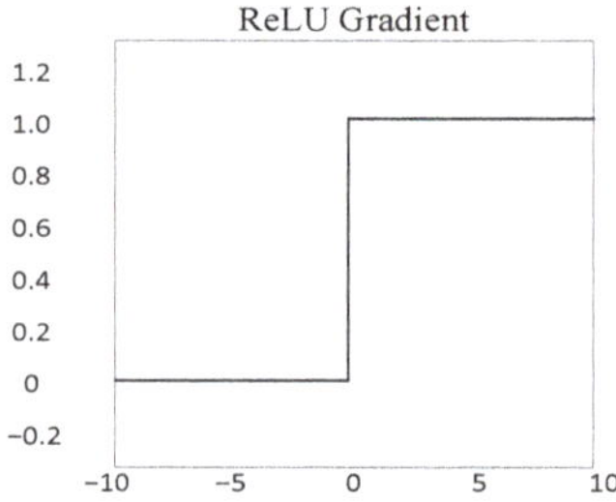

(**b**) ReLU gradient.

Figure 7. ReLU function and gradient.

Because the convolution neural network constructed in this paper is aimed at the mobile end of the forklift AGV equipment and is limited by the computing power of the mobile end, it was considered in order to reduce the computational complexity of the model. Therefore, the H-Swish function was used instead of the ReLU function, which can not only improve the accuracy, but also reduce the computational pressure of the forklift AGV equipment to a certain extent. The Swish [25] function can optimize the effect of the deep convolution network model to a certain extent, which is defined as:

$$swish(x) = x \cdot sigmoid(\beta x) \tag{12}$$

The Swish activation function has no upper bound or lower bound and is smooth and nonmonotonous. However, the calculation of the sigmoid function contained in the function is not friendly to forklift AGV equipment. The value of the H-Swish activation function is similar to that of the Swish activation function, and there is no sigmoid function operation in H-Swish, so it is friendly to the calculation of forklift AGV equipment. The definition is as follows:

$$h\text{-}swish[x] = x \frac{ReLU6(x+3)}{6} \tag{13}$$

5. Model Training and Analysis of Experimental Results

In order to realize the convolution neural network established in this paper, it was used to identify the road scene of forklift AGV equipment in a distribution center warehouse, using the PyTorch deep learning framework, which is the most popular in the field of deep learning and is very suitable for image data processing. We used the GPU computing model, which can improve the computing speed of the convolution neural network. The graphics card used in the model training was 6G NVIDIA GTX1660, video memory, and only one video card was used for training.

5.1. Data Preparation

The convolution neural network established in this paper learns image features from label data, so it needs a distribution center environment image dataset with tags. We used the self-built dataset, including the warehouse environment picture dataset, including 469 logistics distribution center warehouse environment pictures, pictures from Google, Baidu, Bing, and other search engines. The dataset part is as shown in Figure 8. Because the training convolution neural network needs labeled label data, and the dataset built in this paper is only the original picture, we needed to label these pictures manually. The labeling tool was LABELME. In the dataset label, there are six types of tags: shelf, AGV car, pedestrian, ground, pallet, goods. Therefore, these six kinds of objects were segmented in the experiment.

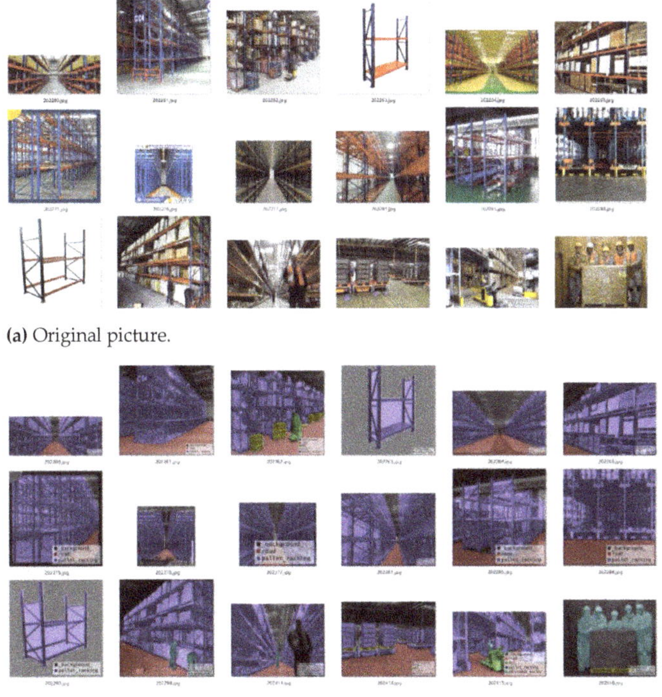

(**a**) Original picture.

(**b**) Label picture.

Figure 8. Distribution center environment image dataset.

5.2. Model Training

The self-built warehouse dataset of the logistics distribution center was used to train the convolution neural network. The dataset includes 375 color training pictures, with a resolution of 224 × 224 and 94 color training pictures, all of which are marked with information. When training the convolution neural network, the random batch processing (Mini-Batch) training mechanism was adopted, the training iteration number of the model was set to 3000, and the training sequence of pictures was randomly generated before each round of training to ensure that the pictures of each round of training were not the same. The generalization ability of the model was increased, and the convergence speed was improved. In the data processing of the convolution neural network, the resolution of the input picture was 224 × 224, but the resolution of the dataset was inconsistent. By compressing the picture, the resolution of the picture was unified to facilitate the input data of the neural network. The objective function of convolution neural network training was the cross-entropy loss function, the learning rate of network training was 10^{-6}, the momentum parameter was 0.9, and every eight pictures were used as a batch of network input (limited by video memory; the larger the number of pictures in batches, the better the training effect).

5.3. Analysis of Experimental Results of Network Model Based on Depthwise Separable Convolution

In order to compare the difference between the network model based on depthwise separable convolution and the original model, the same training set is used to train the model in the environment of the same hardware equipment. In this paper, 375 pictures were used for training and 94 pictures were used for testing. The experiment used six types of tags for training, with a batch size of eight. There were 3000 iterations and 64 epochs of training, and the time consumption in the training process is shown in Figures 9 and 10.

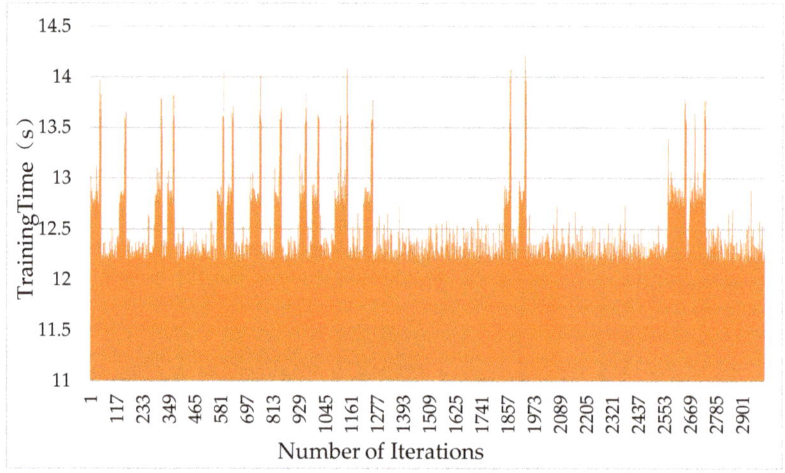

Figure 9. Model training time based on depthwise separable convolution.

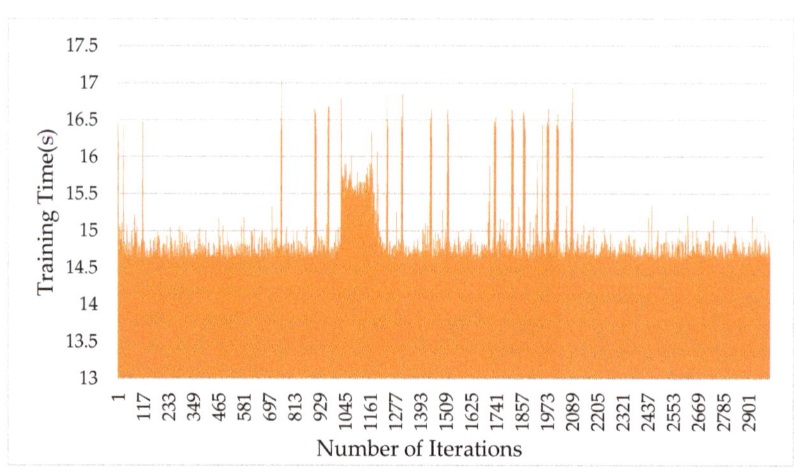

Figure 10. Model training time based on original model.

Among them, batch size is the number of images selected in the dataset for training, usually 8, 16, 32. Iteration completes an iteration for the network model, training a batch-sized picture. Epochs is the number of times the training traverses the dataset. An epoch indicates that each image in the training set participates in the training once. In this experiment, there were 3000 iterations. If the batch size of the training set is eight, then the number of epochs is 64.

As shown in Figure 9, the average training time per round of the network model with depthwise separable convolution was 12.41 s, and the total training time was 10.34 h. In the traditional convolution operation network model, the average training time per round is 14.83 s, and the total training time is 12.36 h. It can be seen that the depthwise separable convolution has an obvious effect on reducing the training time of the model.

After 3000 iterations, both the depthwise separable convolution network model and the original model are basically in a state of convergence, and the loss value is reduced to 2.49. The experimental results show that, in the convolution network model with depthwise separable convolution, although the parameters of the model are reduced, the accuracy of the model is not affected, and the loss value is basically the same as that of the original model, which confirms the effectiveness of the application of depthwise separable convolution.

The number of network model parameters based on depthwise separable convolution is 27.94 million, as shown in Table 3, and the number of parameters of the original model is 29.43 million, which is 1.49 million, and 5% lower than that of the original model. The maximum number of pooled layer and soft connection layer parameters is 0.

Table 3. Network model parameter scale based on depthwise separable convolution.

Name	Input Size	Parameters	Number	Name	Input Size	Parameters	Number
Conv_00	224*224	64,3*3,1,1	1792	Convtr_41d	14*14	512,3*3,1,1	2359808
Conv_10	112*112	64,3*3,1,1	36928	Convtr_40d	14*14	512,3*3,1,1	2359808
Conv_11	112*112	128,1*1,0,1	8320	Convtr_32d	28*28	512,3*3,1,1	2359808
Conv_20	56*56	128,3*3,1,1	147584	Convtr_31d	28*28	512,3*3,1,1	2359808
Conv_21	56*56	256,1*1,0,1	33024	Convtr_30d	28*28	256,3*3,1,1	1180160
Conv_22	56*56	256,3*3,1,1	590080	Convtr_22d	56*56	256,3*3,1,1	590080
Conv_30	28*28	256,3*3,1,1	590080	Convtr_21d	56*56	256,3*3,1,1	590080
Conv_31	28*28	512,1*1,0,1	13312	Convtr_20d	56*56	128,3*3,1,1	295168
Conv_32	28*28	512,3*3,1,1	2359808	Convtr_11d	112*112	128,3*3,1,1	147584
Conv_33	28*28	512,3*3,1,1	2359808	Convtr_10d	112*112	64,3*3,1,1	73856
Conv_40	14*14	512,3*3,1,1	2359808	Convtr_01d	224*224	64,3*3,1,1	36828
Conv_41	14*14	512,3*3,1,1	2359808	Convtr_01d	224*224	8,3*3,1,1	4608
Conv_42	14*14	512,3*3,1,1	2359808	Total			27937564
Convtr_42d	14*14	512,3*3,1,1	2359808				

5.4. Analysis of Network Model Experiment Results Based on H-Swish Activation Function

In order to compare the training results of the network model using the H-Swish activation function with that of the network model using the ReLU activation function, this paper used the same dataset to train the network model. The two models went through 3000 iterations each, and the loss value in the training process is shown in Figure 11 below:

As can be seen from Figure 11, the loss value (loss) of the H-Swish activation function is significantly lower than that of the ReLU activation function: the lowest loss value of the H-Swish activation function is 1.2, and the lowest loss value of using the ReLU activation function is 2.4. The comparison of the experimental data shows that the convolution network model using the H-Swish activation function has higher accuracy. In the convolution network model using the H-Swish activation function, the superiority of the activation function is proven. There is no sigmoid function operation in the activation function, so it is friendly to the calculation of forklift AGV equipment.

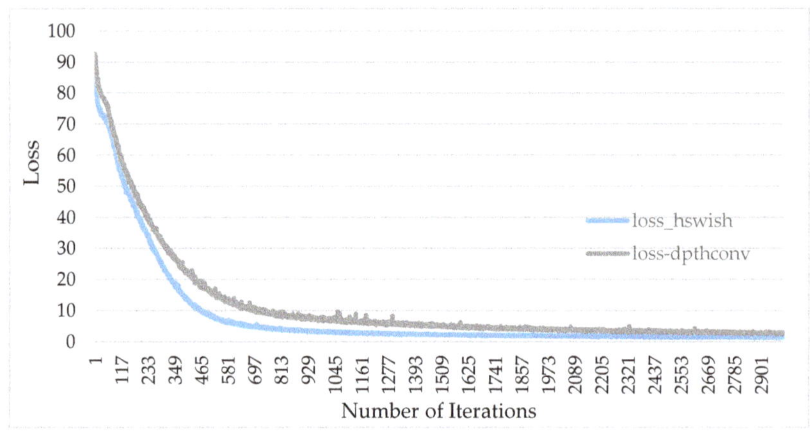

Figure 11. H-Swish and ReLU activation function loss curve.

As shown in Figure 12, 3000 epochs were trained with the network model of the H-Swish activation function and the ReLU activation function, and the segmentation results on the test set image are shown. In the image semantic segmentation, the segmentation effect of using the H-Swish activation function is better, and the segmentation of the target analogy is smoother and more accurate. From the above experimental results, we can see that the segmentation effect of using the H-Swish activation function is better, indicating that using this activation function can improve the performance of the convolution neural network.

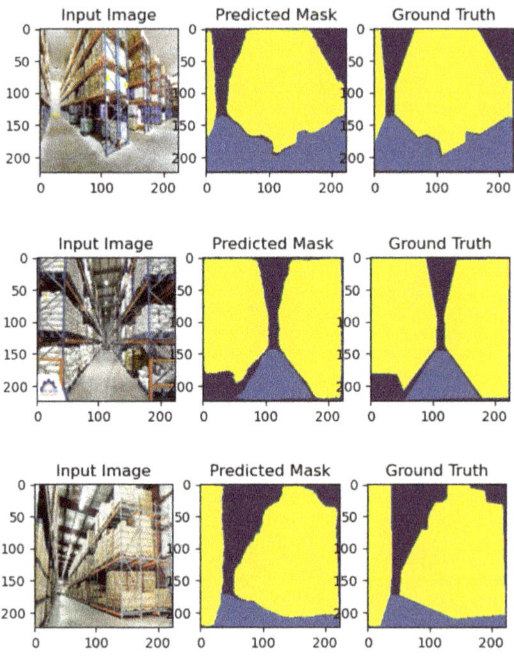

(**a**) Semantic segmentation results of ReLU function.

Figure 12. *Cont.*

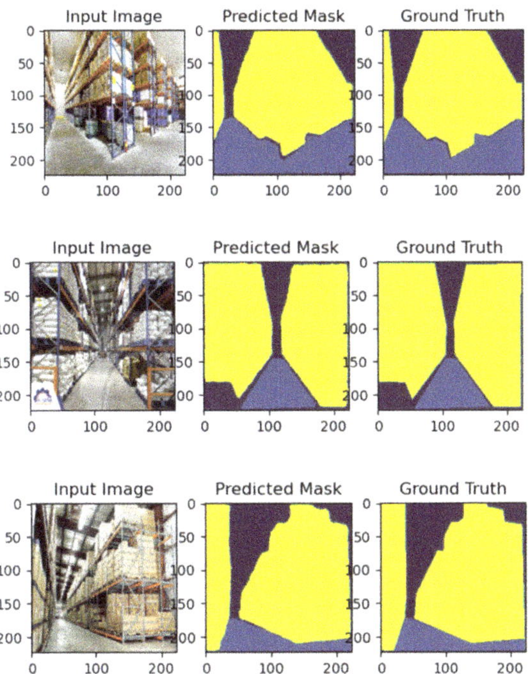

(**b**) Semantic segmentation results of H-Swish function.

Figure 12. Semantic segmentation results of ReLU and H-Swish activation functions.

6. Conclusions

In this paper, the basic principle of depthwise separable convolution is described, and a neural network model based on depthwise separable convolution is proposed on the basis of the SegNet network model. Compared with the original model, the network model established in this chapter has a lower number of parameters and calculations, less memory consumption of the computing equipment in model training, and a shorter convergence time of the training model. On the premise of ensuring the accuracy of the model, the number of parameters of the model proposed in this paper is 1.49 million less than that of the traditional model, and the training time is saved by two hours. The model uses the H-Swish activation function to further improve the accuracy of the model on the basis of reducing the number of model parameters. When using the H-Swish function as the activation function, the loss value is lower than that of ReLU activation function. H-Swish has no sigmoid operation, which reduces the amount of model calculations and is more suitable for the operation of mobile devices, such as forklifts. The network model of this paper was trained by a self-built dataset, which included the data of all kinds of equipment in the logistics distribution center warehouse, which can provide reliable data for the neural network in order to better adapt to the working environment of the warehouse and make the model more suitable for the application of warehouse scene understanding.

This research does not take the small-label objects in the forklift operation environment into account. The classification label types can be improved to make the model suitable for the general industrial environment, which is also a direction for future research.

Author Contributions: Conceptualization, Y.W.; model building, R.Z.; methodology, G.L.; investigation, R.M.; resources, G.L.; experiment, R.Z.; supervision, G.L.; validation, R.M.; writing—original draft, Y.W.; writing—review and editing, R.M. All authors have read and agreed to the published version of the manuscript.

Funding: This research was funded by the Science, Technology and Innovation Commission of ShenZhen Municipality, grant number (No. JCYJ20190807094803721), and Shandong Provincial Natural Science Foundation, grant number (No. ZR2020MF085).

Institutional Review Board Statement: Not applicable.

Informed Consent Statement: Not applicable.

Acknowledgments: We thank other colleagues at the Modern Logistics Research Center of Shandong University for their help.

Conflicts of Interest: The authors declare no conflict of interest.

References

1. Wu, H.; Chi, J.; Tian, G. Instance recognition and semantic mapping based on visual SLAM. *J. Huazhong Univ. Sci. Tech. Nat. Sci. Ed.* **2019**, *47*, 48–54.
2. Wang, M.; Han, B.; Luo, Q. Binocular Visual Navigation and Obstacle Avoidance of Mobile Robots Based on Speeded-Up Robust Features. *CADDM* **2013**, *4*, 18–24.
3. Long, J.; Shelhamer, E.; Darrell, T. Fully Convolutional Networks for Semantic Segmentation. *IEEE Trans. Pattern Anal. Mach. Intell.* **2015**, *39*, 640–651.
4. Noh, H.; Hong, S.; Han, B. Learning Deconvolution Network for Semantic Segmentation. In Proceedings of the IEEE International Conference on Computer Vision (ICCV), Santiago, Chile, 7–13 December 2015; pp. 1520–1528.
5. Badrinarayanan, V.; Kendall, A.; Cipolla, R. SegNet: A Deep Convolutional Encoder-Decoder Architecture for Image Segmentation. *IEEE Trans. Pattern Anal. Mach. Intell.* **2017**, *39*, 2481–24951. [CrossRef] [PubMed]
6. Lin, G.; Milan, A.; Shen, C.; Reid, I. RefineNet: Multi-Path Refinement Networks for High-Resolution Semantic Segmentation. In Proceedings of the 30th IEEE Conference on Computer Vision and Pattern Recognition, Honolulu, HI, USA, 21–26 July 2017; pp. 5168–5177. [CrossRef]
7. Liu, Z. Research of A Local Planner Based on Stereo and 2D Lidar. Master's Thesis, University of Electronic Science and Technology of China, Chengdu, China, 2019.
8. Zhang, X.; Gao, H.; Zhao, J.; Zhou, M. Overview of deep learning intelligent driving method. *J. Tsinghua Univ. Sci. Technol.* **2018**, *58*, 438–444.
9. Mur-Artal, R.; Tardos, J.D. ORB-SLAM2: An opensource slam system for monocular, stereo, and RGB-D cameras. *IEEE Trans. Robot.* **2017**, *33*, 1–8. [CrossRef]
10. Fang, L.; Liu, B.; Wan, Y. Semantic SLAM based on deep learning in dynamic scenes. *J. Huazhong Univ. Sci. Tech. Nat. Sci. Ed.* **2020**, *48*, 121–126.
11. Simonyan, K.; Zisserman, A. Very Deep Convolutional Networks for Large-Scale Image Recognition. In Proceedings of the 3rd International Conference on Learning Representations, ICLR 2015, San Diego, CA, USA, 7–9 May 2015.
12. Chen, L.C.; Papandreou, G.; Kokkinos, I.; Murphy, K.; Yuille, A. DeepLab: Semantic Image Segmentation with Deep Convolutional Nets, Atrous Convolution, and Fully Connected CRFs. *IEEE Trans. Pattern Anal. Mach. Intell.* **2018**, *40*, 834–848. [CrossRef] [PubMed]
13. Chen, C.; Tang, S.; Li, J. Weakly supervised semantic segmentation based on dynamic mask generation. *J. Image Graph.* **2020**, *25*, 1190–1200.
14. Souly, N.; Spampinato, C.; Shah, M. Semi supervised semantic segmentation using generative adversarial network. In Proceedings of the IEEE International Conference on Computer Vision (ICCV), Venice, Italy, 22–29 October 2017; pp. 5689–5697.
15. Wei, Y.; Feng, J.; Liang, X.; Cheng, M.; Zhao, Y.; Yan, S. Object region mining with adversarial erasing: A simple classification to semantic segmentation approach. In Proceedings of the IEEE Conference on Computer Vision and Pattern Recognition, College Park, MD, USA, 25–26 February 2017; pp. 1568–1576.
16. Zhang, X.; Wei, Y.; Feng, J.; Yang, Y.; Huang, T. Adversarial complementary learning for weakly supervised object localization. In Proceedings of the IEEE Computer Society Conference on Computer Vision and Pattern Recognition, Salt Lake City, UT, USA, 18–23 June 2018; pp. 1325–1334.
17. Courbariaux, M.; Bengio, Y.; David, J.P. BinaryConnect: Training Deep Neural Networks with binary weights during propagations. In Proceedings of the Advances in Neural Information Processing Systems, Convention Center, Montreal, QC, Canada, 7–12 December 2015; pp. 3123–3131.
18. Prewitt, J.; Mendelsohn, M.L. The analysis of cell images. *Ann. N. Y. Acad. Sci.* **1966**, *128*, 1035–1053. [CrossRef] [PubMed]
19. Fukunaga, K.; Hostetler, L. The estimation of the gradient of a density function, with applications in pattern recognition. *IEEE Trans. Inf. Theory* **2006**, *21*, 32–40. [CrossRef]

20. Bai, X.; Wang, W. Saliency-SVM: An automatic approach for image segmentation. *Neurocomputing* **2014**, *136*, 243–255. [CrossRef]
21. Chen, L.C.; Papandreou, G.; Kokkinos, I.; Murphy, K.; Yuille, A.L. Semantic Image Segmentation with Deep Convolutional Nets and Fully Connected CRFs. In Proceedings of the 3rd International Conference on Learning Representations, ICLR 2015, San Diego, CA, USA, 7–9 May 2015.
22. Sandler, M.; Howard, A.; Zhu, M.; Zhmoginov, A.; Chen, L.C. MobileNetV2: Inverted Residuals and Linear Bottlenecks. In Proceedings of the IEEE Computer Society Conference on Computer Vision and Pattern Recognition, Salt Lake City, UT, USA, 18–23 June 2018; pp. 4510–4520.
23. Ioffe, S.; Szegedy, C. Batch Normalization: Accelerating Deep Network Training by Reducing Internal Covariate Shift. In Proceedings of the 32nd International Conference on Machine Learning, ICML, Lile, France, 6–11 July 2015; Batch, F., Blei, D., Eds.; pp. 448–456.
24. Nair, V.; Hinton, G.E. Rectified Linear Units Improve Restricted Boltzmann Machines Vinod Nair. In Proceedings of the 27th International Conference on Machine Learning, Haifa, Israel, 21–24 June 2010; pp. 807–814.
25. Ramachandran, P.; Zoph, B.; Le, Q.V. Searching for Activation Functions. In Proceedings of the 6th International Conference on Learning Representations, Vancouver, BC, Canada, 30 April–3 May 2018.

Article

A Data Management Approach Based on Product Morphology in Product Lifecycle Management

Gang Liu [1,2,*], **Rongjun Man** [3] **and Yanyan Wang** [3]

1   School of Mechanical Engineering, Shandong University, Jinan 250061, China
2   Key Laboratory of High Efficiency and Clean Mechanical Manufacture, Ministry of Education, Shandong University, Jinan 250061, China
3   School of Control Science and Engineering, Shandong University, Jinan 250061, China; 201914400@mail.sdu.edu.cn (R.M.); yanyan.wang@sdu.edu.cn (Y.W.)
*   Correspondence: sdulg@sdu.edu.cn; Tel.: +86-531-8839-2269

Abstract: In the product life cycle from conception to retirement, there are three forms: conceptual products, digital products and physical products. The carriers of conceptual products are requirements, functions and abstract structures, and data management focuses on the mapping of requirements, functions, and structures. The carrier of digital products is digital files such as drawings and models, and the focus of data management is the design evolution of product. Physical products are physical entities, and their attributes and states will change over time. Existing data model research often focuses on one or two forms, and it is even impossible to integrate three forms of data into one system. So, a new data management method based on product form is presented. According to the characteristics of the three product form data, a conceptual product data model, a digital product data model, and a physical product data model are established to manage the three forms of data, respectively, and use global object mapping to integrate them into a unified data model. The conceptual product data model has a single data model for a single business stage. The digital product data model uses the core data model as the single data source, and uses one stage rule filter to add constraints to the core data model for each business stage. The physical product data model uses the core data model to manage the public data of the physical phase, and the phase private data model focuses on the private data of each business phase. Finally, a case of Multi-Purpose Container Vessel is studied to verify the feasibility of the method. This paper proposes three product forms of product data management and a unified data management model covering the three product forms, which provides a new method for product life cycle data.

Keywords: product morphology; core data model; phase rule filter; phase private data model

Citation: Liu, G.; Man, R.; Wang, Y. A Data Management Approach Based on Product Morphology in Product Lifecycle Management. *Processes* **2021**, *9*, 1235. https://doi.org/10.3390/pr9071235

Academic Editors: Sergey Y. Yurish and Bhavik Bakshi

Received: 27 May 2021
Accepted: 13 July 2021
Published: 16 July 2021

1. Introduction

With the coming of global economic integration, a deep affinity exists for the manufacturing industry. Customers focus not only on high quality and low cost, but also on innovation, service, environmental protection and the personalization of products. Thus, manufacturers must pay attention to intellectual capital, from a product's initial conception to its retirement.

Product Lifecycle Management (PLM) is defined as "A strategic business approach that applies a consistent set of business solutions that support the collaborative creation, management, dissemination and use of product definition information" [1]. In this decade, PLM systems are recognized for managing the information of the full product lifecycle [2].

A critical aspect of PLM systems is their product information modelling architectures [2,3]. A reasonable and effective product data model can seamlessly integrate and make available all of the product data throughout all of the phases in a full lifecycle, to the right people, who are the manufacturer, suppliers, facilitators, customers and circulation economy enterprises, at the right time.

In the product life cycle, products have three morphologies: conceptual products, digital products, and physical products. The carrier of a conceptual product is an abstract product structure oriented to product requirements and functions. Conceptual product data management focuses on the realization relationship between requirements and functions, and functions and structures. The carrier of digital products is data based on the product design structure, including the attributes of the product and its parts, two-dimensional drawings, three-dimensional models, instructions, analysis reports and other data. The focus of digital product data management is product evolution management. Physical products are physical objects that really exist, and can be used and maintained. Physical product data management focuses on changes in product status and attributes over time. The data management characteristics of these product morphologies determine that there are huge differences in data management modes, and there is a gap.

In this paper, a new product data management method based on product morphology is proposed. The method uses three data models and integrates the three data models into a data management framework to manage product lifecycle data. The first one is a single data model for the conceptual data. The second one uses a core data model and several phase rule filters for the digital data. Additionally, the last one includes a core data model and several phase private data models for the physical data. This approach removes the gaps among the conceptual product, digital product and physical product, which are divided according to product morphology. Thus, all the product data seamlessly integrates into one system.

The structure of this paper is as follows. Section 2 studies the state-of-the-art of a data model in PLM. Section 3 discusses the methodology. Section 4 presents three product morphology types: the conceptual product, the digital product and the physical product. Section 5 achieves a Unified Data Model based on Product Morphology (UDMPM) using Unified Modelling Language (UML). Section 6 studies a case of shipbuilding to validate the data management approach.

2. State-of-the-Art: A Data Model of PLM

In this section, several types of product data modelling methods are reviewed. The researchers studying the product data model focus on different aspects to deal with the problems facing them. Then, different methods of product data description and expression, integration of multi-phase data, single data source and so on are presented. The emphases of them are as follows:

The description and expression of product data describe products in terms of content, attributes, and relationships, which are the basis of data management. There are many related studies, mainly focusing on the following four aspects. (1) Meta-data and meta-model are used for product data modeling and integration. Xia uses meta-models to describe data in the stages of product design, process manufacturing, sales operations, and maintenance [4]. Kreis uses meta-data to describe CAD data and integrates with PDM systems [5]. Tao uses meta-models to describe the basic structure and content of the data set, enables the retrieval and integration of heterogeneous data from different data sources [6]. Dinesh uses meta-models to support the integration of CAD, CAPP, NC tool path verification and MRP systems [7]. (2) Ontology is an explicit and formalised description of common concepts [8,9]. Ontology-based data models are used for product design, production, use, and service data management and integration [10–12], as well as product knowledge [13]. (3) As the most popular data exchange format and data model language, eXtensible Markup Language (XML), Standard for the Exchange of Product Model Data (STEP) and UML were widely adopted in data modelling. XML presented shared data to different information systems [14,15]. STEP was used as a bridge to integrate heterogeneous data [16–18]. UML model administrates product data [4,19]. (4) Product Profiles are multiple benefit models that benefit expected suppliers, customers, and users, and are used for the description and expression of early conceptual product data. Roughly outline the outline of the product, for example, with basic product attributes, customers can experience the functions of future products without completely predicting its shape [20].

The integration of multi-stage data is another research focus. Product plan-BOM (Bill Of Material), design-BOM, craft-BOM, stock-BOM, manufacture-BOM, finance-BOM, customer-BOM, and after service-BOM, is integrated to supply a single data source by developing a data flow [21,22]. Li uses data domains integrating a concept design domain, a structural design domain, a detailed design domain, a process planning domain, a manufacturing domain and a sales and service domain by mapping [3]. Xia puts forward a unified meta-model that supports the data meta models in the stages of product design, process manufacturing, sales and operation, and maintenance services [4]. Cai analyses the main mapping forms of BOM multi-views and designs the mapping rules of BOM multi-view model [23]. A dynamic BOM describing the configuration of a spacecraft assembly indifferent phase of product development is established [24]. Modular instance structure data models of the generalized product are established, and the structure mapping views in different phases of the product life cycle are also presented [25].

A single data source is the next focus. Single data sources guarantee consistency, integrality and security of data to the largest extent. A single data source is obtained by integrating multi-BOM [26]. Sudarsan described a product information-modelling framework, based on the NIST Core Product Model (CPM) and its extensions: The Open Assembly Model (OAM), the Design-Analysis Integration model (DAIM) and the Product Family Evolution Model (PFEM), which can support the full range of PLM information needs [2]. A configurable BOM model abstracts key entities such as product, product attribute, material, material attribute, material brand, is a single data source of product parameter calculation [27].

All the researches above have contributed to information modelling of PLM. However, there are some problems that remain to be solved:

Product data description and expression model one or several stages of product structure, attributes, characteristics and documents. However, it is not a deeper factor, that is, product form. It is the most critical factor for product existence and data evolution. Different product forms have different carriers and data evolution patterns, while the carriers and data evolution patterns at different stages in the same product form are the same. Therefore, a data model that does not take into account the similarities and differences of product form and product stage cannot well manage all product data and its evolution, and support data tracking throughout the life cycle.

Integration of multi-phase data uses complex mapping relations and brings redundant data among multi-BOM, multi-phase data models or multi-domains. The consistency and integrality of product data are usually out of control throughout the course of full lifecycle. Additionally, upward spread of versions is enlarged. Most studies only achieve the integration of conceptual products and digital products or digital products and physical products. BOM or domains of different stages are regarded as the same concepts as each other. The evolution process of "conceptual product", "digital product" and "physical product" has not been described.

Single data sources achieve a single meta-model, a single data model of the design phase or multi-phase data models and multi-BOM, which are not one whole data model. It either cannot express all of the information on a product from concept to retirement, or it is too difficult to implement.

To manage data from the full lifecycle, the paper studies product morphology, and a unified data model is presented. The morphology of a product has three types: a conceptual product, a digital product and a physical product. Every product morphology type has a meaning, a function, and data. According to the characteristics of the product morphology type, the unified data model reveals the three product morphology types and their relations.

3. Methodology

The product life cycle covers the entire process of a product from concept to death, including many business stages. Collecting and analyzing the data of each stage, it is found that the carrier of the product has three forms. Summarizeing the three product forms revealed the differences between them, and we have summarized the characteristics of data evolution.

Facing each product form, studying the characteristics of each phase of the product, it is found that the logical relationship of the data within the product form is similar, and there is an evolutionary relationship between the data in each phase.

Based on the different characteristics of the data evolution of each product form, the data model of each product form is established as a single data source, reducing the mapping relationship between BOMs, reducing data redundancy and the complexity of data management, and it is easy to maintain data consistency. Then, the mapping relationship of global objects is established to achieve the morphological evolution path of "conceptual product" to "digital product" to a "physical product".

UML supports object-oriented technology, can express various objects and their attributes concisely and clearly, can describe rich relationships in data modeling, and is used to establish a unified data model.

4. Study of Product Morphology

A product is a thing that is produced to meet the demands of a customer. There are three product morphology types in a lifecycle: a conceptual product, a digital product and a physical product, which separately are the forms of a concept, digital data and an entity with matter.

A conceptual product is an abstract concept that describes the inherent characteristics of a type of product. It is the result of product conceptual design, and its carrier is the product requirement, function and abstract structure. Conceptual products are the source of product evolution, from which products of different models and specifications can be developed.

A digital product is developed based on conceptual products, which define the detailed functions, structural composition, structural features of parts and components, production processes and materials of the product. Its carrier is product structure information, two-dimensional drawings, three-dimensional models, calculation instructions, simulation analysis data, etc. The product structure is a framework for organizing and managing other data, which exhibits EBOM, PBOM and MBOM. Product data are gradually improved and accurate along with the development process. Versions are used to manage data evolution.

A physical product, also called an instance product, is a physical entity that can be manipulated. It has use value and can truly meet the needs of customers. It exists in a specific time domain and space domain, and its use and management are restricted by time and space. It has a single structure, but its properties and status change over time. The PLM system manages the data extracted from the physical product, including product attributes, status and structure. The above data has time series properties and is unique at a certain moment.

In summary, the conceptual product, digital product and physical product are different morphology types of the same product, and separately involve conceptual, digital and physical things. There are many studies that concern data model of a digital product and a physical product, but, importantly, do not include a conceptual product.

5. Unified Data Model Based Product Morphology

There are large differences in the data of the three product morphology types. The data of a conceptual product is abstract and general, and includes concepts, functions and main structures of multi-series products. The data of a digital product is accurate and changeable, and includes structural data, attributes, requirements, and processes. The structural data of a physical product is usually changeless, and some attributes and states change continually.

According to the differences above, the Unified Data Model based Product Morphology (UDMPM) is made of three layers: Object Layer, Data Model Layer and Document Management Layer, as shown in Figure 1. The Object Layer includes an object pool, which creates and destroys Conceptual Object, Digital Object and Physical Object and manages the evolution among the three types of objects by mapping. The Data Model Layer holds three data models: the Conceptual Product Data Model (CPDM), the Digital Product Data Model (DPDM) and the Physical Product Data Model (PPDM). The three data models separately manage the data of three product forms and support corresponding business applications, acting as single data sources of every product morphology type. Document Management manages all of the unstructured data in the full lifecycle. UDMPM is based on UML. UML is an effective PLM modelling language and supports a Model-Driven Architecture (MDA). Thus, UML can achieve a platform-independent model.

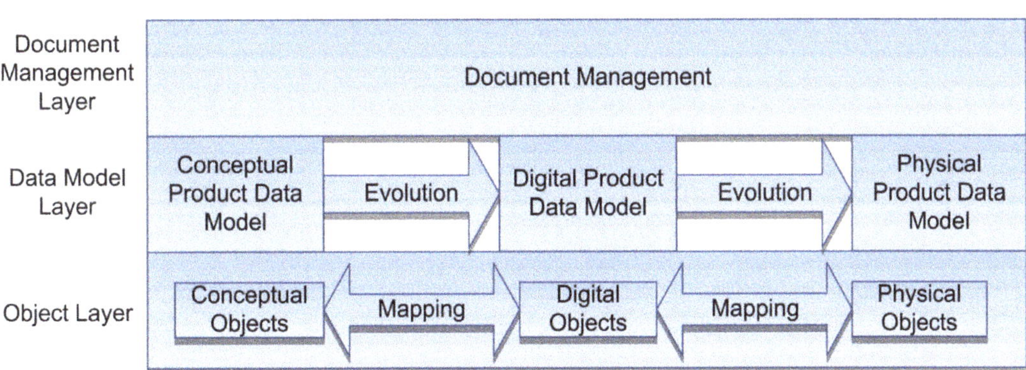

Figure 1. UDMPM Framework.

5.1. Object Layer

The object layer uses an object pool to create and destroy all of the business objects and manages mapping relations among the conceptual object, digital object and physical object. The conceptual object, digital object and physical object are separately the abstract of business objects of a conceptual product, digital product and physical product, including the products, components, parts, functions, requirements, resources and so on. A conceptual object is the source of several digital objects. Many physical objects are derived from the same digital object. Thus, there are mapping relations that describe the evolution of the three objects.

A UML static class structure of all product objects is presented and divided into four groups: Global Objects, Conceptual Product Objects, Digital Product Objects and Physical Product Objects, as shown in Figure 2. As shown in Figure 2, symbol 1 and symbol * represent the one and multi-object, respectively.

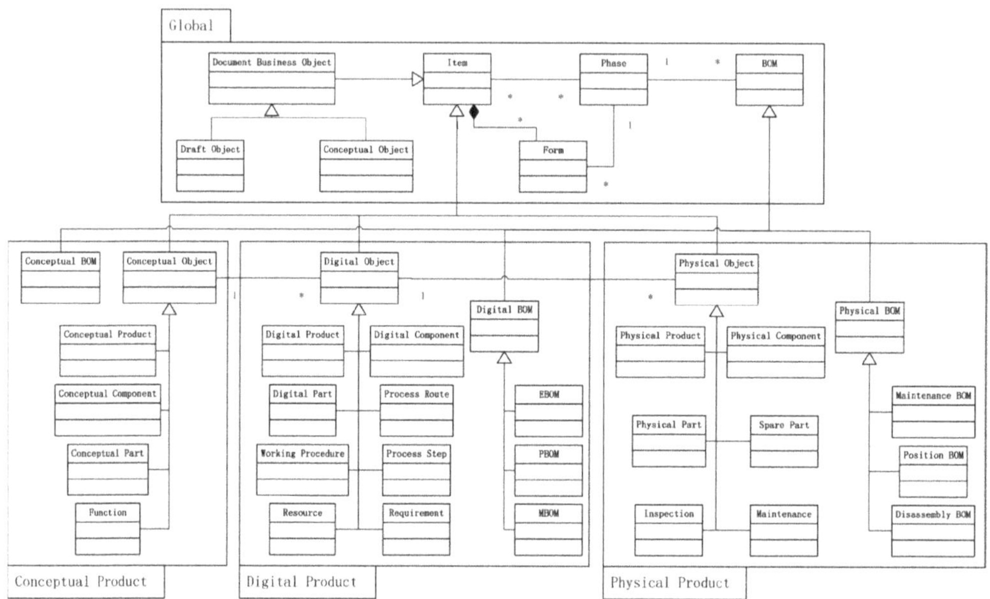

Figure 2. Static Classes Structure of Product Lifecycle Objects (1, * represent the one and multi-object).

Global Objects are the base of all objects. Item is the base class of all the business classes, and BOM is the base class of all the types of BOM objects. They are abstract classes and cannot be instantiated. The subclass of Document Business Object and Document Dataset are used in document management. The main attributes of Item, such as ID, name, revision, vision, security classification, type, owner and time, are managed alone, and other phase-attributes are managed by Form. Phase is the abstract of the business phase. Some subclasses of Item can exist in several Phases, for example that the same products, components and parts are in requirements management, structure design, process design and so on. However, Form and BOM only belong to a Phase. Then, Item can integrate the attributes and structural information of a multi-phase and can hold the access of every phase.

The main subclasses of Item are Conceptual Object, Digital Object and Physical Object, which are basic abstract classes of three product morphology types and cannot be instantiated.

Conceptual Object and Conceptual BOM are major Conceptual Product Objects. Conceptual object is a super-class of all of the conceptual product business objects and has the attributes: state (Working, Modifying, Examining, Released and Invalid) and project. Its major subclasses are Conceptual Product, Conceptual Component, Conceptual Part and Functions, which are the abstracts of corresponding business objects and have attributes, as shown in Table 1. Conceptual BOM is an accessorial class creating the relations of two objects in the conceptual product structure tree. The data model of Conceptual Product is composed of the four subclasses, Conceptual BOM and other Conceptual Product objects, which are derived from Conceptual Object to fit specific needs.

Table 1. Attributes of conceptual object.

Object	Attribute
Conceptual Product	Usage, performance, function, unit, description, mender, modification time
Conceptual Component	Specification, performance, description, mender, modification time
Conceptual Part	Specification, performance, description, mender, modification time
Function	Content, description, mender, modification time

Digital Object and Digital BOM are major base classes of Digital Product Objects. Digital Object is a super-class of all of the digital product business objects and has the attributes: state (Working, Modifying, Examining, Released and Invalid), project, maturity degree and employ degree. Its subclasses are Digital Product, Digital Part, Digital Component, Process Route, Working Procedure, Process Step, Requirement, and Resource, all of which have unique attributes outlined in Table 2. Digital BOM also is an accessorial class, and its subclasses are EBOM, PBOM, and MBOM. All of the subclasses above establish digital product structure.

Table 2. Attributes of digital object.

Object	Attribute
Digital Product	Model, weight, material, design organisation, mender, modification time
Digital Component	Weight, material, mender, modification time
Digital Part	Weight, material, mender, modification time
Process Route	Process type, station, time, description
Working Procedure	Type, description
Process Step	Type, description
Requirement	Content, description
Resource	Resource type, working condition, usage state

Physical Object and Physical BOM are major base classes of Physical Product Objects. Physical Object is a super-class of all physical product business classes. The subclasses of Physical Object are Physical Product, Physical Component, Physical Part, which have attributes as shown in Table 3. Physical BOM is an accessorial class also, and its subclasses are Maintenance BOM, Position BOM and Disassembly BOM. All of the subclasses above construct a physical product structure.

Table 3. Attributes of a physical object.

Object	Attribute
Physical Product	Serial number, batch number, manufacturer, production time, producing area, quality controller, performance, working condition, weight, length, width, height, and other attributes.
Physical Component	Serial number, batch number, producer, production time, producing area, quality controller, weight, length, width, height, material, resource (self-manufacturing, procurement, buy)
Physical Part	Serial number, batch number, producer, production time, producing area, quality controller, weight, length, width, height, material, origin (self-manufacturing, procurement, buy)
Inspection	Inspection type, object, time, place, checker, degree, content, result, standard, tool
Maintenance	Object, failures, state (closed, delay, fixed, new, open, rejected, reopen), maintainer, begin time, finish time, place, description

5.2. Data Model Layer

According to the data characteristics of every product morphology type, Data Model Layer uses corresponding objects when building data models, including CPDM, DPDM and PPDM. The three data models are single data sources of every product morphology type. Mapping relations combine three data models into a whole model and remove the gaps among them.

CPDM, DPDM and PPDM are separately composed of a core data model and phase data models. A core data model is the real single data source of product morphology types and assures the uniqueness and consistency of the data. A phase data model is a logic data model that exhibits the data of a core data model for business applications in a specific phase.

5.2.1. Conceptual Product Data Model

CPDM achieves Function–Structure management, and a Conceptual Core Data Model (CCDM) is the same as a Conceptual Phase Data Model (CPhDM), because there is only one phase. CPhDM is shown in Figure 3. The Conceptual Product and Conceptual component have a Conceptual BOM. At the same time, a Conceptual BOM has several Conceptual Components and Conceptual Parts. With the help of Conceptual BOM, the structure of a Conceptual Product is built. Conceptual Product, Conceptual Component and Conceptual Part have several Functions, and the Function structure is a tree.

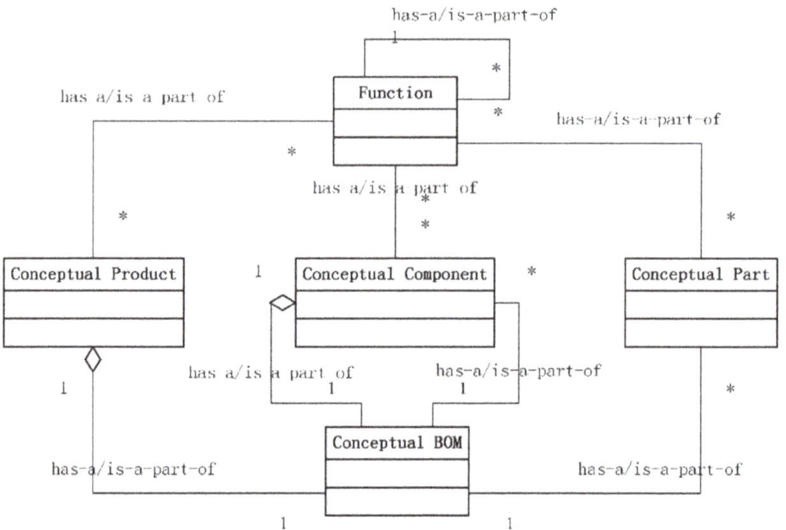

Figure 3. CPhDM (1, * represent the one and multi-object).

5.2.2. Digital Product Data Model

Digital product evolves through several business phases, including requirements management, structure design, process design and so on. DPDM, including a Digital Core Data Model (DCDM), multiple rule filters and phase data model, is presented, as shown in Figure 4. DCDM is the only data source of a digital product, and flexibly manages all of the design data. There is a rule filter that adds constraints to DCDM in every business phase to prevent illogical and unauthorised operations, and then, the phase data models appear. Rule filters are in phase business managers; for example, Structure Rule Filter (SRF) is in Product Structure Editor, which is a tool that manages design structures. In the digital product, there is Requirement Rule Filter (RRF), SRF, Process Rule Filter (PRF) and Requirement Phase Data Model (RPDM), Structure Phase Data Model (SPDM), and Process Phase Data Model (PrPDM).

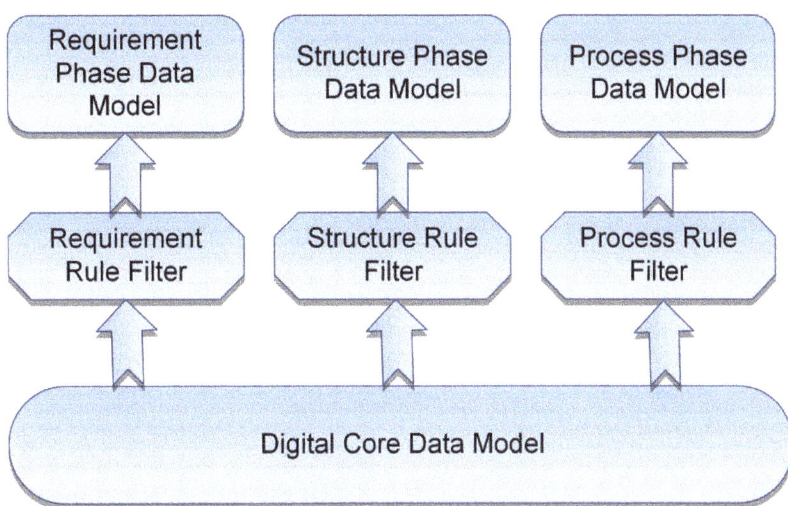

Figure 4. DPDM.

DCDM is very flexible for heterogeneous data and is shown in Figure 5. The digital object contains 0 or more Digital BOMs and 0 or more forms. Digital BOM manages the underlying structure of digital objects, and forms manage the attributes of digital objects at different stages. Digital objects are contained in multiple stages, indicating in which stages are valid. The stage contains multiple digital objects, multiple Digital BOMs, and multiple Forms. A stage has multiple digital objects and their part Digital BOM and Form. Form and Digital BOM belong to a Digital Object and a Phase, which means that it is the Form and Digital BOM of this object at this stage. Digital BOM has multiple Digital Objects and multiple Digital BOMs, which, respectively, indicate which objects it contains and the relationship with the underlying structure.

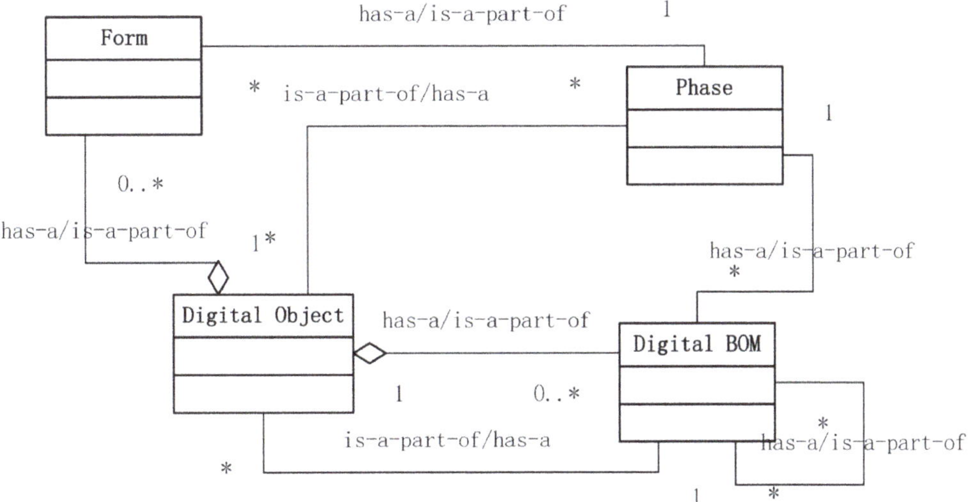

Figure 5. Digital Core Data Model (1, * represent the one and multi-object).

Rule filter uses a set of rules to restrict objects, relations and access. The attributes of the rules are shown in Table 4. In Table 4, symbol 0 represents there is no object. ID is the identification number. The name is used to indicate the meaning of a rule. Phase decides the business phase. Priority means the order of importance. Relation is the relationships between Source Object and Destination Object. Multiplicity represents the quantitative relationship between Source Object and Destination Object in the rule. Access means who has the right to use this rule.

Table 4. Attributes of rules (1, * represent the one and multi-object).

Attribute	Signification	Option
ID	Identify each rule	Number or string
Name	Name of the rule	String
Phase	Phase of the rule	Requirements Management, Structure Design, Process Design or Production
Priority	Right to proceed before another	Number
Relation	Relationship between Source Object and Destination Object	Has-a/is-a-part-of, Specification, Compose, Reference or Requirement
Source Object	The first object of the rule	All types of business objects or data objects
Destination Object	The second object of the rule	
Multiplicity	Number of Source Object and Destination Object in the rule	1/*/0..1/0..*/1..*
Access	Access to operate the rule	

Requirements management perfectly and accurately depicts the product capability that satisfies the demands of customers. The requirements come from a customer's advice and complaints, orders, and development processes. They are managed and tracked by an RPDM.

In the requirements management phase, there are many rules, as shown in Table 5. The Source Objects and Destination Objects have access to this phase, while other objects do not. All of the relations permitted and their multiplicities are shown in Table 5. A Product has a Requirements BOM for managing all of the requirements. The relations between Requirement and Requirement BOM build the requirements tree. Requirements are met by Design Objects, and the relations between them are used for requirements tracking management. Requirement has several subclasses, including Customer Requirement, Usage Requirement, Sale Requirement, Management Requirement, and Technology Requirement.

The rules of the structure design stage are shown in Table 6. Design Object has three subclasses: Design Product, Design Component and Design Part. The subclass of Form is Structure Form. The subclass of Phase is Structure Design Phase. The subclass of Design BOM is EBOM. Both Design Product and Design Component have an EBOM object that manages the design structure. EBOM has multiple EBOM and Design Part, which manage the subordinate structure and the parts contained in this level, respectively. Design Product, Design Component and Design Part each have a Structure Form, which manages the properties of the design phase. There are Specification, Compose, Reference and Requirement relationships between two Design Objects. Design Product, Design Component, Design Part, EBOM and Structure Form all belong to a Structure Design Phase. The relationship between Requirement and Design Object is "Requirement/Function".

Table 5. Set of rules in the requirements management phase (1, * represent the one and multi-object).

Relation	Source Object	Destination Object	Multiplicity
Inherit	Design Object	Requirement	
Inherit	Design Object	Design Product	
Inherit	Form	Requirement Form	
Inherit	Phase	Requirement Management Phase	
Inherit	Design BOM	Requirement BOM	
Has-a/is-a-part-of	Design Product	Requirement BOM	1:1
Has-a/is-a-part-of	Requirement	Requirement BOM	1:0..1
Has-a/is-a-part-of	Requirement	Requirement Form	1:1
Has-a/is-a-part-of	Requirement BOM	Requirement	1:*
Has-a/is-a-part-of	Requirement Management Phase	Requirement BOM	1:*
Has-a/is-a-part-of	Requirement Management Phase	Requirement Form	1:*
Has-a/is-a-part-of	Requirement Management Phase	Requirement	1:*
Has-a/is-a-part-of	Requirement Management Phase	Design Product	1:*
Requirement/Function	Requirement	Design Object	*:*
Has-a/is-a-part-of	Design Product	Requirement Form	1:1
Has-a/is-a-part-of	Phase	Design Object	*:*
Inherit	Requirement	Customer Requirement	
Inherit	Requirement	Usage Requirement	
Inherit	Requirement	Sales Requirement	
Inherit	Requirement	Management Requirement	
Inherit	Requirement	Technology Requirement	

Table 6. Set of rules in the design phase (1, * represent the one and multi-object).

Relation	Source Object	Destination Object	Multiplicity
Inherit	Design Object	Design Product	
Inherit	Design Object	Design Component	
Inherit	Design Object	Design Part	
Inherit	Form	Structure Form	
Inherit	Phase	Structure Design Phase	
Inherit	Design BOM	EBOM	
Has-a/is-a-part-of	Design Product	EBOM	1:1
Has-a/is-a-part-of	Design Component	EBOM	1:1
Has-a/is-a-part-of	EBOM	EBOM	1:*
Has-a/is-a-part-of	EBOM	Design Part	1:*
Has-a/is-a-part-of	Design Product	Structure Form	1:1
Has-a/is-a-part-of	Design Component	Structure Form	1:1
Has-a/is-a-part-of	Design Part	Structure Form	1:1
Specification, Compose, Reference, Requirement	Design Object	Design Object	*:*
Has-a/is-a-part-of	Structure Design Phase	Design Product	1:*
Has-a/is-a-part-of	Structure Design Phase	Design Component	*:*
Has-a/is-a-part-of	Structure Design Phase	Design Part	*:*
Has-a/is-a-part-of	Structure Design Phase	EBOM	1:*
Has-a/is-a-part-of	Structure Design Phase	Structure Form	1:*
Requirement/Function	Requirement	Design Object	*:*

The rules of the process design stage are shown in Table 7. Design Object has eight subclasses: Design Product, Design Component, Design Part, Process Route, Working Procedure, Process Step, Resource and Work Area. The subclasses of Process Route are Assembly Process and Manufacturing Process. The subclass of Form is Process Form. The subclass of Phase is Process Phase. Process BOM (PBOM), Bill Of Process (BOP), Bill Of Resource (BOR) and Bill Of Work Area (BOWA), which separately manage structures of process, resource and work area, are subclasses of Design BOM. PBOM is subordinate to Design Product, Design Component and PBOM, and manages a hierarchical process structure. PBOM contains Design Part, and is subordinate to Process Phase. Process Form belongs to Design Product, Design Component, Design Part and Process Phase to manage the process attributes of product, component and part. Process Phase has Process Form, Design Product, Design Component, Design Part, Process Route, Working Procedure, Process Step, Resource, Work Area, Assembly Process, Manufacturing Process, BOP, BOR and BOWA.

5.2.3. Physical Product Data Model

In contrast to a conceptual product and digital product, a physical product has individualities. The data of physical products, which are from one digital product, are not the same as each other. The structures and most of the attributes of physical products, components and parts are changeless, but the state and other attributes change frequently. Physical components and parts are replaced often. The objects of inspection, MRO and recycling are the physical object, such as the physical product, physical components and physical parts.

Thus, PhDM is composed of the Physical Core Data Model (PCDM) and several private data models, including the Inspection Private Data Model (IPrDM), Service Private Data Model (SPrDM) and Recycling Private Data Model (RPrDM). PCDM describes the objects, structure and attributes of the physical product. Private data models manage the private data of each phase. Phase data models are the sum of the PCDM and private data models, and there are three phase data models: the Inspection Phase Data Model (IPDM), Service Phase Data Model (SPDM) and Recycling Phase Data Model (RPDM), as shown in Figure 6.

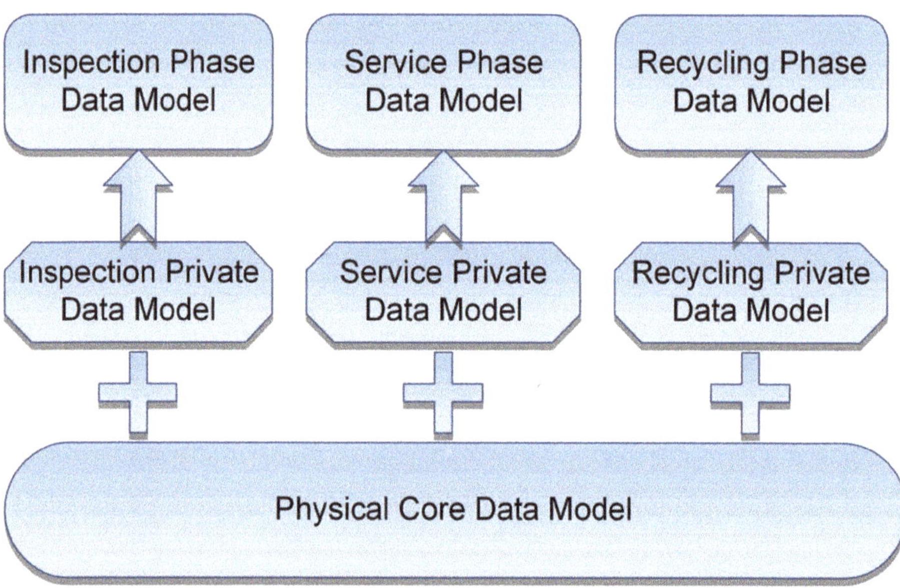

Figure 6. PPDM.

Table 7. The set of rules in the process phase (1, * represent the one and multi-object).

Relation	Source Object	Destination Object	Multiplicity
Inherit	Design Object	Design Product	
Inherit	Design Object	Design Component	
Inherit	Design Object	Design Part	
Inherit	Design Object	Process Route	
Inherit	Design Object	Working Procedure	
Inherit	Design Object	Process Step	
Inherit	Design Object	Resource	
Inherit	Design Object	Work Area	
Inherit	Process Route	Assembly Process	
Inherit	Process Route	Manufacturing Process	
Inherit	Form	Process Form	
Inherit	Phase	Process Phase	
Inherit	Design BOM	PBOM	
Inherit	Design BOM	BOP	
Inherit	Design BOM	BOR	
Inherit	Design BOM	BOWA	
Has-a/is-a-part-of	Design Product	PBOM	1:1
Has-a/is-a-part-of	PBOM	PBOM	1:*
Has-a/is-a-part-of	Design Component	PBOM	1:1
Has-a/is-a-part-of	PBOM	Design Part	1:*
Has-a/is-a-part-of	Process Phase	PBOM	1:*
Has-a/is-a-part-of	Design Product	Process Form	1:1
Has-a/is-a-part-of	Design Component	Process Form	1:1
Has-a/is-a-part-of	Design Part	Process Form	1:1
Has-a/is-a-part-of	Process Phase	Process Form	1:*
Has-a/is-a-part-of	Process Phase	Design Product	1:*
Has-a/is-a-part-of	Process Phase	Design Component	*:*
Has-a/is-a-part-of	Process Phase	Design Part	*:*
Has-a/is-a-part-of	Process Phase	Process Route	*:*
Has-a/is-a-part-of	Process Phase	Working Procedure	*:*
Has-a/is-a-part-of	Process Phase	Process Step	*:*
Has-a/is-a-part-of	Process Phase	Resource	*:*
Has-a/is-a-part-of	Process Phase	Work Area	*:*
Has-a/is-a-part-of	Process Phase	Assembly Process	*:*
Has-a/is-a-part-of	Process Phase	Manufacturing Process	*:*
Has-a/is-a-part-of	Process Phase	BOP	*:*
Has-a/is-a-part-of	Process Phase	BOR	*:*
Has-a/is-a-part-of	Process Phase	BOWA	*:*
Design/Assembly	Design Product	Assembly Process	*:*
Design/Assembly	Design Component	Assembly Process	*:*
Design/Manufacture	Design Part	Manufacturing Process	*:*
Has-a/is-a-part-of	Design Process	BOP	1:1
Process/Resource	Design Process	Resource	*:*
Process/Work Area	Design Process	Work Area	*:*
Has-a/is-a-part-of	BOP	BOP	1:*
Has-a/is-a-part-of	Working Procedure	BOP	1:1
Process/Resource	Working Procedure	Resource	*:*
Process/Work Area	Working Procedure	Work Area	*:*
Has-a/is-a-part-of	BOP	Process Step	1:*
Process/Resource	Process Step	Resource	*:*
Process/Work Area	Process Step	Work Area	*:*
Has-a/is-a-part-of	Resource	BOR	1:1
Has-a/is-a-part-of	BOR	BOR	1:*
Has-a/is-a-part-of	BOR	Resource	*:*
Has-a/is-a-part-of	Work Area	BOWA	1:1
Has-a/is-a-part-of	BOWA	BOWA	1:*
Has-a/is-a-part-of	BOWA	Work Area	*:*

As shown in Figure 7, PCDM comprises Neutral Structure, Position Structure and Physical Structure. A digital product will produce thousands of physical products. In the maintenance phase, the attributes and status of physical products are different and will change over time. Each physical product requires a physical structure to manage attributes and status data separately. However, the design and process data of these physical products are the same, and the maintenance planning and solutions, failure modes and other data of the MRO stage are also the same. If the same data are managed with a physical structure, it will bring thousands of instances of data redundancy and easily lead to data inconsistency. Therefore, a neutral structure is used to manage the common data of physical products of the same design. Position Item inherits structural information, and Physical Object obtains other attributes from Design Object. Time Stamp stores the time relative to when Physical Object is installed on Position Item.

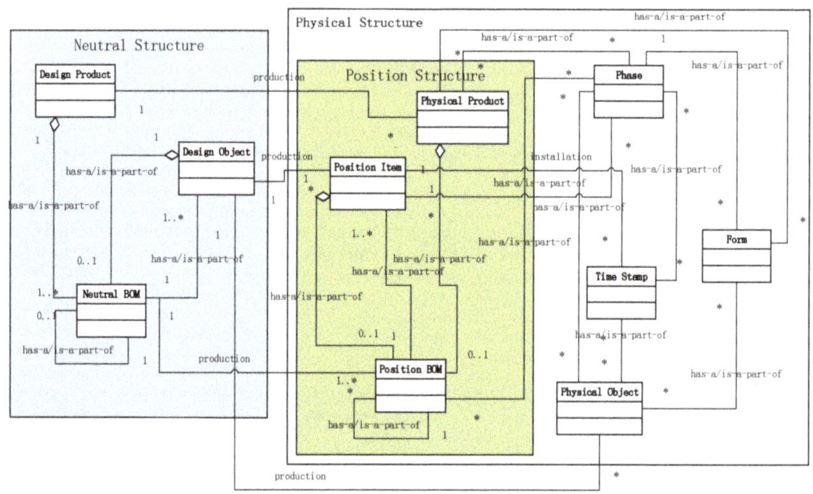

Figure 7. Physical Core Data Model (1, * represent the one and multi-object).

Neutral Structure bridges the gap between the design product and physical product and manages common structures of physical products from one batch, and it is a simplified mirror of the design structure with respect to the granularity and assembly structures in the processes of use and repair. It manages instructions, operations process, maintenance methodology and necessary design information. Design Product, other Design Object and Neutral BOM construct the Neutral Structure, such as an SPDM.

Position Structure, which is made up of Position Item and Position BOMs, mirrors the Neutral Structure and defines configuration data and position meta-data, which manage maintenance and repair documents that relate to the assembly position. Position Structure represents a virtual physical product that is independent of specific physical components and parts. Every Physical Product has a Position BOM to manage its sub-Item—Position Item. Position Item is a type of placeholder for a physical component and part in Position Structure and builds the structure with the help of Position BOM.

Based on Position Structure, Physical Structure manages the data of disassembly and assembly of specific physical components and parts with the help of Time Stamp. At any time, the structure of a physical product is extracted from Physical Structure, because one Physical Object can be installed on different Position Items, and one Position Item can use different Physical Objects at different times. However, the relations between Position Item and Physical Object are 1:0, 0:1 and 1:1 every moment. Physical Product and other Physical Objects have Forms to manage their attributes in different business phases. The phase decides where the Form, Physical Product and other Physical Objects are utilised.

In the inspection phase, according to quality standards, a strict working process and methods are used to supervise and control the quality of a physical product and to avoid providing defective physical products to the market. The aims of inspection are physical products, components and parts. The data of inspection are queues of inspections and bugs. IPrDM is built as shown in Figure 8. Physical Product, Physical Component and Physical Part inheriting from Physical Object exist in the whole PPDM. They can be inspected several times and use Inspection to store the data. The information of defects is described by Bug. Inspection and Bug are subclasses of Physical Object, too, and they are mainly the business object of the inspection phase. Physical Product, Physical Component, Physical Part and Bug have Inspection Form to address their attributes in the phase.

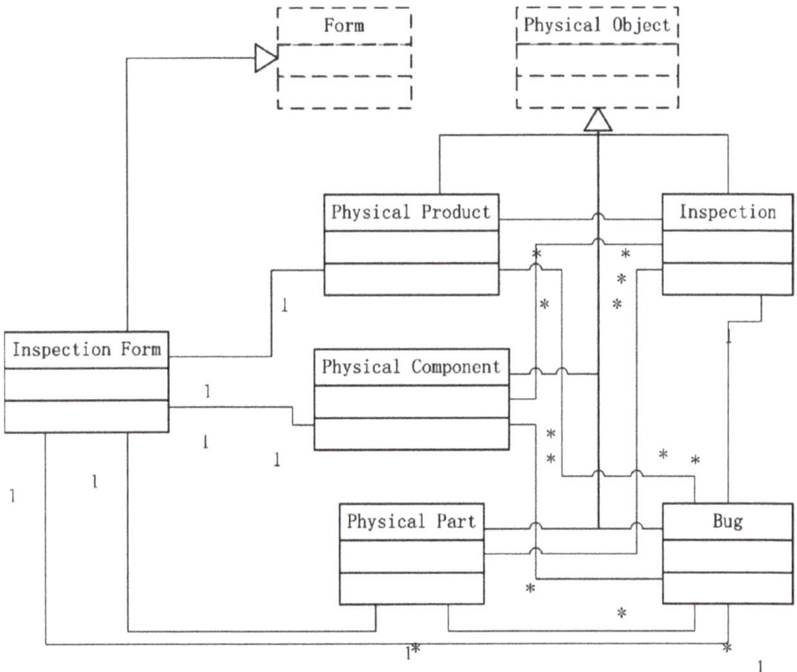

Figure 8. Inspection Private Data Model (1, * represent the one and multi-object).

Service contains the use, maintenance, repair, and overhaul. The data contain the working condition, failure, maintenance, repair, overhaul, and complaint. Based on PCDM, SPrDM is built, as shown in Figure 9. The Physical Product is owned by a Customer, who has Gripe, and has many uase data, which involved Working Condition and brings Failure. It has MRO, including Maintain, Repair and Overhaul. MRO addresses the Failure of Physical Product, Physical Component and Physical Part. Physical Product, Physical Component and Physical Part have Service Form managing their attributes in the Service phase.

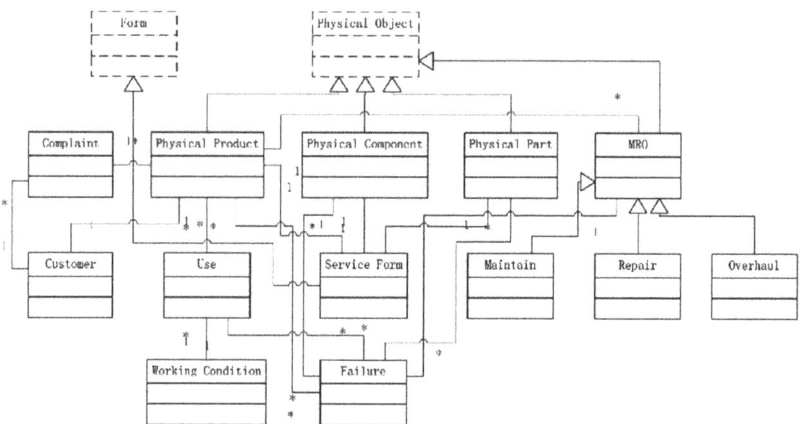

Figure 9. Service Private Data Model (1, * represent the one and multi-object).

Recycling contains disassembly, cleanout, detection, reuse, remanufacturing, regeneration, burning, and Landfill. When out of service, physical products are taken to circulation economy enterprises. Physical products and components are disassembled, and the physical parts are cleaned. After detection, some are reused directly in other products, and others are remanufactured to recover the performance, regenerated for the material then burned and landfilled. Based on PCDM, RPrDM is built, as shown in Figure 10. Physical Product and Physical Component are dismantled, and we obtain Physical Component, Physical Part, Material and Waste. Some can be reused. Blank, which is machined to a new Physical Part, comes from Material. Waste receives Innocent Treatment. A new Physical Product is presented by Remanufacture and Reassembly of Physical Component and Physical Part. All Physical Objects in the recycling phase have a Recycling Form for their recycling attributes. The content and management of the Process Route are the same as that for the Process Phase.

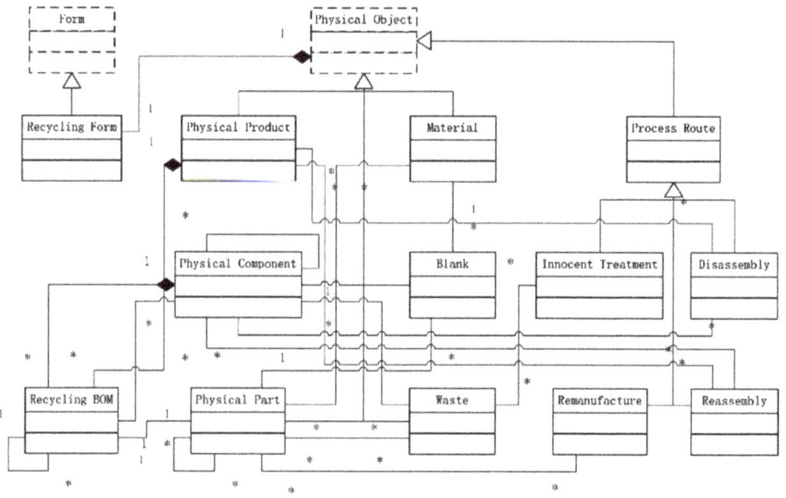

Figure 10. Recycling Private Data Model (1, * represent the one and multi-object).

5.3. Document Management

Document management is the basic function of PLM and is the first requirement of customers. It manages unstructured data, including two-dimensional drawings, three-dimensional models, calculations, specifications, and simulation analysis data. Unstructured data are transferred and stored as a whole one, not considering the logic and stored structure.

Document management uses meta-data to manage all of the types of documents, as shown in Figure 11. DoDataset is meta-data to manage the file. Item is related to Document and DoDataset by citation, specification, and representation. Document, which is a superclass of Model and Draft, organises DoDataset of several files for the logic file, for example, a manual is composed of files in Chinese and in English, Model contains UG NX files and CATIA files, and Draft includes a metric drawing and an imperial drawing of the same content.

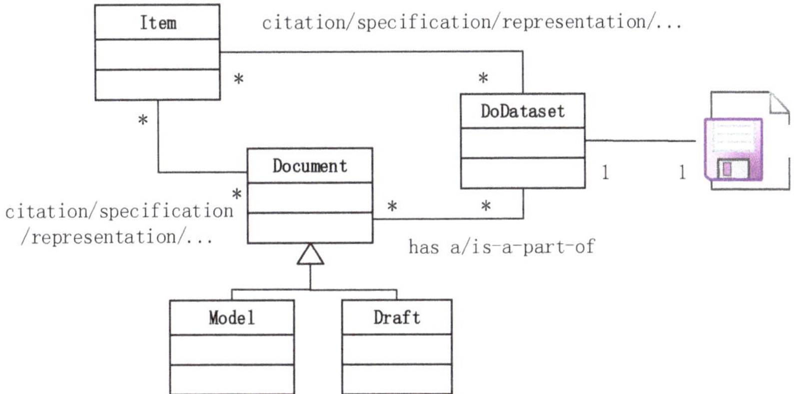

Figure 11. Document Management Data Model (1, * represent the one and multi-object).

6. Case Study

An example using UDMPM to manage data on shipping in Huanghai Shipbuilding Co. Ltd. is studied. A sship is a complex product system, involving several majors. There are complex relations among phases, system, sections, components, parts and documents. The ship's data management has obvious characteristics, and a unified data model for a conceptual ship, digital ship and physical ship is built. Because the ship has too much data, the case focuses on the marine engine system.

In a shipyard, a conceptual ship manages a ship type, such as a Multi-Purpose Container (MPC) Vessel, a Ro-Ro Passenger Vessel, or an Oil Tanker. MPC is a type of ship to transfer containers, and its CCDM is shown in Figure 12. Conceptual MPC uses a function tree and a conceptual structure to manage all the conceptual data. The function of a marine engine system includes voyage power, cooling, air exchange and separate oil. The Marine Engine System has a Diesel Engine System, a Steam Turbine System, a Ventilation System and an Oil Separator. Diesel Engine Systems and Steam Turbine Systems both achieve voyage power and are redundant units. However, a conceptual MPC can select one or two types. Function–Structure mapping describes the relation between the function and structure units.

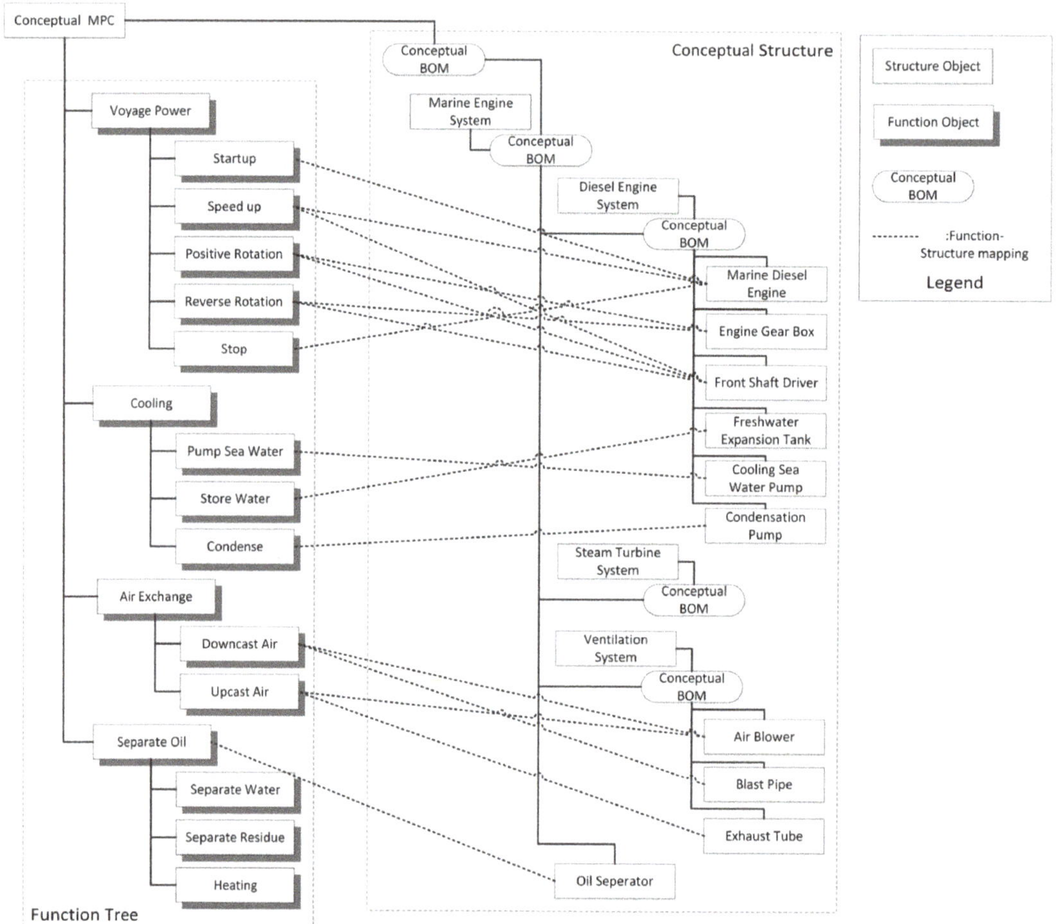

Figure 12. Conceptual Ship—Multi Purpose Container vessel.

Digital MPC originates from conceptual MPC and manufactures one or several physical MPCs. The DCDM of digital MPC is shown in Figure 13. Some digital objects can cover several business phases for uniqueness, such as the digital MPC existing in the requirements management phase, the structure design phase and the process design phase. Thus, digital MPC has Requirement Form, Structure Form and Process Form, which separately contain the attributes of corresponding phases. Requirement Form owns attributes, such as the seaworthy area, dead weight ton and seating capacity. Structure Form includes the length, width, height, and draught. The Process Form has a section number, max section weight, and welding requirements. Digital MPC uses Requirement BOM, Structure BOM and Process BOM to construct the backbone of the data. The black line, red line and green line separately express the Requirement Management Rule of RRF, the Structure Design Rule of SRF and the Process Design Rule of PRF, and RPDM, SPDM and PrPDM. In SPDM, sub-Items of the Marine Engine Room are the Diesel Engine System and the Ventilation System, which disappear in PrPDM because they are virtual parts, and they appear only for planning processes of structure design management. The long broken line is Requirement–Structure Mapping. The short dotted line is the Citation between design parts. The Exhaust Tube has a single structure, and Exhaust Tube A, B and C have different installation sites. They separately represent structure information and process information on the same part. Digital MPC, Marine Engine Room and Exhaust Tube separately own Joining Process Route, Fixing Process Route and Machining Process Route. The Machining Process Route is shown in Figure 13, and others are omitted. The bending process step uses the Bending Machine in the Cold Working Shop.

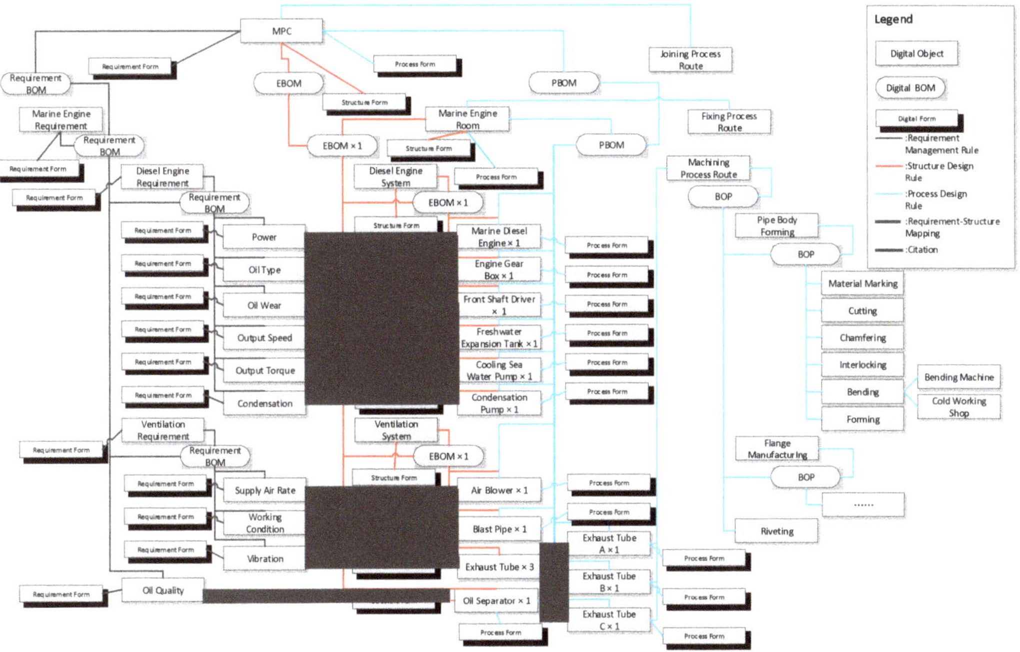

Figure 13. Digital Ship—Multi Purpose Container vessel.

Every physical MPC has a unique ID, such as HCY-93, which has 8000 Dead Weight Ton (DWT), as shown in Figure 14. In PCDM, Neutral Structure is a mirror of PBOM, and the object comes from PBOM, also. The Position Structures of HCY-93 and HCY-94 are copies of Neutral Structure, using position objects. Timestamp manages the installation relation of position objects and physical objects. Before 6/19/11, the Air Blower uses CZ-80A, and after 6/19/11, it uses JCZ-50B. All of the physical objects are produced according to their corresponding design objects and obtain design attributes from them, such as the design performance and working condition, for example, the KYDH204SD-23 Oil Separator. The CLH100-125 Sea Water Pump has Inspection Form for inspection information, Service Form for use and MRO information and Recycling Form for disposal and recycling information. In the Inspection phase, HCY-93-08 Marine Engine Room experiences Air Tight Inspection and Water Tight Inspection, in which two bugs are found. In service, the KYDH204SD-23 Oil Separator is maintained, because of failures tracked by the system such as Oil Leak Accident and Shake Seriously. In recycling, 8DKM-28 Diesel Engine has Remanufacture Process Route to recover performance, and Stainless Steel is obtained from the HCY-93-08-02 Blast Pipe.

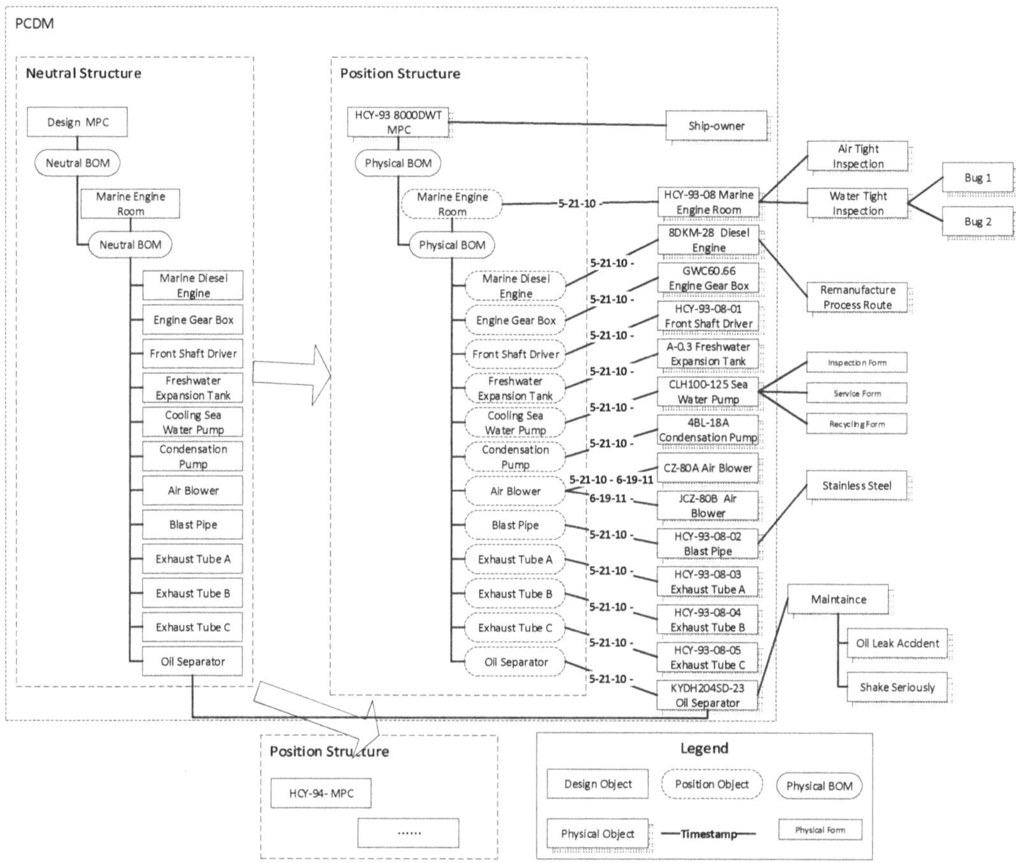

Figure 14. Physical Ship—Multi Purpose Container vessel.

7. Discussion

There are three forms in the product life cycle: conceptual products, digital products and physical products. The carrier and data evolution characteristics of these three product forms are different, so it is difficult to efficiently integrate product life cycle data.

In the product life cycle, there are several business stages such as conceptual design, structural design, process design, manufacturing, and service use. Many previous studies established a data model or BOM for each stage, and integrated them using mapping. Other studies established a model for all stages as the only data source. The former abandons the common characteristics of data in multiple stages of a product form and weakens the evolutionary relationship between them. The latter cannot manage the data of each product form according to its characteristics and distinguishes the evolution mode of each product form.

The paper summarizes the connotation of the three forms of products, and proposes a unified product data management model that covers the differences in data management among three product forms. Based on previous researches, the data model is constructed at two levels: product form and business stage. We have inherited the business-oriented modeling method and the idea of a single data source, and combined the two organically.

8. Conclusions

In this paper, we present a data management approach that is based on three product morphology types for concept data, digital data and physical data over the course of a full lifecycle. Three product morphology types and UDMPM are the core of the approach. Product morphology is classified into three types: conceptual product, digital product and physical product, by the carrier and the evolutionary characteristics of the product. According to the evolutionary characteristics of every product morphology type, UDMPM makes a corresponding management mode. A conceptual product has one business phase, and a single data model, and CCDM is the best approach. Because designing a product involves requirements management, structure design and process design, and because it evolves continually, a flexible DCDM allows for consistency and integrality. In addition, business applications are supported by phase data models, which are founded by adding phase rule filters on the DCDM. The structure and features of a physical product are not changeless, but properties and status change over time. Then, PCDM manages changeless data, and the phase private data model describes the changed data. In general, the approach integrates three data management modes into a whole system, which addresses all of the data in the full lifecycle, eliminating data redundancy and holding consistency and integrality.

Previous research focused on the integration of multi-stage BOM or data model with data mapping, or a single data source of multiple stages. Based on the above research, the paper summarizes the three forms of product connotation and proposes a unified product data management model, establishing a single data source for each product form, manage multiple stages of data and the data evolution between them, using global object mapping to integrate three data models, and providing a new method for integrated management of product life cycle data.

In the current research, big product data are of great significance to product operation and maintenance and design and manufacturing. Digital twin technology is also a current research hotspot. The data characteristics of big data are different from the data studied in this article. Digital twins are another carrier of products, and their evolutionary characteristics are also different from the data studied in this article. Due to limited conditions, the paper did not study the management model of big data and digital twins. This is an area for future research.

Author Contributions: Conceptualization, Y.W.; Data curation, R.M.; Formal analysis, G.L.; Funding acquisition, Y.W.; Investigation, R.M.; Methodology, G.L.; Project administration, G.L.; Resources, G.L.; Supervision, G.L.; Validation, R.M.; Writing—original draft, Y.W.; Writing—review and editing, R.M. All authors have read and agreed to the published version of the manuscript.

Funding: This research was funded by the National Key Research and Development Program of China, grant number 2018YFB1702601, the Major Scientific and Technological Innovation Projects in Shandong Province, grant number 2019JZZY010442 and Natural Science Foundation of Shandong Province, grant number ZR2020MF085.

Acknowledgments: We thank managers and engineers of Shandong Hoteam Software Co. Ltd. and Huanghai Shipbuilding Co. Ltd. for their kind help.

Conflicts of Interest: The authors declare no conflict of interest.

References

1. Gecevska, V.; Chiabert, P.; Anisic, Z.; Lombard, F.; Cus, F. Product lifecycle management through innovative and competitive business environment. *J. Ind. Eng. Manag.* **2010**, *3*, 323–336. [CrossRef]
2. Sudarsan, R.; Fenves, S.J.; Sriram, R.D.; Wang, F. A product information modeling framework for product lifecycle management. *Comput. Aided Des.* **2005**, *37*, 1399–1411. [CrossRef]
3. Li, Y.; Wan, L.; Xiong, T. Product data model for PLM system. *Int. J. Adv. Manuf. Technol.* **2011**, *55*, 1149–1158. [CrossRef]
4. Zihang, X.; Jiping, Y.; Sheng, Y.; Tao, H. Research on Product Lifecycle Model-Based on UML. *IOP Conf. Ser.* **2021**, *632*, 052078.
5. Kreis, A.; Hirz, M.; Stadler, S.; Salchner, M.; Rossbacher, P. Convenient connection technology data model supporting optimized information exchange between CAx-systems. *Comput. Aided Des. Appl.* **2018**, *15*, 771–778. [CrossRef]

6.	Tao, J.; Yu, S. A Meta-model based Approach for LCA-oriented Product Data Management. *Procedia CIRP* **2018**, *69*, 423–428. [CrossRef]

7.	Dhamija, D.; Koonce, D.A.; Judd, R.P. Development of a unified data meta-model for CAD-CAPP-MRP-NC verification integration. *Comput. Ind. Eng.* **1997**, *33*, 19–22. [CrossRef]

8.	Studer, R.; Benjamins, V.R.; Fensel, D. Knowledge engineering: Principles and methods. *Data Knowl. Eng.* **1998**, *25*, 161–197. [CrossRef]

9.	Sanfilippo, E.M. Feature-based product modelling: An ontological approach. *Int. J. Comput. Integr. Manuf.* **2018**, *31*, 1097–1110. [CrossRef]

10.	Ebrahimipour, V.; Rezaie, K.; Shokravi, S. An ontology approach to support FMEA studies. *Expert Syst. Appl.* **2010**, *37*, 671–677. [CrossRef]

11.	Gim, D.M.; Vegetti, M.; Leone, H.P.; Henning, G.P. PRoduct ONTOlogy: Defining product-related concepts for logistics planning activities. *Comput. Ind. Prod. Lifecycle Model. Anal. Manag.* **2008**, *59*, 231–241.

12.	Lu, Y.; Wang, H.; Xu, X. ManuService ontology: A product data model for service-oriented business interactions in a cloud manufacturing environment. *J. Intell. Manuf.* **2019**, *30*, 317–334. [CrossRef]

13.	Dori, D.; Shpitalni, M. Mapping Knowledge about Product Lifecycle Engineering for Ontology Construction via Object-Process Methodology. *CIRP Ann. Manuf. Technol.* **2005**, *54*, 117–122. [CrossRef]

14.	Choi, S.S.; Yoon, T.H.; Noh, S.D. XML-based neutral file and PLM integrator for PPR information exchange between heterogeneous PLM systems. *Int. J. Comput. Integr. Manuf.* **2010**, *23*, 216–228. [CrossRef]

15.	Hu, X.; Xu, Y. Research on semi-structured and unstructured data storage and management model for multi-tenant. *J. Inf. Technol. Res.* **2019**, *12*, 49–62. [CrossRef]

16.	Zhang, Y.; Zhang, C.; Wang, H. An Internet based STEP data exchange framework for virtual enterprises. *Comput. Ind.* **2000**, *41*, 51–63. [CrossRef]

17.	Mehli-Qaissi, J.; Coulibaly, A.; Mutel, B. Product data model for production management and logistics. *Comput. Ind. Eng.* **1999**, *37*, 27–30. [CrossRef]

18.	Behera, A.K.; McKay, A.; Earl, C.F.; Chau, H.H.; Robinson, M.A.; de Pennington, A.; Hogg, D.C. Sharing design definitions across product life cycles. *Res. Eng. Des.* **2019**, *30*, 339–361. [CrossRef]

19.	Eynard, B.; Gallet, T.; Roucoules, L.; Ducellier, G. PDM system implementation based on UML. *Math. Comput. Simul. Comput. Eng. Syst. Appl.* **2006**, *70*, 330–342. [CrossRef]

20.	Albert, A.; Jonas, H.; Benjamin, W.; Gustav, N.B.; Nicolas, R.; Nicolas, H.; Sascha, O.; Nikola, B. Product Profiles: Modelling customer benefits as a foundation to bring inventions to innovations. *Procedia CIRP* **2018**, *70*, 253–258. [CrossRef]

21.	Zhou, C.; Cao, Q. Design and implementation of intelligent manufacturing project management system based on bill of material. *Clust. Comput.* **2019**, *22*, 8647–8655. [CrossRef]

22.	McKay, A.; Chau, H.H.; Earl, C.F.; Behera, A.K.; de Pennington, A.; Hogg, D.C. A lattice-based approach for navigating design configuration spaces. *Adv. Eng. Inform.* **2019**, *42*, 100928. [CrossRef]

23.	Shuyu, C.; Die, H.; Yang, F.; Jianxin, X. Research on BOM multi-view mapping method orient to aircraft in-service data management. *J. Phys. Conf. Ser.* **2020**, *1550*, 032124.

24.	Jing, F.; Yang, W.; Zhang, Y.; Zhou, R.; Gao, X. Dynamic BOM Construction Technology and Application Based on MBD. *J. Phys. Conf. Ser.* **2020**, *1486*, 072050.

25.	Li, H.; Ji, Y.; Luo, G.; Mi, S. A modular structure data modeling method for generalized products. *Int. J. Adv. Manuf. Technol.* **2016**, *84*, 197–212. [CrossRef]

26.	Wu, W.; Fan, X.; Zhou, G.; Huang, Y.; Cao, Y.; Lin, G. A Material Recommendation System for Enterprise Manufacturing Integrated Systems. *Zhongguo Jixie Gongcheng/China Mech. Eng.* **2019**, *30*, 1856–1865.

27.	Xia, L.; Ri, N.J.; Han, T.Z. Product parameter calculation based on configurable BOM model. *J. Phys. Conf. Ser.* **2020**, *1486*, 032036.

Article

Condition Monitoring of Drive Trains by Data Fusion of Acoustic Emission and Vibration Sensors [†]

Oliver Mey [1,*], André Schneider [1], Olaf Enge-Rosenblatt [1], Dirk Mayer [1], Christian Schmidt [2], Samuel Klein [2] and Hans-Georg Herrmann [2]

[1] Fraunhofer Institute for Integrated Circuits IIS, Division Engineering of Adaptive Systems, 01069 Dresden, Germany; andre.schneider@eas.iis.fraunhofer.de (A.S.); olaf.enge@eas.iis.fraunhofer.de (O.E.-R.); dirk.mayer@eas.iis.fraunhofer.de (D.M.)

[2] Fraunhofer IZFP Institute for Nondestructive Testing, 66123 Saarbrücken, Germany; christian.schmidt@izfp.fraunhofer.de (C.S.); Samuel.Klein@izfp.fraunhofer.de (S.K.); hans-georg.herrmann@izfp.fraunhofer.de (H.-G.H.)

[*] Correspondence: oliver.mey@eas.iis.fraunhofer.de

[†] This paper is an extended version of our paper published in Automation, Robotics & Communications for Industry 4.0: 1st IFSA Winter Conference (ARCI' 2021).

Abstract: Early damage detection and classification by condition monitoring systems is crucial to enable predictive maintenance of manufacturing systems and industrial facilities. Data analysis can be improved by applying machine learning algorithms and fusion of data from heterogenous sensors. This paper presents an approach for a step-wise integration of classifications gained from vibration and acoustic emission sensors in order to combine the information from signals acquired in the low and high frequency ranges. A test rig comprising a drive train and bearings with small artificial damages is used for acquisition of experimental data. The results indicate that an improvement of damage classification can be obtained using the proposed algorithm of combining classifiers for vibrations and acoustic emissions.

Keywords: condition monitoring; vibration; acoustic emission; drive train; data fusion; machine learning

Citation: Mey, O.; Schneider, A.; Enge-Rosenblatt, O.; Mayer, D.; Schmidt, C.; Klein, S.; Herrmann, H.-G. Condition Monitoring of Drive Trains by Data Fusion of Acoustic Emission and Vibration Sensors. *Processes* **2021**, *9*, 1108. https://doi.org/10.3390/pr9071108

Academic Editor: Sergey Y. Yurish

Received: 1 June 2021
Accepted: 22 June 2021
Published: 25 June 2021

1. Introduction

Increasing demand for efficient, reliable and available industrial production systems has led to the development of systems for continuous condition monitoring (CM). By the analysis of sensor signals, failures and wear can be detected in an early stage, enabling cost-effective predictive maintenance and prohibiting severe damages [1–3]. A widely implemented application is CM of bearings. These are critical parts of many industrial drives. Established fields of application include wind energy [4] or process industry [5].

A common monitoring approach includes vibration measurements, usually conducted with accelerometers [6]. Damage detection is traditionally carried out by the extraction of features from the time domain signals or the respective frequency domain representation. An overview of commonly used features such as RMS, Variance or Kurtosis, especially with application to low speed bearings, is given in [7]. Since bearings exist in a wide range of geometries and are driven at different speeds and loads, CM systems have to be adapted individually for a sufficient damage detection capability [8]. Still, the performance of vibration-based CM systems suffers when dynamic acceleration levels become low due to slow rotation speeds, when disturbance levels are high due to coupling with noisy gearboxes or other drive train components, and in case of varying speeds.

In many practical cases of vibration analysis for CM, the relevant signals are deteriorated by background noise or interference. The Wiener filter technique is suitable for these tasks, for slowly varying disturbances, and adaptive digital filters can also be applied [9]. Originating from the removal of interference from signals in the medical diagnostic sector, the removal of noise and interference from bearing vibration signals has been implemented [10]. One option is the utilization of a reference correlated to the interfering signal, a technique also applied in active noise control systems [11]. Another option is the so-called adaptive line enhancer, which does not need a dedicated reference. This method has already been tested in real industrial environments [12].

Measurement of vibration at higher frequencies between 35 and 40 kHz was tested decades ago, and a higher sensitivity to bearing wear could be proven [13]. Additionally, CM based on ultrasonic structure-borne noise or acoustic emissions has been evaluated [14,15]. The method is based on the detection of impulsive events (bursts) sent out from the rolling contact of damaged bearing elements. One of the first applications were the large and slowly rotating main bearings in off-shore cranes [16]. Several works investigated the performance of vibration and ultrasonic CM in comparison to each other, demonstrating the high sensitivity of Acoustic Emission (AE), e.g., for monitoring of gearboxes [17] and bearings [18]. While being very sensitive, disadvantages of the method include limited range of the impulsive waves in the structure and being insensitive to global mechanical faults such as misalignment of drive shafts, imbalances or loose parts [19]. Another challenge in comparison to vibration signals is the high data rate resulting from the needed sampling frequency of AE signals [20].

For both methods, the vibration approach and ultrasonic approach, a number of signal analysis algorithms have been developed, either in the time or frequency domain. More recent approaches use the corresponding sensor data as an input to powerful machine learning (ML) algorithms [21–24]. ML has been successfully applied in a previous work to this paper for the detection of mechanical faults such as imbalances [1]. Several works investigated how data from multiple vibration sensors can be jointly processed to obtain diagnostic information about a rotating machine [25–27].

AE signatures can estimate the remaining useful life of a slowly rotating bearing [28]. A fatigue test introduced flat spots to the bearing rings. However, the AE signatures were not analyzed in depth to distinguish between different damage types. Hase investigated the influence of damages on the AE signatures in a fatigue test. Different damages could be distinguished both in the time and frequency domains. However, an attempt for automation of classification by machine learning was not reported [29]. An application of machine learning to the classification of AE signatures of a sliding bearing in a fatigue test was presented in [30]. A deep neural network classified different damages, but a limited accuracy was reached. Saufi et al. applied a classification to several features of vibration and AE signals generated from rotating bearings. Different artificial defects could be distinguished; however, no attempt for data fusion was described [31]. Support Vector Machines (SVMs) were applied to classify AE signals from artificially damaged bearings. The input features of the SVM were skewness and kurtosis calculated from the measured signals. Classified damages included cracks in the outer ring of the bearing and debris [32].

Additionally, fusion of heterogeneous sensors has been investigated. The authors of [33] give an overview and a taxonomy of the approaches: with respect to the analysis layer that implements the fusion method, data fusion, feature fusion, and decision fusion can be distinguished. For an example of decision fusion, in [3], an approach based on fusion of AE and vibration signals is presented. Classifiers for both signals are trained and afterwards merged by a random forest network to detect gearbox faults. Further approaches to detect gearbox faults use information from multiple motor current sensors [34] or from vibration sensors and rotary encoders [35] for a neural network-based classification. In [36], features from various heterogenic sensor time series are extracted, decorrelated using principle component analysis and afterwards classified using a naïve Bayes classifier to detect motor faults. The fusion of two-dimensional image data from an infrared camera with one-dimensional vibration signals using convolutional neural networks and t-distributed stochastic neighbor embedding is shown in [37].

CM with integrated sensors or by acquisition of operation parameters supports the continuous analysis of the system's state. A CM system, thus, can improve resilience, if a feedback loop automatically switches operation modes according to the detected state of the system components. On the other hand side, the CM itself is a system component and has to provide a resilient behavior. Especially for critical systems, the CM function should still provide a defined minimum of functionality in case some of its components fail. This is also supported by a heterogeneous sensor architecture: a common mode failure, i.e., several sensors failing due to the same cause being prevented [38]. If one sensor fails, the remaining sensors can still provide data for CM. In the best case, both sensors are operating, and their data can be fused for an improved analysis.

The aim of this work is to provide a novel approach for data fusion of vibration and AE signals with a realistic test bench for bearing damage detection. In particular, several damage types should be classified, even if the setup is dismantled and reassembled between the measurements of the data used for model training and validation. The system concept comprises an interference canceling at the input stage, and also a parallel implementation of a feature fusion approach and isolated classifiers for both sensor systems. This enables an analysis of the benefit with respect to the system complexity. In addition, the resulting CM system is resilient against a single sensor failure due to the remaining isolated classifier which is based only on the remaining sensor. The source code used for this work is available at Github [39].

2. Methods

In this section, the used drive train setup will be explained first (Section 2.1). Specifications of the conducted measurements will be given in the same section. The conducted data preprocessing steps will be described in Section 2.2. The used approach, the data from the AE and vibration sensors as well as the rotation speed information are used to calculate a combined classification, as shown in Section 2.3. Its actual implementation follows in Section 2.4.

2.1. Experimental Setup and Measurement

The proposed approach is demonstrated with a drive train setup, at which multiple bearing failures can be simulated (see Figure 1). It follows the basic concept documented in [1] and is extended by an instrumentation with AE sensors. The setup comprises a cage rotor induction motor (WEG V3.5111) connected to a bearing shaft by a flexible coupling, driven by an electronic motor inverter that allows the output rotation speed to be set by the operator. The velocity of the shaft was determined using a consumer-grade digital laser tachometer (Akozon DT2234C+) and was controlled via a potentiometer connected to the inverter unit. The inverter uses the feedback of the motor to ensure a constant rotational speed after setting it with the aforementioned potentiometer.

The simulated bearing failures include damaged outer rings, damaged inner rings and damaged rolling elements. All the aforementioned failures are tested in two severities; this means that one small slot and one larger slot are inserted into the inner ring and the outer ring. Instead of a slot, the defect at the rolling element is a flat spot. The damaged parts as well as their counterparts without damage are depicted in Figure 2. The bearings are precision magneto ball bearings of type E12 (DIN 615) (12 × 32 × 7 mm), which are essentially deep groove ball bearings with an outer ring with only one shoulder, providing a separable outer ring from the cage and inner ring. The cage allows removal from the inner ring and individual bearing balls within the cage can be swapped. Thus, with this type of bearing it is possible to selectively mount individual defects or defect combinations. Damages of different sizes are introduced to the inner rings, the outer rings and the rolling elements by wire electrical discharge machining (EDM). The defect sizes were chosen in relation to the diameter of the rolling elements of 4.7 mm. For the cases of inner and outer rings, slots with widths of 8% and 40% in relation to the ball diameter were inserted (corresponding to 376 and 1880 µm each). A flattening was introduced to the rolling elements with depths of 2% and 8% of the rolling element diameter (corresponding to 94 and 376 µm each). The precision of the wire EDM amounts to ±10 µm. The relative dimensions of the bearing damages are are thus in a similar range to those reported in [31], where damages in the range between 9.6% and 28.8% of the rolling element diameter were used.

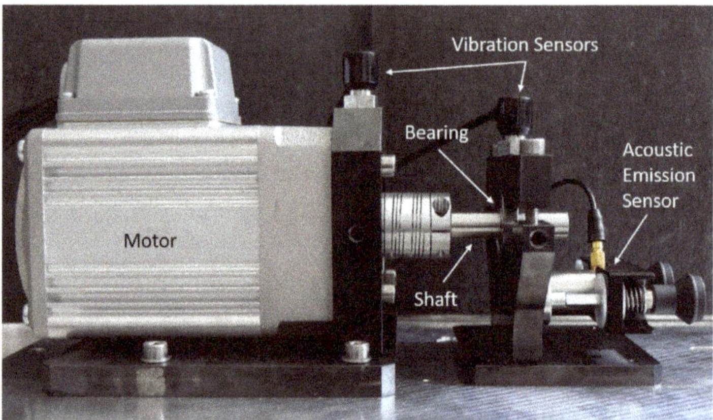

Figure 1. Drive train setup with vibration and acoustic emission sensors.

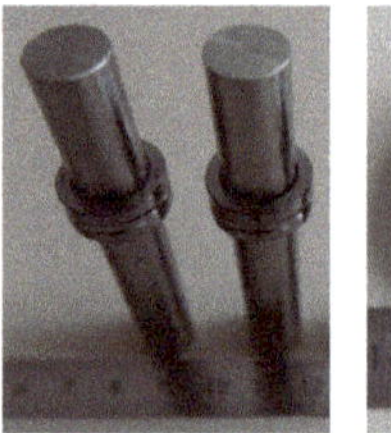

(**a**) Inner ring   (**b**) Outer ring   (**c**) Rolling element

Figure 2. Artificially damaged bearings, each without damage (**left**) and small damage (**right**).

For vibration measurements, the setup is instrumented with ICP accelerometers (PCB 607A11, 100 mV/g, 0.5 Hz–10,000 Hz) at the bearing holder frame of the motor and the bearing holder frame of the shaft. Both vibration sensors are digitized by a DT 9837A USB-DAQ from Measurement Computing GmbH. This 4-channel signal analyzer is equipped with a 24-bit ADC, which supports sample rates of up to 100 kSPS and is well suited for the vibration measurements. AE signals were acquired by a piezo transducer (Vallen VS30V, 20 kHz–80 kHz) connected to an AEP5 preamplifier for signal conditioning. These signals were digitized with an USB oscilloscope (PicoScope 2204A: 2-channel, 8-bit ADC, sample rate up to 100 MSPS, analog bandwidth of 10 MHz). Both sensors were connected to a single board computer (Raspberry Pi 3B+) running custom software to acquire both data streams simultaneously. Data were stored in two separate files. However, all data could be synchronized afterwards by frequently incorporating precise time-stamps into the files. This synchronization was carried out using the system clock of the single board computer and thus did not include USB-related delays. Therefore, the overall synchronization performance was estimated to be ±10 ms.

In order to separate the effect of mounting a damaged bearing from variability of the setup when demounting and mounting due to uncertainties in torque of screws, orientation of damage at the bearing, etc., measurements were repeated several times after demounting and mounting the bearing. For the case of the outer ring defect, the orientation of the defect was varied within the angles −45°, 0° and 45° with respect to the vertical, which resulted in a higher number of measurements for this defect type. In total, data from 29 measurements were collected and used to train the neural network-based classifiers, as explained in the following sections. Measurements were conducted at five different speeds (600/1000/1400/1800/2200 rpm) and with different damaged bearings mounted. Table 1 lists the number of measurements obtained together with their respective specifications. There is one measurement for each specification, which is not used for model training but holds out for the validation of the trained classifiers. These measurements represent the validation dataset, while the other measurements are referred to as the development dataset.

Table 1. Parameters of used datasets.

Defect Type	No Defect	Inner Ring Defect		Outer Ring Defect		Rolling Element Defect	
Defect Size	-	376 μm	1880 μm	376 μm	1880 μm	94 μm	376 μm
Number of measurements development dataset	3	3	3	4	3	3	3
Number of measurements validation dataset	1	1	1	1	1	1	1

2.2. Data Preprocessing

For the classification, vibration signals from the sensor mounted close to the bearing were used. The acquired vibration data contain a significant DC offset, since the accelerometers were powered over the signal line. A Bessel high pass filter of order 4 and a cut-off frequency were applied to the vibration signal. With a roll-off of 80 dB per decade, it allowed for a sharp separation of the information contained in the vibration from the offset. By applying the filter to time series twice, forwards and backwards, the effective order of the filtering operation doubled, while the phase lag was compensated. Details on the design of digital filters can be found in [40]. The algorithms are well established and part of standard libraries such as Scipy [41], which was used here.

Further, the signal contains interference from power mains, i.e., it is deteriorated by harmonics of the mains frequency of 50 Hz. Thus, an interference canceling has to be implemented (see Figure 3). The sensor signal $d(k)$ is composed of a useful signal $x(k)$ and the interference $v(k)$. A reference signal $u(k)$ of the same frequency as the interference is generated. By applying an appropriate filter W with an impulse response $\underline{w}$, a signal can be obtained that cancels the interference in the sensor signal. The remaining signal $\hat{x}(k)$ thus provides a good estimation of the useful signal:

$$\hat{x}(k) = d(k) - \sum_{n=1}^{N} \underline{w}_n u(n-k).$$

(1)

To obtain the filter coefficients of the filter W, the time-discrete Wiener–Hopf equation can be solved:

$$\mathbf{R}_{uu}\underline{w} = \underline{R}_{ud},$$

(2)

where $\mathbf{R}_{uu}$ denotes the autocovariance matrix of the reference signal $u(k)$ and $\underline{R}_{ud}$ the vector of cross-covariance values of $u(k)$ and $d(k)$. For a single harmonic interference, a filter of order $N = 2$ is sufficient.

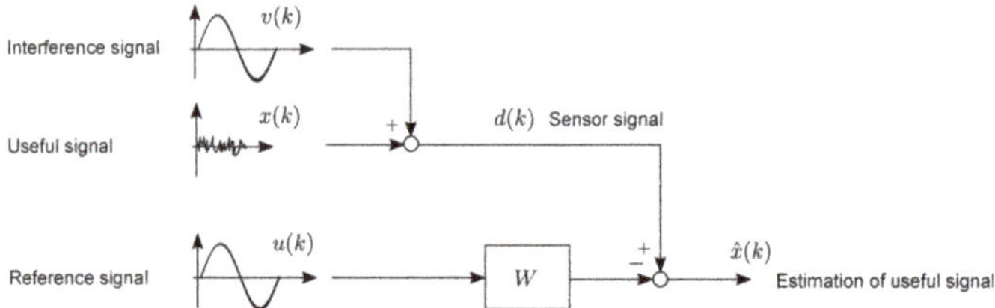

Figure 3. Interference reduction with a Wiener filter.

In a further preprocessing step, data at constant rotation speed were extracted to not include the ramping up and down effects in the training data. Afterwards, measurements from both sensors were divided into snippets to create distinct samples for training and evaluation of the classifiers. An AE sample consists of 8000 consecutive values measured at a sampling rate of 390,625 values per second (20.48 ms per sample), while a vibration sample consists of 4096 values, which corresponds to one second of measuring. The resulting datasets with the vibration and AE snippets were complemented by the extracted rotation speed as an additional column. For each vibration or AE sample, the mean and standard deviation were calculated and were afterwards standardized using these values. From each standardized sample, the Fast Fourier Transformation (FFT) was calculated. For AE data, the frequency range between 512 and 50,000 Hz was extracted, and for the vibration data, the frequency range between 5 and 512 Hz was used. The previously calculated mean and standard deviation values were appended to the FFT-transformed samples and thereby created a feature vector for each sample. An additional scaling of all datasets to a feature range between 0 and 1 was conducted based on the range parameters of the stage 1 training dataset. As a result of the described preprocessing steps, the feature vector for each AE snippet comprises 1014 values while the feature vector of each vibration snippet comprises 2045 values.

2.3. Data Fusion Approach

In this paper, a combined system is presented, which utilizes low-frequency vibration signals from accelerometers and high-frequency acoustic emission (AE) signals from ultrasonic transducers. By feeding the signals into a multi-stage classification algorithm (see Figure 4a), the capabilities of the detection should be enhanced to a broad range of faults and damages. The classifier architecture thereby consists of 3 parts, which are all trained to detect the defect state of the system:

- A classifier which receives data from the acoustic emission sensor (AE classifier);
- A classifier which receives data from the acceleration sensor (vibration classifier);
- A classifier which receives activations of the AE and vibration classifiers at intermediate layers (combined classifier) as well as the rotation speed.

The intention for this approach was to increase the sensitivity and to improve the resilience with respect to sensor failures. The proposed classifier architecture provides classifications based only on single sensor streams on the one hand (using the data from one sensor), and classifications based on fused sensor information on the other hand (using all available sensors). Thereby, sensitivity of the respective sensor data for the detection of different defect states can be evaluated and the performance gain due to the conducted sensor fusion can be assessed. Additionally, faults can even be detected if a single sensor stream is missing, albeit not with the maximum prediction accuracy.

In order to improve the training of the combined classifier, a stage 1 dataset and a stage 2 dataset were defined from the development dataset. An illustration of the splitting of the whole set of measurements into the stage 1 and 2 datasets as well as the validation dataset is depicted in Figure 4b. Like the validation dataset, the stage 2 dataset comprises one measurement per defect type and size. In an optimal configuration, the stage 2 dataset is only used to train the combined classifier, while the AE classifier and the vibration classifier are trained with the stage 1 dataset.

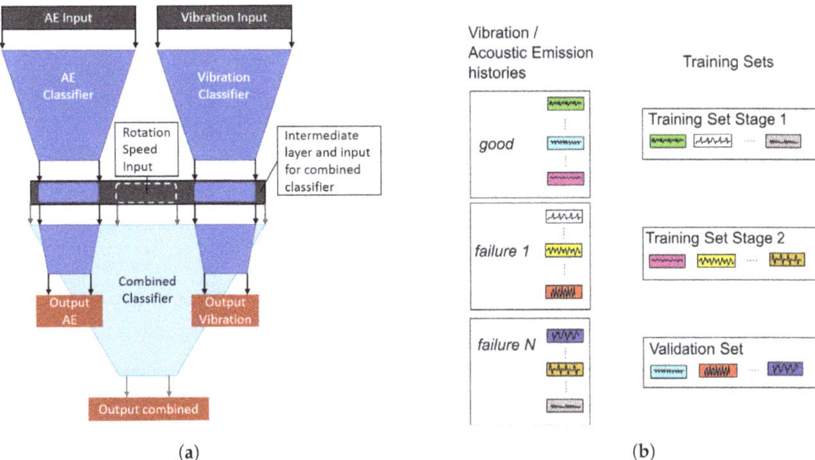

Figure 4. (**a**) Splitting of the measurements into stage 1 and stage 2 datasets as well as a validation dataset; (**b**) sketch of the proposed classifier architecture.

2.4. Classifier Setup

All the subclassifiers mentioned in Section 2.3 are multi-layer perceptrons. This type of neural network together with the frequency transformed input samples was chosen, since for vibration measurements it was shown that it can lead to classifiers with better generalization ability compared to, e.g., convolutional neural networks, which receive the time domain signals as an input [1]. The AE classifier has three hidden fully connected (dense) layers, each with 1024 nodes, and the vibration classifier has two hidden layers, also consisting 1024 nodes, respectively. Dropout was applied to all hidden layers of both classifiers to reduce overfitting. The combined classifier uses the activations of the last hidden layer of both the AE classifier and the vibration classifier as well as the scaled rotation speed as one concatenated input vector. In addition, it has two hidden layers, each with 128 nodes. No dropout was applied to the combined classifier since enough variability was induced by the intermediate layer values of the AE and vibration classifiers. The neural network was implemented using Tensorflow [42] and the used source code is documented at the corresponding Github repository [39]. Its layer architecture is depicted in Figure 5a.

During the training phase of the combined classifier, the weights of the AE classifier and the vibration classifier are frozen to only fit the weights of the combined classifier.

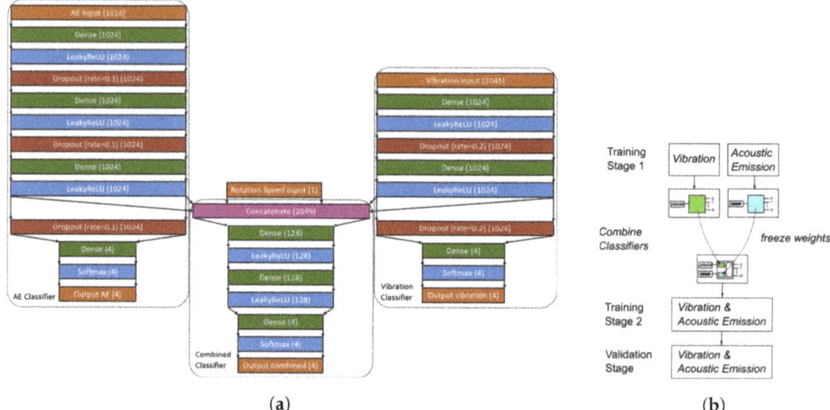

(a)

(b)

Figure 5. (**a**) Neural network implementation of the proposed classifier architecture; (**b**) training procedure of the stacked classifiers.

3. Results

All vibration data were treated with the preprocessing and filtering as described above. The interference canceling was implemented successively for the frequencies 50, 100, 200, 300, 400 and 500 Hz. Figure 6 shows an example for a time series acquired in the undamaged state. Obviously, the Wiener filter with the reference signal suppressed only the interference while leaving the vibration signals from the bearing unaffected. An overview of the measured and preprocessed sensor time series as well as their transformation into the frequency domain are provided in Figure 7.

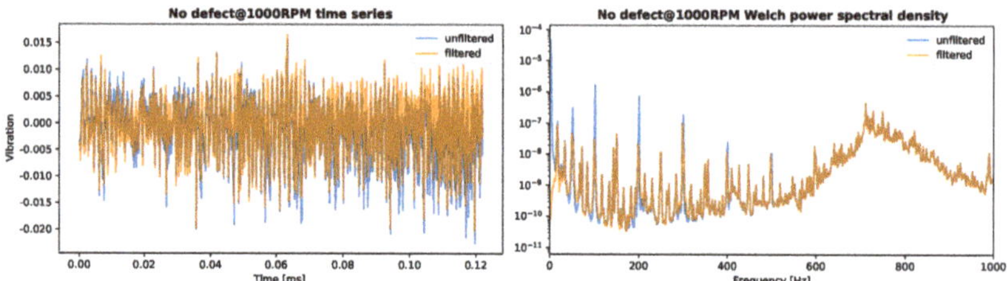

Figure 6. Result of the signal preprocessing: left: time domain; right: frequency domain.

At first and as a baseline, both the AE and the vibration classifiers were trained on the full training dataset (including stage 1 and stage 2 training datasets). Each one acquired the scaled feature vectors from the respective sensor. Both classifiers achieved accuracies close to 100% on the training dataset. Classification results of the trained classifiers on the unseen validation dataset are visualized in Figure 8a,b. While both classifiers correctly or almost correctly classified the defect free measurement and the measurement with the damaged rolling element, they both failed to classify the measurement of the outer ring defect. The inner ring defect can be recognized by the AE classifier, while the vibration classifier is only partially able to correctly classify this measurement.

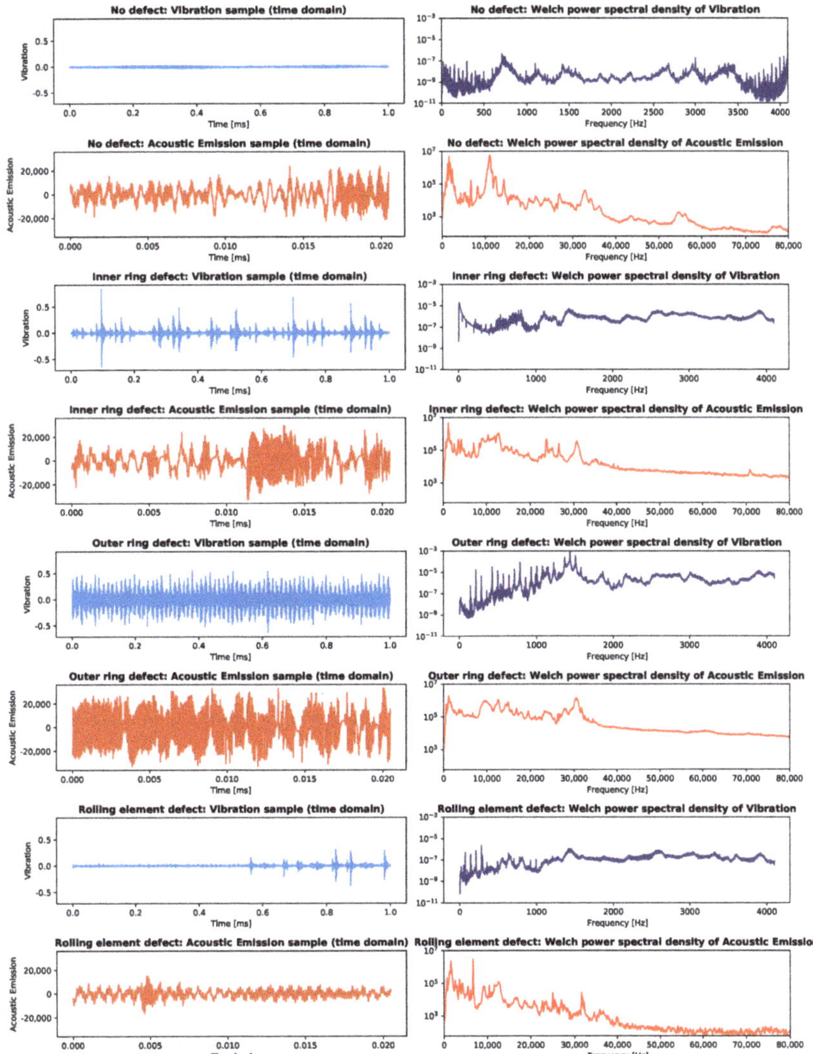

Figure 7. Examples from the measurement data. Left column: samples from all four used defect states in time domain alternating between vibration and acoustic emission. Right column: power spectral density calculated by Welch's method of the samples in the left column.

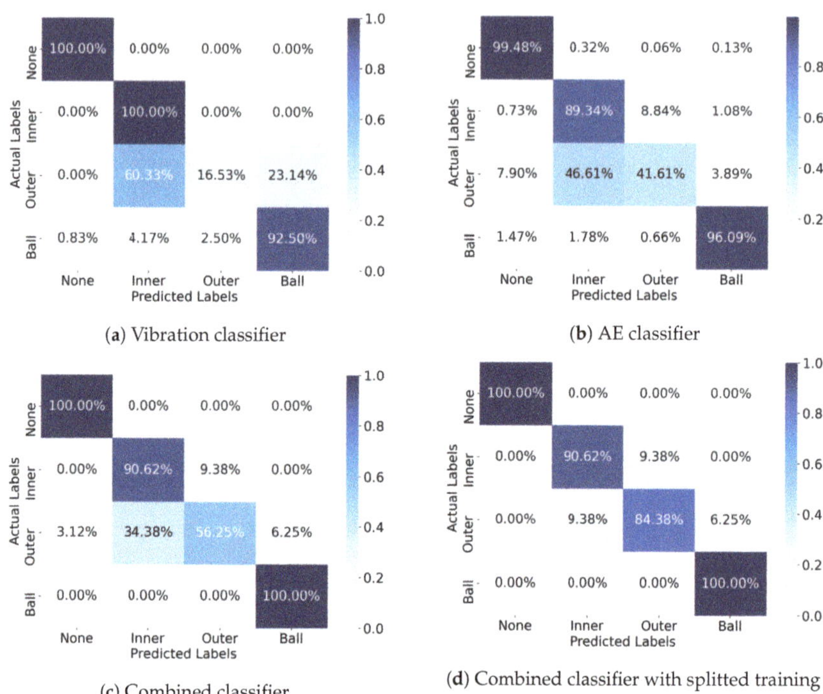

(**a**) Vibration classifier

(**b**) AE classifier

(**c**) Combined classifier

(**d**) Combined classifier with splitted training dataset

Figure 8. Confusion matrices of the classification results on unseen validation measurements.

Here, we see a typical weak point of drive train CM systems: since the frequency spectra of single samples inside one measurement differ only slightly, it is a relatively easy task for any machine learning algorithm to recognize other samples from the same measurement. However, there are comparatively huge differences in frequency spectra between samples from different measurements, even when they belong to the same defect class and especially when the setup is dismantled and reassembled between the measurements, as was carried out here. It is therefore easier for the classifier to learn characteristic features of each measurement rather than to learn characteristic features of each defect state. A much larger number of measurements would be necessary to compensate for these effects and to achieve robust classifiers. An approach to increase the robustness of the defect state classification is to combine the information from both used sensor modalities. This is achieved by the proposed combined classifier. In a first experiment, the combined classifier is trained on the full training dataset. The pretrained vibration and AE classifiers thereby provide the input for the combined classifier. Their weights, however, remain unchanged during this procedure. In addition, the rotation speed is also an input to the combined classifier. The classification accuracies of the resulting classifier on the validation dataset are depicted in Figure 8c. Still, the outer ring defect cannot be classified correctly and the accuracy for the inner ring defect is not perfect. Again, the reason for this result is the overfitting of the vibration and AE classifiers. Since all classifiers receive input data from the same measurements, the combined classifier is not able to learn when to trust more the vibration classifier and in which cases the AE classifier is more reliable, because both of them are almost perfectly able to correctly classify samples from the dataset they are trained on.

To let the combined classifier learn when to trust which of the preclassifiers more, it needs to be trained with data from measurements unseen to the vibration and the AE classifiers. For this purpose, the splitting of the training dataset in the stage 1 and the stage 2 training datasets, as described in Section 2.3, was conducted. At first, the vibration classifier and the AE classifier were trained on data from the stage 1 training dataset. The combined classifier was afterwards fitted to data from the stage 2 training dataset. As with the validation data, both preclassifiers were not able to perfectly classify the measurements from the stage 2 dataset and therefore the combined classifier could learn when to trust which one more or less. The validation result of the combined classifier trained with the described procedure is shown in Figure 8d. A remarkable improvement in the classification accuracy is apparent, which indicates that the resulting classifier is much more robust to changing vibration behavior caused by a remounting of the setup compared to the classifier approaches discussed before.

4. Discussion and Outlook

Using the presented approach, bearing damages at drive trains can be detected early, precisely and robustly. While both sensor types enable a defect state classification separately, their data fusion further improves the fault detection sensitivity when they are trained on data from different measurements. On the other hand, in case of a sensor failure, a defect state classification can still be obtained since classifications are also calculated solely depending on one of the sensors at the same time. This is beneficial compared to the approaches reported in the literature (e.g., [3,34,37]), which only work when all sensor information is available. Still, it has to be considered that the predictions with partially missing sensor data lead to a lower accuracy compared to the case without sensor failure.

Comparisons in terms of fault detection accuracy are generally difficult to conduct since defect severity differs and also the experimental setups are not identical. Additionally, the division of the datasets into training, test and validation data differs significantly between the individual works. Many studies conduct a random train–test split between all the available data [3,22,27] as this is a common practice in the machine learning field. However, as stated above, vibration characteristics vary only slightly once a setup is completely assembled. As a result, almost perfect recognition accuracies can be achieved, while the classification algorithm used has only learned how a certain measurement can be identified, but does not have a correct internal representation of the error state. In an actual production facility, maintenance works are still necessary even when sophisticated CM systems are in place. During maintenance, screws may be replaced, lubricants may be refilled or adoptions due to changing products are necessary. These real-world conditions were simulated in this study by using independent measurements for model development and validation. In addition, the setup was dismantled and reassembled between each measurement, as explained in Section 2.1. The used algorithms for combining information, on the other hand, could be compared when applied to identical datasets. This work paves the way for a future comparative study by publishing the used dataset [43] as well as the used source code [39].

In a next step, the concept needs to be evaluated on a broader and continuous rotation speed range as well as with differently sized drive trains. A further limitation is that no load was used in this study. Investigations with varying loads could make the obtained dataset even more realistic. Additional sensor modalities (e.g., motor current, thermography) can increase the capabilities of the CM system further. Using a higher number of sensors in turn requires the development of algorithms with which the computational complexity can be reduced. Investigations need to be conducted that quantify the diagnostic information gain of each sensor for the defect state classification so that sensors that do not improve the detection capability of the system can be removed. Still, the data fusion concept presented here is not limited to AE and vibration data but is applicable for the fusion of many kinds of sensor time series, such as sound, temperature or pressure sensors. In particular, it is beneficial when a CM system needs to be resilient against sensor failures.

Author Contributions: Conceptualization, all authors; construction of the experimental setup, C.S.; measurements, S.K.; software, O.M. and A.S.; data curation, A.S., O.M., and S.K.; writing—original draft preparation, O.M., and D.M.; writing—review and editing, O.E.-R.; funding acquisition, D.M. and H.-G.H. All authors have read and agreed to the published version of the manuscript.

Funding: This research received internal funding by the Fraunhofer Cluster of Cognitive Internet Technologies. The open access publication was funded by Fraunhofer-Gesellschaft zur Förderung der angewandten Forschung e.V.

Data Availability Statement: The measured vibration and acoustic emission data are available at the Fraunhofer Fordatis database [43] under CC-BY 4.0 license.

Conflicts of Interest: The authors declare no conflict of interest. The funders had no role in the design of the study; in the collection, analyses, or interpretation of data; in the writing of the manuscript, or in the decision to publish the results.

References

1. Mey, O.; Neudeck, W.; Schneider, A.; Enge-Rosenblatt, O. Enge-Rosenblatt, Machine Learning-Based Unbalance Detection of a Rotating Shaft Using Vibration Data. In Proceedings of the 2020 25th IEEE International Conference on Emerging Technologies and Factory Automation (ETFA), Vienna, Austria, 8–11 September 2020; pp. 1610–1617
2. Dadouche, A.; Rezaei, A.; Wickramasinghe, V.; Dmochowski, W.; Bird, J.W.; Nitzsche, F. Sensitivity of Air-Coupled Ultrasound and Eddy Current Sensors to Bearing Fault Detection. *Tribol. Trans.* **2008**, *51*, 310–323. [CrossRef]
3. Li, C.; Sanchez, R.-V.; Zurita, G.; Cerrada, M.; Cabrera, D.; Vásquez, R.E. Gearbox fault diagnosis based on deep random forest fusion of acoustic and vibratory signals. *Mech. Syst. Signal Process.* **2016**, *76*, 283–293. [CrossRef]
4. Yang, W.; Tavner, P.J.; Crabtree, C.J.; Feng, Y.; Qiu, Y. Wind turbine condition monitoring: Technical and commercial challenges. *Wind. Energy* **2014**, *17*, 673–693. [CrossRef]
5. Wu, W.; Zheng, Y.; Chen, K.; Wang, X.; Cao, N. A Visual Analytics Approach for Equipment Condition Monitoring in Smart Factories of Process Industry. In Proceedings of the 2018 IEEE Pacific Visualization Symposium, PacificVis 2018, Kōbe, Japan, 10–13 April 2018; IEEE: Piscataway, NJ, USA, 2018; pp. 140–149. ISBN 978-1-5386-1424-2.
6. Renwick, J.T.; Babson, P.E. Vibration Analysis—A Proven Technique as a Predictive Maintenance Tool. *IEEE Trans. Ind. Applicat.* **1985**, *IA-21*, 324–332. [CrossRef]
7. Caesarendra, W.; Tjahjowidodo, T. A Review of Feature Extraction Methods in Vibration-Based Condition Monitoring and Its Application for Degradation Trend Estimation of Low-Speed Slew Bearing. *Machines* **2017**, *5*, 21. [CrossRef]
8. Randall, R.B.; Antoni, J. Rolling element bearing diagnostics—A tutorial. *Mech. Syst. Signal Process.* **2011**, *25*, 485–520. [CrossRef]
9. Widrow, B.; Stearns, S.D. *Adaptive Signal Processing*; 8. Print; Prentice-Hall: Englewood Cliffs, NJ, USA, 1985; ISBN 9780130040299.
10. Khemili, I.; Chouchane, M. Detection of rolling element bearing defects by adaptive filtering. *Eur. J. Mech.-A/Solids* **2005**, *24*, 293–303. [CrossRef]
11. Lee, S.K.; White, P.R. The enhancement of impulsive noise and vibration signals for fault detection in rotating and reciprocating machinery. *J. Sound Vib.* **1998**, *217*, 485–505. [CrossRef]
12. Zimroz, R.; Bartelmus, W. Application of Adaptive Filtering for Weak Impulsive Signal Recovery for Bearings Local Damage Detection in Complex Mining Mechanical Systems Working under Condition of Varying Load. *Solid State Phenom.* **2011**, *180*, 250–257. [CrossRef]
13. Broderick, J.J.; Burchill, R.F.; Clark, H.L. Design and fabrication of prototype system for early warning of impending bearing failure. NASA CR 123717. 1972.
14. Rezaei, A.; Dadouche, A.; Wickramasinghe, V.; Dmochowski, W. A Comparison Study Between Acoustic Sensors for Bearing Fault Detection Under Different Speed and Load Using a Variety of Signal Processing Techniques. *Tribol. Trans.* **2011**, *54*, 179–186. [CrossRef]
15. Kang, M.; Kim, J.; Kim, J.-M.; Tan, A.C.C.; Kim, E.Y.; Choi, B.-K. Reliable Fault Diagnosis for Low-Speed Bearings Using Individually Trained Support Vector Machines With Kernel Discriminative Feature Analysis. *IEEE Trans. Power Electron.* **2015**, *30*, 2786–2797. [CrossRef]
16. Rogers, L.M. The application of vibration signature analysis and acoustic emission source location to on-line condition monitoring of anti-friction bearings. *Tribol. Int.* **1979**, *12*, 51–58. [CrossRef]
17. Elasha, F.; Greaves, M.; Mba, D.; Fang, D. A comparative study of the effectiveness of vibration and acoustic emission in diagnosing a defective bearing in a planetry gearbox. *Appl. Acoust.* **2017**, *115*, 181–195. [CrossRef]
18. Al-Ghamd, A.M.; Mba, D. A comparative experimental study on the use of acoustic emission and vibration analysis for bearing defect identification and estimation of defect size. *Mech. Syst. Signal Process.* **2006**, *20*, 1537–1571. [CrossRef]
19. Mba, D. Development of Acoustic Emission Technology for Condition Monitoring and Diagnosis of Rotating Machines: Bearings, Pumps, Gearboxes, Engines, and Rotating Structures. *Shock Vib. Dig.* **2006**, *38*, 3–16. [CrossRef]
20. Mo, Z.; Wang, J.; Zhang, H.; Zeng, X.; Liu, H.; Miao, Q. Vibration and Acoustics Emission Based Methods in Low-Speed Bearing Condition Monitoring. In Proceedings of the 2018 Prognostics and System Health Management Conference (PHM-Chongqing), Chongqing, China, 26–28 October 2018; pp. 871–875.
21. Janssens, O.; Slavkovikj, V.; Vervisch, B.; Stockman, K.; Loccufier, M.; Verstockt, S.; van de Walle, R.; van Hoecke, S. Convolutional Neural Network Based Fault Detection for Rotating Machinery. *J. Sound Vib.* **2016**, *377*, 331–345. [CrossRef]

22. Shao, H.; Jiang, H.; Wang, F.; Zhao, H. An enhancement deep feature fusion method for rotating machinery fault diagnosis. *Knowl.-Based Syst.* **2017**, *119*, 200–220. [CrossRef]
23. Liu, R.; Yang, B.; Zio, E.; Chen, X. Artificial intelligence for fault diagnosis of rotating machinery: A review. *Mech. Syst. Signal Process.* **2018**, *108*, 33–47. [CrossRef]
24. Lei, Y.; Yang, B.; Jiang, X.; Jia, F.; Li, N.; Nandi, A.K. Applications of machine learning to machine fault diagnosis: A review and roadmap. *Mech. Syst. Signal Process.* **2020**, *138*, 106587. [CrossRef]
25. Liu, J.; Hu, Y.; Wang, Y.; Wu, B.; Fan, J.; Hu, Z. An integrated multi-sensor fusion-based deep feature learning approach for rotating machinery diagnosis. *Meas. Sci. Technol.* **2018**, *29*, 55103. [CrossRef]
26. Yang, J.; Bai, Y.; Wang, J.; Zhao, Y. Tri-axial vibration information fusion model and its application to gear fault diagnosis in variable working conditions. *Meas. Sci. Technol.* **2019**, *30*, 95009. [CrossRef]
27. Nath, A.G.; Udmale, S.S.; Raghuwanshi, D.; Singh, S.K. Improved Structural Rotor Fault Diagnosis using Multi-sensor Fuzzy Recurrence Plots and Classifier Fusion. *IEEE Sens. J.* **2021**, 1. [CrossRef]
28. Elforjani, M.; Shanbr, S. Prognosis of Bearing Acoustic Emission Signals Using Supervised Machine Learning. *IEEE Trans. Ind. Electron.* **2018**, *65*, 5864–5871. [CrossRef]
29. Hase, A. Early Detection and Identification of Fatigue Damage in Thrust Ball Bearings by an Acoustic Emission Technique. *Lubricants* **2020**, *8*, 37. [CrossRef]
30. König, F.; Jacobs, G.; Stratmann, A.; Cornel, D. Fault detection for sliding bearings using acoustic emission signals and machine learning methods. *IOP Conf. Ser. Mater. Sci. Eng.* **2021**, *1097*, 12013. [CrossRef]
31. Saufi, S.R.; Ahmad, Z.A.B.; Leong, M.S.; Lim, M.H. Low-Speed Bearing Fault Diagnosis Based on ArSSAE Model Using Acoustic Emission and Vibration Signals. *IEEE Access* **2019**, *7*, 46885–46897. [CrossRef]
32. Omoregbee, H.O.; Heyns, P.S. Fault Classification of Low-Speed Bearings Based on Support Vector Machine for Regression and Genetic Algorithms Using Acoustic Emission. *J. Vib. Eng. Technol.* **2019**, *7*, 455–464. [CrossRef]
33. Duan, Z.; Wu, T.; Guo, S.; Shao, T.; Malekian, R.; Li, Z. Development and trend of condition monitoring and fault diagnosis of multi-sensors information fusion for rolling bearings: A review. *Int. J. Adv. Manuf. Technol.* **2018**, *96*, 803–819. [CrossRef]
34. Azamfar, M.; Singh, J.; Bravo-Imaz, I.; Lee, J. Multisensor data fusion for gearbox fault diagnosis using 2-D convolutional neural network and motor current signature analysis. *Mech. Syst. Signal Process.* **2020**, *144*, 106861. [CrossRef]
35. Jiao, J.; Zhao, M.; Lin, J.; Ding, C. Deep Coupled Dense Convolutional Network With Complementary Data for Intelligent Fault Diagnosis. *IEEE Trans. Ind. Electron.* **2019**, *66*, 9858–9867. [CrossRef]
36. Stief, A.; Ottewill, J.R.; Baranowski, J.; Orkisz, M. A PCA and Two-Stage Bayesian Sensor Fusion Approach for Diagnosing Electrical and Mechanical Faults in Induction Motors. *IEEE Trans. Ind. Electron.* **2019**, *66*, 9510–9520. [CrossRef]
37. Yuan, Z.; Zhang, L.; Duan, L. A novel fusion diagnosis method for rotor system fault based on deep learning and multi-sourced heterogeneous monitoring data. *Meas. Sci. Technol.* **2018**, *29*, 115005. [CrossRef]
38. Verhulst, E.; Sputh, B.; van Schaik, P. Antifragility: Systems engineering at its best. *J. Reliable Intell. Environ.* **2015**, *1*, 101–121. [CrossRef]
39. Supplementary Information: Source Code Documentation of This Paper at Github. Available online: https://github.com/deepinsights-analytica/mdpi-arci2021-paper (accessed on 24 June 2021).
40. Antoniou, A. *Digital Signal Processing: Signals, Systems and Filters*; McGraw-Hill: New York, NY, USA, 2005.
41. Virtanen, P.; Gommers, R.; Oliphant, T.E.; Haberland, M.; Reddy, T.; Cournapeau, D.; Burovski, E.; Peterson, P.; Weckesser, W.; Bright, J.; et al. SciPy 1.0: Fundamental algorithms for scientific computing in Python. *Nat. Methods* **2020**, *17*, 261–272. [CrossRef] [PubMed]
42. Abadi, M.; Agarwal, A.; Barham, P.; Brevdo, E.; Chen, Z.; Citro, C.; Corrado, G.S.; Davis, A.; Dean, J.; Devin, M.; et al. TensorFlow: Large-Scale Machine Learning on Heterogeneous Systems. 2015. Available online: https://www.tensorflow.org/ (accessed on 31 May 2021).
43. Mey, O.; Scheider, A.; Enge-Rosenblatt, O.; Mayer, D.; Schmidt, C.; Klein, S.; Herrmann, H.-G. Condition Monitoring of Drive Trains by Data Fusion of Acoustic Emission and Vibration Sensors. *Fraunhofer Fordatis* **2021**. [CrossRef]

 *processes*

Article

Magnetic Particle Inspection Optimization Solution within the Frame of NDT 4.0

Andreea Ioana Sacarea, Gheorghe Oancea and Luminita Parv *

Department of Manufacturing Engineering, Transilvania University of Brasov, 29 Eroilor Boulevard, 500036 Brasov, Romania; andreea.sacarea@unitbv.ro (A.I.S.); gh.oancea@unitbv.ro (G.O.)
* Correspondence: luminita.parv@unitbv.ro; Tel.: +40-740-851-396

Abstract: The quality of product and process is one of the most important factors in achieving constructively and then functionally safe products in any industry. Over the years, the concept of Industry 4.0 has emerged in all the quality processes, such as nondestructive testing (NDT). The most widely used quality control methods in the industries of mechanical engineering, aerospace, and civil engineering are nondestructive methods, which are based on inspection by detecting indications, without affecting the surface quality of the examined parts. Over time, the focus has been on research with the fourth generation in nondestructive testing, i.e., NDT 4.0 or Smart NDT, as a main topic to ensure the efficiency and effectiveness of the methods for a safe detection of all types of discontinuities. This area of research aims at the efficiency of methods, the elimination of human errors, digitalization, and optimization from a constructive point of view. In this paper, we presented a magnetic particles inspection method and the possible future directions for the development of standard equipment used in the context of this method in accordance with the applicable physical principles and constraints of the method for cylindrical parts. A possible development direction was presented in order to streamline the mass production of parts made of ferromagnetic materials. We described the methods of analysis and the tools used for the development of a magnetic particle inspection method used for cylindrical parts in all types of industry and NDT 4.0; the aim is to provide new NDT 4.0 directions in optimizing the series production for cylindrical parts from industry, as given in the conclusion of this article.

Keywords: NDT; magnetic particle inspection; optimization; Industry 4.0

Citation: Sacarea, A.I.; Oancea, G.; Parv, L. Magnetic Particle Inspection Optimization Solution within the Frame of NDT 4.0. *Processes* **2021**, *9*, 1067. https://doi.org/10.3390/pr9061067

Academic Editor: Sergey Y. Yurish

Received: 31 May 2021
Accepted: 16 June 2021
Published: 18 June 2021

Publisher's Note: MDPI stays neutral with regard to jurisdictional claims in published maps and institutional affiliations.

1. Introduction

As digitalization evolves from year to year, many processes or industries are unable to cope with this progress [1,2]. Technological processes precede the progress of digitalization, and thus Industry 4.0 was later developed, which also led to the development of processes such as NDT, where we can now see a lot of digitalization, improvements and optimizations in terms of Industry 4.0 [2–4]. Despite this, the development of Industry 4.0 comes with changes in terms of the user–equipment interface, and any change involves the unknown, which can bring about reluctance or lack of confidence in the user in the displayed result. Digitalization comes primarily with the improvement and development of new processes and methods, but it also comes with the disposal of human errors, especially in the field of NDT methods where the final interpretation and final decision involves the human factor [5–7]. NDT, which stands for the term "nondestructive testing", does not affect the surface quality of the parts, the material, or the geometry of the tested components; this term is found in various publications, books, and media, and it is known as NDT, NDI, or NDE using different endings (e.g., testing, inspections, examinations, evaluations). The abbreviation used in this manuscript is NDT and includes all inspection processes by nondestructive methods. From Industry 4.0, which involves smart, digitized, and optimized nondestructive inspection processes, the development of the NDT 4.0 industry is currently facing challenges such as infrastructure limitations, lack of presence of sensors for transmitting measured values, recorded parameters, etc. The development of NDT 4.0 involves firstly ensuring the direct connectivity of the devices to displays in the laboratories, to analyzing the parameters or data in real time. It is an advantage to develop DAS (data acquisition system) systems by taking values and signals and processing them at a PC through connection cables and private intranet platforms or even wireless connections.

A shortcoming of these NDT methods for Industry 4.0 development and digitalization is the recent development of new equipment and new design proposals [1,8–10]. One of the main challenges of this industry is, first of all, to digitize and make smart connections between already used equipment/installations and PCs for displaying and processing the data recorded through sensors or connection cables, or more precisely, to form DAS systems. If the installations are old and do not have Internet options or other smart applications, then it is necessary to firstly implement electronics and "host" computers for these DAS connections. Along with their implementation, it is necessary to develop IT applications to ensure cooperative infrastructures and also intelligible and intuitive interfaces, but they also need to have the role of supporting security and limiting access. Both Industry 4.0 and the transition and approach to the concept of NDT 4.0 are sensory areas based on data acquisition and processing.

The main purpose of the NDT 4.0 approach and development is to have real-time access to all the obtained results and the used parameters [11,12]. Many papers have been published on the development of the NDT industry, addressing its concepts in terms of the design, production, maintenance, lifespan, examinations, etc. [11–15]. In this manuscript, we adopt an approach to the NDT 4.0 concept that involves optimizing inspection parameters and eliminating the human factor. Specifically, this paper addresses the concept of NDT 4.0 by optimizing and developing the magnetic particle inspection method for cylindrical parts, which we discuss later.

Industry 4.0 and NDT have undergone several stages of revolutions, starting from NDT 1.0. As the earliest generation, NDT 1.0 appeared with the development of the Industry 1.0 concept through the need to monitor and ensure safety by performing periodic checks through a prescribed and probable process, using tools that are not complex in order to increase human capacity. Examples of such developments are tactile devices, improved visual inspection through magnifying glasses, and other auxiliary devices. NDT 2.0 represents the second stage of process development, represented by the use of electricity and the use of analog devices by using waves or frequencies outside the sensory domain perceived by the human body. During this stage, NDT methods such as ultrasound, eddy currents, X-rays, fluorescent penetration substances in the PT method, and fluorescent particles for the MT method were developed, as well as techniques developed for visual examination and inspection by developing devices such as cameras. NDT 3.0, which represents the third stage of development in industry, includes the development in terms of digital technology, through the acquisition, storage, and processing of data from inspection processes, the development of displays, and the reproduction of 2D and 3D defects; its aim is to provide an efficient, safe, and faster interpretation of the obtained results. There have been developments in the inspection chambers for methods using probes and also sensors such as ET, UT, thermography, visual inspections, X-rays, and even industrial tomography.

NDT 4.0 represents the latest generation of the development of this industry and includes the combination of generation 3.0 and generation 2.0 by eliminating human errors and improving inspection systems and the continuous development of digital technologies and physical methods used, with this development being achieved independently or even simultaneously. It aims at an improvement and optimization in this generation of both repair processes and overhauls and the maintenance throughout the life of the equipment. Another need for the development of the NDT 4.0 industry is the cause of depletion of natural resources and environmental pollution; here, we refer to the sustainable development of NDT processes that develops digitalization while also eliminating human error [1–4]. Sustainability is the field that in recent year has attracted the most attention in production research from various industries and is an important factor not only in technological development but also in the innovation of sustainable systems. The role of production in sustainability is important because in many processes, methods are consuming natural resources or resources that affect the quality of the environment and the health of human factors. The biggest confrontation in the world is to face a land with finite resources and limited natural resources.

NDT 4.0 is characterized by the elimination of human errors and the transmission of information and physical parameters in real time through connections made with laboratories (i.e., Connected Lab). Figure 1 describes the evolution of nondestructive testing methods in terms of digitization and optimization according to the evolution of Industry 4.0. It is observed that with the evolution of these NDT methods, the influence of the human senses is diminished with the development of new technologies (sensors, probes, devices), and four important stages are highlighted in the figure, i.e., from NDT 1.0 to NDT 4.0.

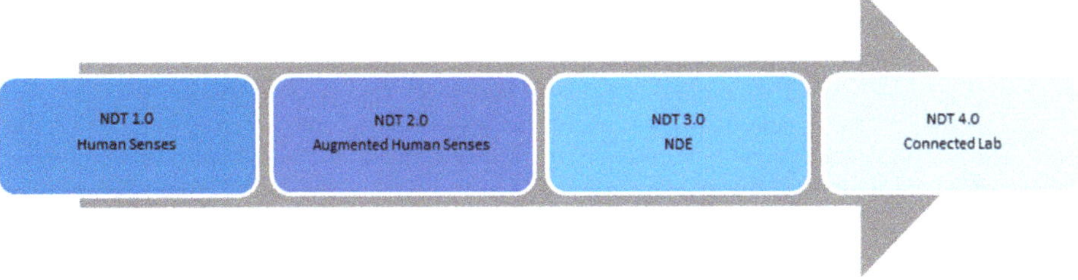

Figure 1. NDT concept evolution.

Nondestructive inspection methods, as specified above, are characterized in particular by the fact that they can assess and characterize the internal or surface defects of the mounted or unmounted parts, without causing any damage to their surface. Each method has evolved over time, and in recent decades, the field of NDT has been increasingly used due to the speed of examination and efficient operation. Each method is based on its own physical principles and is specific to a certain location of defects or certain groups of materials or classes of products. The methods are divided according to the industries in which they operate and are based on the materials that needs to be examined. In addition to the advantages of these NDT methods, there is a disadvantage for each method because there are limitations, especially in terms of the material examined in relation to the physical parameters involved in the inspection. We can speak of limitations in terms of effectiveness and precision in cases where they are not applied correctly on specific product/material groups. The effectiveness of the methods is also influenced by the human factor that establishes the chosen method and the parameters used for different industries; this fact may be due to the choice of using an inefficient method and with maximum yield for a group of materials. For example, regarding the method used with magnetic particles, it is characterized in particular by the fact that it can be applied only to ferromagnetic materials; if the material concerns alloys with poorly magnetizable substances, then there is a risk that the method will not yield or even have suboptimal inspection results [16,17]. The magnetic particle method (often with the abbreviation MPI—magnetic particle inspection or MT—magnetig testing) is the most common method used in inspecting steel parts, especially in the aerospace industry, from small components (such as washers) to large parts of great importance that are sometimes critical in the operating cycle of an aircraft, such as landing gear, engine components, etc. The method is also found in the automotive industry and in the railway industry for the inspection of bearings and their components, often made of special steels, and also in the civil engineering industry to characterize cracks in cement concrete specimens [18–23]. Although it seems easy to understand and apply, the method is complex and is based on many theories such as the effects of metallurgical materials, fluid movement dynamics, human factors and electromagnetic theories—with all these factors integrated into one method and one applicability.

In this manuscript, the magnetic particle method is approached together with the analysis of the physical parameters used for the cylindrical parts and their optimization by developing an auxiliary device that reduces the inspection time and eliminate human errors, as well as ensure an ergonomic position of the inspector during the execution of the method while using standard control equipment. There are very few studies and auxiliary tools that ensure a correct determination of the tests and parameters used, and therefore in this manuscript, we address this concern to ensure a uniform magnetic field of the cylindrical parts subjected to examination.

The values established for the electric current intensity have an importance in ensuring an optimal magnetic field that can highlight the existing discontinuities on the surface of the parts, and so attention is given to the values of the tangential magnetic field and the remaining magnetism in the part. The biggest risk in the wrong choice of parameters is to not highlight the discontinuities or to mark them with false indications, which has a high risk factor for any type of examined part. The magnetic field is a property with high sensitivity in any field, and it has unique properties in terms of impact on components/technologies. Magnetic phenomena are very important in any field of science, and the remaining magnetism has become an area of interest, as it is a critical characteristic in certain functional ensembles. The magnetic particle method is the method most often used to detect microdefects (usually linear cracks) in the case of materials with ferromagnetic properties. The principle of examination with this method is quite simple, namely that for the cylindrical parts, standard control equipment is used, with a central conductor mounted, through which electric current passes [5,7]. The parts are magnetized, and then the applied magnetic particles are attracted to the defective areas; however, it is important that the orientation of the defects is about 90 degrees from the direction of the field lines, as this is one of its basic characteristic methods [24–26]. The smaller the angle, the lower the sensitivity and efficiency of the method, and the lower the probability of detecting defects. A main advantage of this method—which is agreed by many industries, especially the car and aircraft manufacturers—is that as long as the method is applied correctly and the chosen parameters are optimal, there is a probability that defects are detected from the beginning.

In this manuscript, we aim to optimize the MT method with a central conductor by developing a device that ensures a uniform magnetic field on the surface of cylindrical parts, thereby ensuring constant parameters during the process and the same probability of detecting defects on the entire surface. The design of the device is chosen first, with the help of a multicriteria analysis, taking into account both the examination conditions and working environments used and also the design conditions for manufacturing and the physical parameters to be obtained. In the current control stage, the parts are positioned directly on the central conductor, and they are in direct contact with it, which implies a nonuniform magnetization on the surface of the part, i.e., a nonuniform magnetic field with values gradually decreasing from the contact area with the mandrel piece to the opposite surface of the part, where it is no longer in contact with the mandrel. The value of the magnetic field is one of the most important features of this inspection process because if the CM value is too low, there is a risk that fine or even large defects are not highlighted, and if the CM value is too high, each imprint can be highlighted, which leads to a difficult interpretation from a visual point of view.

2. Materials and Methods

The materials and methods used in this manuscript are known methods in research, i.e., data acquisition systems and manufacturing design; these are applied here for the development of a new device to help with the improvement and optimization of the magnetic particle method. Data acquisition systems are used for measuring and recording physical parameters, and manufacturing design is used not only for the correct choice of device design and thinking, but also for the constraint and importance of physical parameters, as well as the effective design of the constructive solution, chosen in the shadow of an applied multicriteria analysis.

The data acquisition system (DAS) in this work measures the values of physical parameters used in magnetic particle inspection. The data acquisition system here consists of a current generating source, Hall probes for measuring magnetic field values, and a digital screen that allows for recording and displaying measured values. Hall probes are based on the effect of the same name and are used mainly to measure a tangential magnetic field or the remaining magnetic field values. Hall effect sensors are characterized by the measuring range as well as the working environment temperatures. In the case of the magnetic particle method, the working temperature is controlled, most often the equipment placed in industrial halls with a controlled temperature or in testing laboratories. With the help of a data acquisition system consisting of a generating source and devices for measuring the magnetic field, provided with Hall sensors, tangential magnetic field values can be recorded.

The manufacturing design in this work looks at designing a device that aims to optimize the NDT inspection process with magnetic particles for cylindrical parts. To do this, the design conditions were taken into account so that its realization process can be easily implemented [1,3]. It is used conceptually to design the parts in pairs, trying as much as possible to use the same types of parts, both in one part of the device and in the other. Choosing the design of parts with simple configuration and sections helps to facilitate costs, time, and manufacturing processes, especially if users opt for additive manufacturing by the 3D realization of certain components of the device. In our work, these were used as main elements in the construction of the proposed model; the parts consisted of two cylindrical drums, moving arms, the sole of the device, all with sections with simple geometries. A second condition that was taken into account in the design of the variant proposed for the device is that of the use of symmetrical parts. One of the important characteristics of the symmetrical parts is their fast and correct assembly, as no sensors or other auxiliary devices are needed in order to orient them. It is important that the device has as few constructive elements and with geometries as simple as possible and can be disassembled, in order to ensure safe transport, easy packaging conditions, and a quick and correct assembly.

Assembly must be done as easily as possible, both manually and automatically—it is much easier to program assembly robots with simple and short movements than large and complex movements. We opted for a device that has few main constructive elements that require assembly, such as: support/seating soles, fixing, moving arms, drums for rotating parts and height actuation device. We took into account our desire to have a device with a utility for a wider range of cylindrical products with different diameters, and thus it was intended that the disassembly/assembly of the two drive drums be performed as simple and fast as possible. Depending on the diameters of the examined parts, the two provided drums must be interchangeable so as to ensure the coaxiality of the central conductor and the part. Both the drums and the support soles and movable arms were designed symmetrically, and their symmetry is considered the sign of quality assurance, following the manufacturing process of the device. From the point of view of additive manufacturing, the configuration of symmetrical parts is easier to achieve; moreover, a symmetrical part is easier to assemble with another part that is also symmetrical, so there is no risk of confusing the positions or directions of placement or assembly.

Regarding the design concept, the proposed device model resembles a jack, possibly with a hydraulic drive. We approached the design from a classic configuration of the hydraulic table assembly, where the upper part is arranged with two drums (made of soft material without degrading the surface of the part during rotation) to ensure a constant rotation with a controlled speed of the part. With the rotation of the cylindrical part, the risk of false indications or the risk of deleting the relevant indications is also eliminated.

Without the presence of an auxiliary device, the parts are handled manually during the examination. The device is made of the following component parts (see Figure 2):

- Supporting legs/movable arms that ensure sufficient height;
- Interchangeable drums for parts with different diameters made of soft material such as plastic, silicone, or silicone-coated plastic;
- A support sole, provided with supports to facilitate its transfer from one piece of equipment to another.

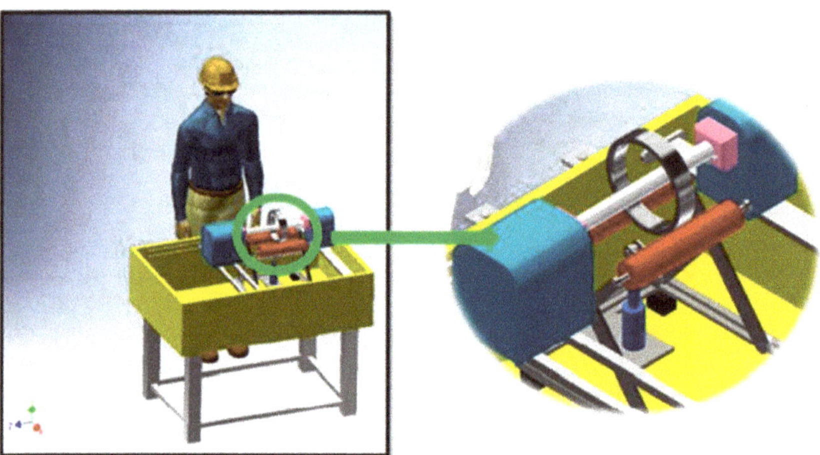

Figure 2. Main view of the device and the optimal position of the part in which the coaxiality is ensured.

Figure 2 shows the design proposal for such a device, provided in the tank of a standard equipment for examination with magnetic particle inspection. In accordance with the manufacturing design as described above, it is very important that the device can be easily assembled and has a configuration with simple geometry and that we use symmetrical parts. The proposed design has a simple configuration; it does not fill the volume of the tank, and it provides the necessary space for handling parts. In addition, it has a very small footprint at the base of the basin and is sealed so that magnetic particles does not remain accumulated on the inner surfaces of the device. The access to the part to be examined is not affected, but the possibility of inspection the parts is facilitated, given that the parts are closer to the visual field of the operator.

In order to build the device, we took into account that this method generates a magnetic field, and that the materials chosen in its construction must not be ferromagnetic, especially in the area near the central conductor/magnetic poles. If the supporting elements of the parts are ferromagnetic, they will be magnetized, and in turn they can magnetize the already demagnetized parts following the control. With the realization of this new type of device, we propose an optimization of the inspection process from the point of view of ensuring the constant working parameters, as well as from the point of view of the working time of the entire inspection process.

Regarding the physical principles and relationships, it can be seen that this NDT method—like any other method in this category—is based on physical methods and principles, as well as constraints in this regard; therefore, one of the most important constraints is that the parts must be made of ferromagnetic material, and the ideal inspection case is to ensure a uniform magnetic field on all surfaces of the parts subject to inspection [24,25]. A common problem in this type of inspection is the remaining magnetism left in the part, which can affect other parts during operation in the assembly of which it is part; this is a critical feature even for human safety if it is part of an engine assembly from an aircraft for example, or other areas with high sensitivity in use [20,26,27]. The ideal case of physical parameters implies a constant value of the magnetic field, which also means a uniform control and a similar highlighting for all inspection areas. Thus, the ideal case is to magnetize simultaneously both the upper part and the lower part, without the need for a 180 degrees rotation (in the case of supporting the part on the central conductor) of the part to ensure a magnetic field also on the lower surface.

The value of the magnetic field usually decreases in proportion to the distance from the mandrel of the part; the role of the mandrel in this type of assembly is to ensure the passage of electric current in order to highlight the discontinuities, both at the inner surface of the part and its outer surface. There is a linear decrease with the increase of the distance between the central conductor and the part to be examined.

One of the most important physical characteristics is represented by the magnetic field and the assurance that it is evenly distributed on the surface of the parts. The approximate values must be between 30 and 60 A/cm, as this is an interval with values that guarantee the highlighting of possible indications (such as cracks). If the values are lower than this interval, then there is the possibility that the defects are not highlighted, meaning that the part is not magnetized enough. If the values are too high, then the magnetism of the part is too high, and the particles of either part becomes attracted even in the indications of fine scratches. Thus, we can talk about irrelevant indications, which make it difficult to examine the surfaces.

Another important feature is the choice of the type of magnetization, depending on the orientation of the defects to be highlighted; there are two types of magnetization, the circumferential tangential magnetization and the longitudinal tangential magnetization. The importance of this characteristic is given by one of the constraints of the method—the orientation of the defects must always be perpendicular to the direction of the magnetic field lines.

Demagnetization of parts after inspection is also an important factor to consider in this method because it is important that the parts are to be demagnetized evenly on their entire surface so that there are no high values of magnetism remaining in the part. It is preferable for the values of the magnetism after inspection to be close to zero because a possible magnetism on or in the part can affect the functional state of the system of which it is a part, or even the part of other neighboring components.

The multicriteria analysis method was chosen in this manuscript in order for us to choose the optimal variant from a constructive point of view, not only of the device to ensure the optimization of physical parameters but also of inspection. Within the multicriteria analysis, several criteria were determined from a constructive point of view, namely the criteria of the materials used, the chosen components, and the characteristics they must meet. It is very important that the multicriteria analysis is performed in relation to the chosen criteria.

Case Study

In order to carry out a case study highlighting the importance of ensuring constant physical parameters, we used an example of a cylindrical piece with a diameter of approximately 100 mm, positioned on a central conductor with a diameter of 55 mm. It is considered the ideal case in ensuring a magnetization in both directions, having a magnetic field between 30 and 60 A/cm on the entire surface of the part, as well as in ensuring that at the area of contact with the central conductor, the magnetic field is at the limit (higher than 30 A/cm) and that in the lower part (180° to central conductor), it is below the minimum required value.

The way of obtaining the parameters consisted of the effective application of the inspection method on standard equipment. Thus, the part was positioned on the central conductor, where a magnetization cycle of at least 5 s is ensured, and then the tangential magnetic field in both directions—longitudinal and circular, respectively—was checked in turn. The current values were entered depending on the performance of the equipment used, so as to ensure the optimal tangential magnetic fields in order to ensure the detection of possible relevant indications. After establishing the magnetization parameters, a device was used to measure the magnetic field, and its display should show the magnetic field value on different areas of the part, starting from the contact area of the central conductor and the part to the lower end of the part (Figure 3).

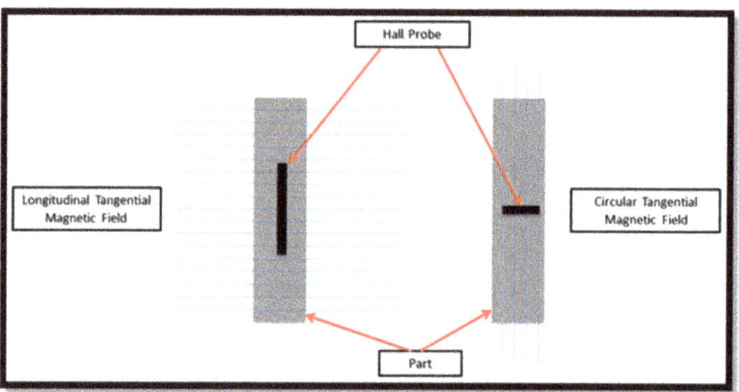

Figure 3. Hall probe positioning for magnetic field measurement.

Figure 3 illustrates the orientations of detectable discontinuities according to the orientation of the magnetic field lines. On the left side of the image, the field lines are longitudinal, in the horizontal direction, which implies the possibility of detecting discontinuities perpendicular to them. The indication is highlighted on the part in the vertical direction, with a black line. For the circular tangential magnetic field, it is the opposite case, in which the field lines have a vertical direction, and the discontinuity is oriented horizontally, perpendicular to the direction of the field lines. Because the detection of defects is characterized by the direction of the field lines, most of the times we chose to use the equipment capable of using a combined magnetization in both directions, in order to achieve a short inspection time and an efficient examination. For a longitudinal tangential magnetic field, it is possible to detect circumferential defects (the left side), and for a circular tangential magnetic field, it is possible to detect longitudinal defects (the right side). After ensuring the optimal parameters, the part was magnetized in both directions simultaneously, having a combined magnetization, followed by a demagnetization with the same values of current intensity. Demagnetization was followed by measuring the remaining magnetism on the entire surface of the part.

The case study was performed on standard control equipment, where the part is in contact with the central conductor (current control situation). For example, one cylindrical part made of ferromagnetic material (stainless steel) with an outer diameter equal to 100 mm was chosen as a reference. The part was positioned on the central conductor; the physical parameters were chosen such that the tangential magnetic field on the surface of the part was between 30 and 60 A/cm. The magnetization time was also chosen, and the demagnetization of the parts was activated. The total magnetization time of the part in both directions was at least 5 s, and the demagnetization was performed with the same amperage values as those used for magnetization. For example, for a part with the outer diameter equal to 100 mm and a central conductor with a 55 mm diameter, the optimal amperage values are 1.5 kA (circular direction) and 5.5 kAT (longitudinal direction).

Figure 4 presents the case in which the parts are in direct contact with the central conductor and thus the lack of coaxiality of the central conductor and the part can be observed. In this case, the examination was not efficient because the examination of the contact area between part and mandrel cannot be ensured completely.

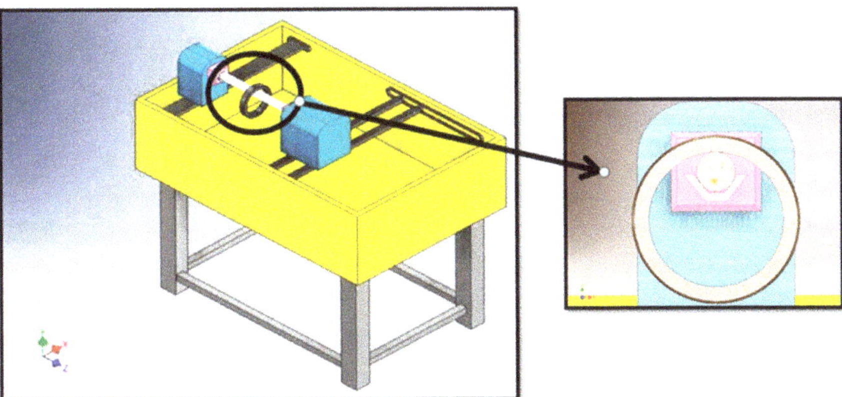

Figure 4. The cylindrical part position on the standard equipment.

Figure 5 presents the basic elements of standard magnetic particle inspection equipment.

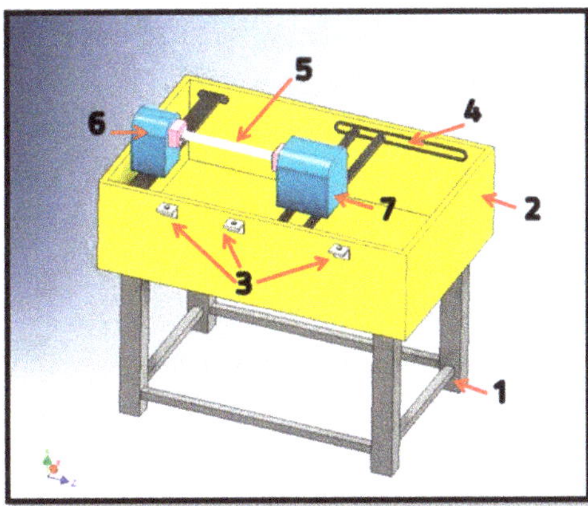

Figure 5. Equipment components: 1—support soles; 2—basin; 3—electrically operated buttons; 4—rail for mobile pole; 5—central conductor; 6—fixed pole–coil assembly; 7—mobile pole–coil central conductor assembly.

Data processing began with the introduction of current intensity parameters for two magnetization directions (longitudinal and circumferential) until magnetic field values were obtained between specified ranges. The values were measured with a data acquisition system composed of a display with the measured values and a probe for measuring the values of the magnetic field, provided with a Hall sensor. After the values of the intensities were established in accordance with the optimal magnetic field for each direction of magnetization separately (circumferential, longitudinal), the effective inspection took place on both directions of magnetization.

At the end of this case study, we performed a multicriteria analysis in order to propose an optimal design of the device, namely a design that ensures a constant magnetic field distributed on the surface of the part. Following the multicriteria analysis, the final construction form of the device was verified and given approval in order to ensure the optimal parameters. In this way, the part can be considered positioned concentrically with the central conductor, and in this case, the coaxially of the central conductor and the part leads to the measurement of a uniform tangential magnetic field distribution on the entire circumference of the part to be inspected. The same values were measured on the entire circumference on the ring from 0 to 180 degrees, and at the same time the values of the remaining magnetic field were constant regardless of the positioning of the Hall probe on the surface of the part.

3. Results

As described in the case study, a graphical distribution of the values of the circumferential and longitudinal tangential magnetic field was performed to observe and analyze the variation of these values, in case the cylindrical part was not positioned coaxially with the central conductor (Table 1). We considered as an example a part with a 100 mm diameter and a central conductor with a diameter of 55 mm. The Hall effect probe was used to measure the values of the magnetic field, in different positions that signify different angles on the diameter of the circle part. The following angles are chosen: $0°$, $15°$, $30°$, $45°$, $60°$, $75°$, $90°$, $105°$, $120°$, $135°$, $150°$, $165°$, and $180°$, with $180°$ representing at the same time the farthest point of the part from the central conductor. The values are centralized in the Table 1 in order for us to observe the influence of the part–conductor distance during the inspection, the influence of the variation of the physical parameters, and the safety of the detection of possible relevant indications (e.g., cracks).

Table 1. Changes in the tangential magnetic field parameters.

Positions [Degrees]	Longitudinal Magnetic Field (A/cm)	Circumferential Magnetic Field (A/cm)
0	39.8	40.2
15	39.8	40.2
30	39.8	40.2
45	39.6	40
60	39.6	40
75	39.5	39.1
90	39.2	38.8
105	39	38
120	38.9	37.9
135	38.5	37.5
150	38.4	37.4
165	38	37
180	37.5	36.5

In Figure 6, we observed the change of magnetic field for the first case in which the part is positioned in contact with the mandrel. We observed a decrease of the magnetic field proportional to the distance of the part from the central conductor. A greater distance of the part from the mandrel implies a decrease of the magnetic field. The values of this magnetic field decreases with the part–conductor distance. In the area of the part and central conductor, the contact surface where the graphic value on the x-axis lies is almost 0; this is the area where the discontinuities can be best detected if at that point the value of the magnetic field is at least 30 A/cm. If the value measured to ensure the parameters is equal to, for example, 56 A/cm in the area of $180°$ (cf. x-axis), there is a risk of irrelevant indications in the maximum contact area of the part and central conductor, where the values are very high in that area. However, if on the contact area the values of the magnetic field are at the limit of the range, i.e., equal to 30 A/cm, in this case there is the risk that in the area furthest from the central conductor, i.e., $180°$ according to the x-axis, the field values are too small for discontinuities to be detected.

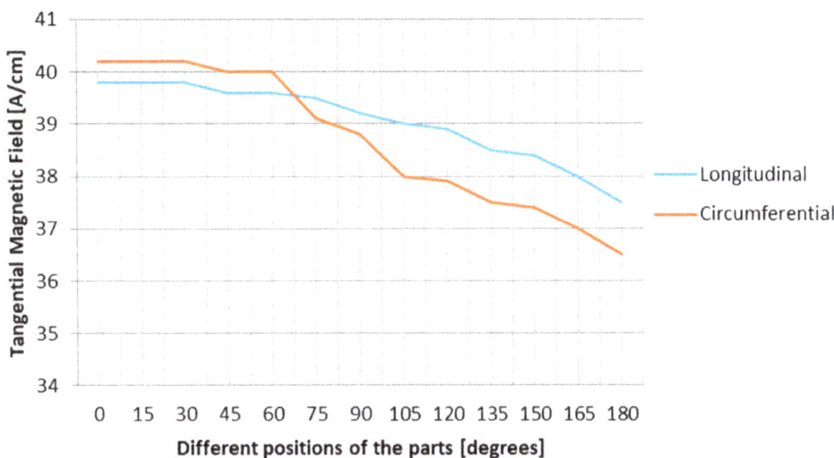

Figure 6. Changes in the magnetic field for the part with central conductor.

The current positioning of the part with the proposed new model, mounted on the standard equipment, is coaxial with the center of the central conductor, which implies a uniform magnetization, meaning that a uniform magnetic field is distributed on the surface of the part to be examined. In the current situation, where having uniform magnetization implies a uniform magnetic field on the surface, the highlighting of the indications is the same regardless of the surface of the part because the part–conductor distance is constantly kept with the entire diameter of the part, and so the proposed coaxially is ensured (Table 2).

Table 2. Changes of tangential magnetic field parameters using an auxiliary device.

Positions [Degrees]	Longitudinal Magnetic Field [A/cm]	Circumferential Magnetic Field [A/cm]
0	38	39
15	38	39
30	38	39
45	38	39
60	38	39
75	38	39
90	38	39
105	38	39
120	38	39
135	38	39
150	38	39
165	38	39
180	38	39
195	38	39
210	38	39
225	38	39
240	38	39
255	38	39
270	38	39
285	38	39
300	38	39
315	38	39
330	38	39
345	38	39
360	38	39

Because the design of the device is provided with two rotating drums, it is no longer necessary to involve the human factor to ensure the rotation of the part and thus reduces the risk of deleting the indications or masking them. The values of the magnetic field are this time constant on 360 degrees around the central conductor. In Figure 7, we can see a linear change of the parameters on the entire circumference of the part, i.e., from 0 to 360 degrees, and the values of the tangential magnetic field are constant.

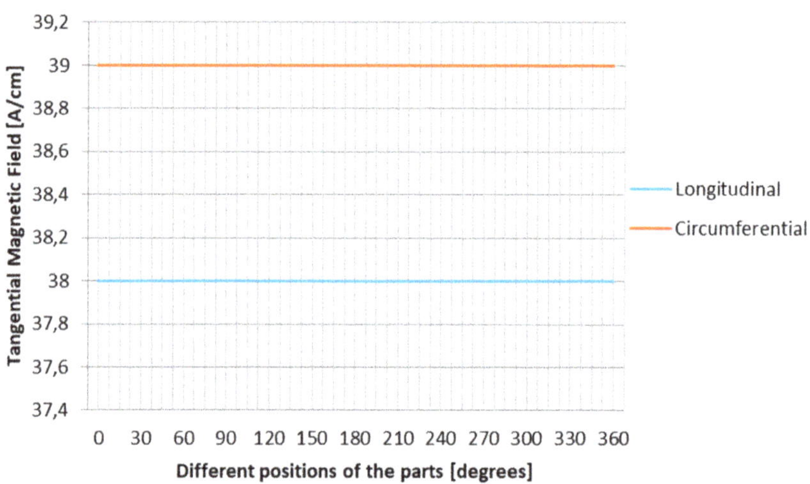

Figure 7. Tangential magnetic field parameters vs. different positions of the parts, measuring using an auxiliary device.

When the part was positioned coaxially with the central conductor, the lines of the magnetic field remain constant on the graph regardless of the part–conductor distance, as these were in coaxiality. In both directions, longitudinally and circumferentially, the values remain constant, and this implies a correct and efficient choice of the physical parameters, meaning that the defect can be highlighted without irrelevant indications and that the risk of not highlighting relevant discontinuities is no longer present.

The Figures 6 and 7 describe the change in the magnetic field for two cases: the part in contact with the central conductor and the part in coaxiality with the central conductor. In both cases, the change in the magnetic field remaining in the part is similar to the tangential magnetic field (i.e., values also vary), and this implies an uneven demagnetization. It can be considered a disadvantage especially in the case of parts that no longer go through other demagnetization stations and are part of an assembly mounted directly on cars, planes, missiles, etc. If the remaining magnetic field is not measured properly and the parts still have magnetism values above the accepted limit, there is a risk that they fail during assembly, or in the worst case, during the running of an aircraft, for example.

The 3D model of the proposed device, as well as the current positioning of the parts during the inspection, can be seen in Figure 8.

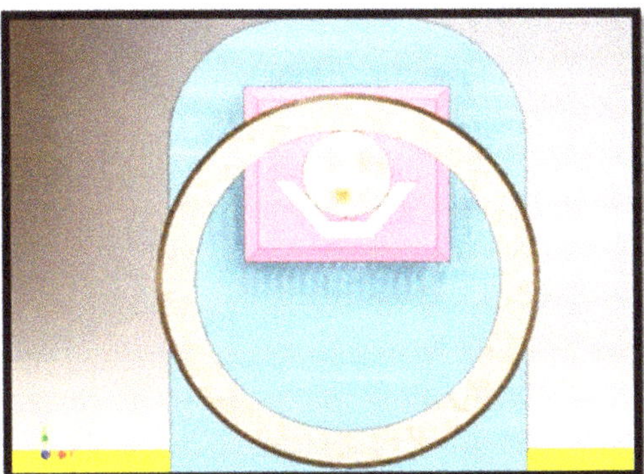

Figure 8. Current position of the parts on the central conductor.

In the multicriteria analysis that was performed (in accordance to [27]), the following criteria were taken into account (Table 3):

Table 3. Multicriteria analysis.

	1	2	3	4	5	6	7	8	9	10	Points	Level	γi
1	1/2	0	0	0.5	0.5	0	1	1	0	0	3.5	8	0.85
2	1	1/2	1	1	1	0	1	1	1	0	7.5	2	3.75
3	1	0	1/2	0.5	1	0	1	1	0.5	0.5	6	5	2.2
4	0.5	0	0.5	1/2	0	0	1	1	0.5	0	4	7	1.1
5	0.5	0	0	1	1/2	0	1	1	0	0.5	4.5	6	1.38
6	1	1	1	1	1	1/2	1	1	1	0	8.5	1	5.1
7	0	0	0	0	0	0	1/2	1	0	0	1.5	9	0.29
8	0	0	0	0	0	0	0	1/2	0	0.5	1	10	0.12
9	1	0	0.5	0.5	1	0	1	1	1/2	1	6.5	4	2.64
10	1	1	0.5	1	0.5	1	1	0.5	0	1/2	7	3	3.15

γi—weighting coefficient (calculated with the FRISCO formula).

- Criterion 1: Assembling/disassembling the device elements
- Criterion 2: Height adjustment according to the diameters of the parts to be examined
- Criterion 3: Ease of operation of orders by operating staff
- Criterion 4: Cycle time of the entire process being under 30 s
- Criterion 5: Fixing mode on several standard control equipment
- Criterion 6: Possibility of interchanging the drums with the support plate for the examination of other configurations of parts, other than the cylindrical ones,
- Criterion 7: Possibility of use for other configurations of parts
- Criterion 8: Establishing the lifting time of the part until ensuring the coaxiality
- Criterion 9: The need to develop a software product for automated operation based on existing calculations
- Criterion 10: Resistance to different weights of the examined parts

In order to perform the analysis, values were assigned to the criteria as follows: 1 being "more important"; 0.5 being "just as important"; and 0 being "less important".

The FRISCO formula for calculating the weighting coefficients γi is presented in Equation (1):

$$\gamma i = \frac{p + |\Delta p| + m + 0.5}{\frac{N_{crt}}{2} + |\Delta p'|} \tag{1}$$

where p is the score obtained along a row by the respective element; Δp is the difference between considered element and the element from the last level; m is the number of outclassed criteria; N_{crt} is the number of considered criteria; $\Delta p'$ is the difference between the score of the considered element and the score of the first element.

In multicriteria analyses, four variants of the device were proposed as follows:

- Device with hydraulic actuation for height adjustment, made of plastics (A);
- Device with electric actuation for height adjustment, made of plastics (B);
- Device with hydraulic actuation for height adjustment, made of composite materials (C);
- Device with electric actuation for height adjustment, made of composite materials (D).

The values of the elements for the calculation of the weighting coefficients were proposed in Table 4.

Table 4. The values of the elements for the calculation of the weighting coefficients.

Δp		m		p		$\Delta p'$	
$\Delta p1$	2.5	$m1$	2	$p1$	3.5	$\Delta p'1$	−5
$\Delta p2$	6.5	$m2$	8	$p2$	7.5	$\Delta p'2$	−1
$\Delta p3$	5	$m3$	5	$p3$	6	$\Delta p'3$	−2.5
$\Delta p4$	3	$m4$	3	$p4$	4	$\Delta p'4$	−4.5
$\Delta p5$	3.5	$m5$	4	$p5$	4.5	$\Delta p'5$	−4
$\Delta p6$	7.5	$m6$	9	$p6$	8.5	$\Delta p'6$	0
$\Delta p7$	0.5	$m7$	1	$p7$	1.5	$\Delta p'7$	−7
$\Delta p8$	0	$m8$	0	$p8$	1	$\Delta p'8$	−7.5
$\Delta p9$	5.5	$m9$	6	$p9$	6.5	$\Delta p'9$	−2
$\Delta p10$	6	$m10$	7	$p10$	7	$\Delta p'10$	−1.5

The score for the established variants are described in Table 5.

Table 5. Notes (Ni) offered to the established variants.

Criteria	Variant A	Variant B	Variant C	Variant D
	Ni	Ni	Ni	Ni
1	8	7	8	7
2	8	9	8	8
3	9	9	8	8
4	10	10	9	9
5	10	7	7	7
6	10	7	9	9
7	10	8	9	9
8	8	9	7	7
9	9	9	8	8
10	10	8	8	8

4. Discussion

In accordance with the Table 6, we found that the largest score is that of Variant A, i.e., hydraulically operated height, with adjusting device made of plastic materials. It is observed that in the absence of a device by positioning the part in direct contact with the central conductor, the values of the magnetic field decrease with the increase of the part–conductor distance. If the part was positioned in direct contact with the central conductor, according to Figure 5, it is observed that the values of the circumferential and longitudinal magnetic fields decrease with the distance between the part and the central conductor. Along with the variation of the tangential magnetic field, there was a variation of a remaining magnetic field left in the part after inspection, and a value of a rest magnetism outside the maximum imposed values can affect during the operation of the part.

Table 6. Consequence of weight.

Criteria	γ_i	Variant A Ni	Variant A Ni × γ_i	Variant B Ni	Variant B Ni × γ_i	Variant C Ni	Variant C Ni × γ_i	Variant D Ni	Variant D Ni × γ_i
1	0.85	8	6.8	7	5.95	8	6.8	7	5.95
2	3.75	8	30	9	33.75	8	30	8	30
3	2.2	9	19.8	9	19.8	8	17.6	8	17.6
4	1.1	10	11	10	11	9	9.9	9	9.9
5	1.38	10	13.8	7	9.66	7	9.66	7	9.66
6	5.1	10	51	7	35.7	9	45.9	9	45.9
7	0.29	10	2.9	8	2.32	9	2.61	9	2.61
8	0.12	8	0.96	9	1.08	7	0.84	7	0.84
9	2.64	9	23.76	9	23.76	8	21.12	8	21.12
10	3.15	10	31.5	8	25.2	8	25.2	8	25.2
Total			191.52		168.22		169.63		168.78

During the optimization of the magnetic particle inspection of the cylindrical parts on standard control equipment, according to Figure 6, it is observed that the existence of an auxiliary device that allows for coaxiality of the central conductor and the part ensures a graphic linearity of the working parameters used during the inspection. This also involves a uniform demagnetization of the part, meaning that a constant magnetic field remains on the entire surface of the part.

From the ergonomic point of view, it can be seen in Figure 9 that the inspector does not have to rotate the parts; instead, the inspector can keep the right body position intact and can focus only on the visual examination of the part in establishing its conformity or not. The inspector should not need to perform additional body movements that from an ergonomic point of view may distract their attention during the inspection.

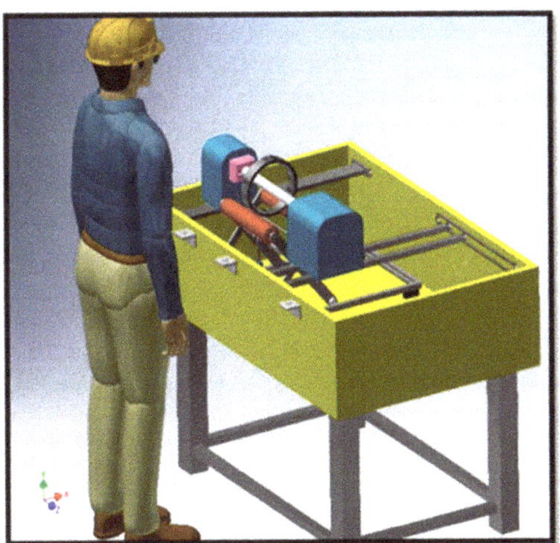
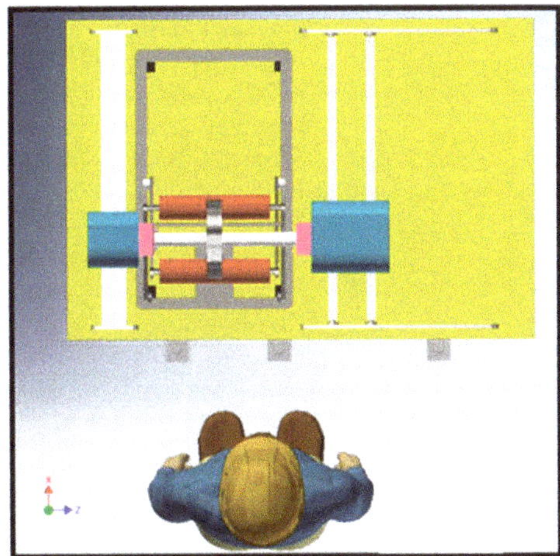

Figure 9. The operator position for a standard equipment using an auxiliary device.

The cycle time decreased by 50% of the total examination time. For example, for a part with a diameter of 200 mm, if magnetization–demagnetization and inspection would take about 30 s on a portion of 180 degrees and another 30 s on the other remaining portion by grinding the part for a new successive magnetization, and if inspection by the use of such a device spans a time of 30 s for the entire surface instead of 60 s and is done only once with constant values for 360 degrees—during which the inspector examines the part without the human factor acting in its handling—we can say we have a decrease of 50%

of the total time, as well as a higher safety in the process and in ensuring the optimal control parameters.

Figure 9 presents the positioning of the operator in relation to the inspection equipment, and it can be seen that in this case, the inspection of the part can be easily done on the entire surface inside the cylindrical part by driving the two drums that will rotate the part. Human errors are eliminated because the risk of deleting the indications is reduced when the operator takes the part manually in order to be able to conduct inspection on all the surfaces. The coaxiality and the constant maintenance of the determined physical parameters are ensured.

5. Conclusions

The design of the device proposed to be used during the inspection of cylindrical parts on standard equipment—an assembly with a central conductor—has a number of advantages, both in terms of working parameters used and in terms of process safety and ergonomics in terms of the inspector's position at work, and it also involves a reduction in control time. Usually, without this device, in order to have a uniform distribution of the magnetic field over the entire surface of the part to be examined, it is necessary to rotate the part 180 degrees around the central conductor in order to have magnetism on the lower surface in addition to the upper surface, and thus there are two magnetizations performed. By using the proposed device, we observed that ensuring coaxiality between the central conductor and the part is a main advantage, as this implies that the magnetization is done only once during the rotation of the piece on the two drums of the device, without the need for two successive magnetizations. A decrease of the inspection time by approximately 50% compared to the initial time means an optimization both from the constructive point of view and from the examination time point of view in the case of series production, as well as the elimination of human errors.

At the same time, both uniform magnetization and uniform demagnetization are ensured, and thus the risk of retaining the remaining magnetism on certain areas of the piece is eliminated. In the classic examination situation, we can speak of the successive magnetization in two stages and of the successive demagnetization in two stages and this one. In order to automate the inspection process by the method described above, it is desirable to make a physical prototype for such a device and the software application to automatically establish (according to calculations) the distance at which we should operate the device with the two drums, so that there is coaxiality between the central conductor and the part to be examined. The height at which the part will be relative to the reference surface (the sole of the device) is determined by two factors: the inner diameter of the cylindrical part and the diameter of the two drums, which are permanently connected to each other to ensure a stable position. The parts also provide strength depending on their weight. It is possible to automate the process as future development guidelines, considering that the advantage of automation would be the optimization of the method, as well as the transition to NDT 4.0 by eliminating the human factor.

Some important advantages are the elimination of the contact area between the inner diameter of the part and central conductor, the ability to remove the risk of not having that area examined and correctly subjected to the control process, and the ensuring of a constant magnetic field on the entire circumference of the part. For example, in the case of a cylindrical part with a large diameter subject to nondestructive testing on equipment without a fastening device, in the moment of checking the intensity of the two magnetic fields, if the Hall sensor is positioned at the top of the ring (part in contact with central conductor), the value indicated by the device will be larger than the one at the bottom coming out of the area of the central conductor. In this case, the magnetization of the part is not performed evenly over its entire surface; to have a correct magnetization, it is necessary to rotate the part by 180 degrees. This difference in the intensity of the magnetic field flux is obvious because in the area of the central conductor, the value of the field is always higher than that at the extremities.

One of the disadvantages is that the device is limited only to the examination of cylindrical parts on standard control equipment, and the limitation is for these part configurations. If the geometry of the part is different and the thickness/radius varies greatly, then depending on the values of the magnetic field, it will have to be magnified several times, but without this device, the part may need to be in direct contact with the central conductor to ensure optimal physics parameters. If the values of the magnetic field from one thickness to another do not differ much and are in the optimal range (30–60 A/cm), then the inspection must be in good condition. There is the possibility of positioning flow indicators (with artificial cracks of different dimensions) on the surface of the part on the area of small thickness and large, and during magnetization in addition to the parameters of the magnetic field, it can be seen how well highlighted or not certain cracks are.

A possible future direction of research is to expand the range of products that can be examined to ensure a constant magnetic field, possibly for parts with variable geometry. Another possible development directive can involve mounting Hall sensors on the two drive drums so that the values of the magnetic field and of the remaining magnetism are measured and shown on a display. At the same time, this measurement of the value of the magnetic field of the remaining magnetism could be transmitted wirelessly to an electronic application to monitor the process and its parameters from reference to reference. All these represent future directions of improvement and optimization that can lead to the development of manufacturing processes and control processes.

Author Contributions: Conceptualization, A.I.S., G.O. and L.P.; methodology, A.I.S. and L.P.; resources, A.I.S. and L.P.; writing—original draft preparation, A.I.S.; writing—review and editing, A.I.S., G.O., and L.P.; visualization, A.I.S., G.O. and L.P.; supervision, G.O. and L.P. All authors have read and agreed to the published version of the manuscript.

Funding: This research received no external funding.

Institutional Review Board Statement: Not applicable.

Informed Consent Statement: Not applicable.

Conflicts of Interest: The authors declare no conflict of interest.

References

1. Chakraborty, D.; McGovern, M.E. NDE 4.0: Smart NDE. In Proceedings of the 2019 IEEE International Conference on Prognostics and Health Management (ICPHM), San Francisco, CA, USA, 17–20 June 2019.
2. Im, K.-H.; Kim, S.-K.; Jung, J.-A.; Cho, Y.-T.; Woo, Y.-D.; Chiou, C.-P. NDE Detection Techniques and Characterization of Aluminum Wires Embedded in Honeycomb Sandwich Composite Panels Using Terahertz Waves. *Materials* **2019**, *12*, 1264. [CrossRef]
3. He, M.; Matsumoto, T.; Takeda, S.; Uchimoto, T.; Takagi, T.; Miki, H.; Chen, H.-E.; Xie, S.; Chen, Z. Nondestructive evaluation of plastic damage in a RAFM steel considering the influence of loading history. *J. Nucl. Mater.* **2019**, *523*, 248–259. [CrossRef]
4. Miwa, T. Non-Destructive and Quantitative Evaluation of Rebar Corrosion by a Vibro-Doppler Radar Method. *Sensors* **2021**, *21*, 2546. [CrossRef] [PubMed]
5. Kostroun, T.; Dvořák, M. Application of the Pulse Infrared Thermography Method for Nondestructive Evaluation of Composite Aircraft Adhesive Joints. *Materials* **2021**, *14*, 533. [CrossRef]
6. Vrana, J.; Singh, R. NDE 4.0—A Design Thinking Perspective. *J. Nondestruct. Eval.* **2021**, *40*, 8. [CrossRef] [PubMed]
7. Abell, J.A.; Chakraborty, D.; Escobar, C.A.; Im, K.H.; Wegner, D.M.; Wincek, M.A. Big data driven manufacturing—Process-monitoring-forquality philosophy. *J. Manuf. Sci. Eng.* **2017**, *139*, 1–12. [CrossRef]
8. He, M.; Shi, P.; Xie, S.; Chen, Z.; Uchimoto, T.; Takagi, T. A numerical simulation method of nonlinear magnetic flux leakage testing signals for nondestructive evaluation of plastic deformation in a ferromagnetic material. *Mech. Syst. Signal Process.* **2021**, *155*, 107670. [CrossRef]
9. Shi, P.; Jin, K.; Zheng, X. A magnetomechanical model for the magnetic memory method. *Int. J. Mech. Sci.* **2017**, *124–125*, 229–241. [CrossRef]
10. Usarek, Z.; Augustyniak, B. Evaluation of the impact of geometry and plastic deformation on the stray magnetic field around the bone-shaped sample. *Int. J. Appl. Electromagn. Mech.* **2015**, *48*, 195–199. [CrossRef]
11. Shi, P.; Su, S.; Chen, Z. Overview of Researches on the Nondestructive Testing Method of Metal Magnetic Memory: Status and Challenges. *J. Nondestruct. Eval.* **2020**, *39*, 1–37. [CrossRef]

12. Boller, C.; Altpeter, I.; Dobmann, G.; Rabung, M.; Schreiber, J.; Szielasko, K.; Tschuncky, R. Electromagnetism as a means for understanding materials mechanics phenomena in magnetic materials. *Mater. Werkst.* **2011**, *42*, 269–278. [CrossRef]
13. Dovbov, A. The method of metal magnetic memory—The new trend in engineering diagnostics. *Weld. World* **2005**, *49*, 314–319.
14. Vrana, J. NDE perception and emerging reality: NDE 40 value extraction. *Mater. Eval.* **2020**, *78*, 835–851. [CrossRef]
15. Vrana, J.; Kadau, K.; Amann, C. Smart data analysis of the results of ultrasonic inspections for probabilistic fracture mechanics. *VGB PowerTech* **2018**, *7*, 38–42.
16. Escobar, C.A.; Wincek, M.A.; Chakraborty, D.; Morales-Menendez, R. Process-monitoring-for-quality—Applications. *Manuf. Lett.* **2018**, *16*, 14–17. [CrossRef]
17. Bray, D.E.; Stanley, R.K. *Nondestructive Evaluation: A Tool in Design, Manufacturing and Service*; CRC Press: Boca Raton, FL, USA, 1998.
18. Sola, M.; Lagüela, S.; Riveiro, B.; Lorenzo, H. Non-destructive testing for the analysis of moisture in the masonry arch bridge of Lubians (Spain). *Struct. Control Health Monit.* **2013**, *20*, 1366–1376. [CrossRef]
19. Diamanti, N.; Redman, D. Field observations and numerical models of GPR response from vertical pavement cracks. *J. Appl. Geophys.* **2021**, *81*, 106–116. [CrossRef]
20. Rasol, M.A.; Pérez-Gracia, V.; Fernandes, F.M.; Pais, H.C.; Solla, M.; Santos, C. NDT assessment of rigid pavement damages with ground penetrating radar: Laboratory and field tests. *Int. J. Pavement Eng.* **2020**, *21*, 1–16. [CrossRef]
21. Rasol, M.A.; Pérez-Gracia, V.; Fernandes, F.M.; Pais, J.C.; Santos-Assunçao, S.; Santos, C.; Sossa, V. GPR laboratory tests and numerical models to characterize cracks in cement concrete specimens, exemplifying damage in rigid pavement. *Measurement* **2020**, *158*, 107662. [CrossRef]
22. Rasol, M.A.; Pérez-Gracia, V.; Solla, M.; Pais, H.C.; Fernandes, F.M.; Santos, C. An experimental and numerical approach to combine Ground Penetrating Radar and computational modeling for the identification of early cracking in cement concrete pavements. *NDT E Int.* **2020**, *115*, 102293. [CrossRef]
23. Fernandes, F.; Pais, J. Laboratory observation of cracks in road pavements with GPR. *Constr. Build. Mater.* **2017**, *154*, 1130–1138. [CrossRef]
24. Schimleck, L.; Dahlen, J.; Apiolaza, L.A.; Downes, G.; Emms, G.; Evans, R.; Moore, J.; Pâques, L.; Van den Bulcke, J.; Wang, X. Non-Destructive Evaluation Techniques and What They Tell Us about Wood Property Variation. *Forests* **2019**, *10*, 728. [CrossRef]
25. Meyendorf, N.G.; Bond, L.J.; Curtis-Beard, J.; Heilmann, S. NDE 4.0—NDE for the 21st century—The internet of things and cyber physical systems will revolutionize NDE. In Proceedings of the 15th Asia Pacific Conference for Non-Destructive Testing (APCNDT2017), Singapore, 13–17 November 2017.
26. Liu, Y.; Hu, B.; Lan, X.; Fu, P. Micromagnetic characteristic changes and mechanism induced by plastic deformation of austenitic stainless steel. *Mater. Today Commun.* **2021**, *27*, 102188. [CrossRef]
27. Zavadskas, E.K.; Antucheviciene, J.; Chatterjee, P. Multiple-Criteria Decision-Making (MCDM) Techniques for Business Processes Information Management. *Information* **2019**, *10*, 4. [CrossRef]

processes

Article

Multifunctional Technology of Flexible Manufacturing on a Mechatronics Line with IRM and CAS, Ready for Industry 4.0

Adriana Filipescu [1], Dan Ionescu [1,2], Adrian Filipescu [1,2,*], Eugenia Mincă [2] and Georgian Simion [1]

[1] Department of Automation and Electrical Engineering, "Dunărea de Jos" University of Galați, 800008 Galați, Romania; adriana.filipescu@ugal.ro (A.F.); dan.ionescu@ugal.ro (D.I.); georgian.simion@ugal.ro (G.S.)

[2] Doctoral School of Fundamental Sciences and Engineering, "Dunărea de Jos" University of Galați, 800008 Galați, Romania; eugenia.minca@ugal.ro

[*] Correspondence: adrian.filipescu@ugal.ro; Tel.: +40-724-537-594

Abstract: A communication and control architecture of a multifunctional technology for flexible manufacturing on an assembly, disassembly, and repair mechatronics line (A/D/RML), assisted by a complex autonomous system (CAS), is presented in the paper. A/D/RML consists of a six-work station (WS) mechatronics line (ML) connected to a flexible cell (FC) equipped with a six-degree of freedom (DOF) industrial robotic manipulator (IRM). The CAS has in its structure two driving wheels and one free wheel (2 DW/1 FW)-wheeled mobile robot (WMR) equipped with a 7-DOF robotic manipulator (RM). On the end effector of the RM, a mobile visual servoing system (*eye-in-hand* VSS) is mounted. The multifunctionality is provided by the three actions, assembly, disassembly, and repair, while the flexibility is due to the assembly of different products. After disassembly or repair, CAS picks up the disassembled components and transports them to the appropriate storage depots for reuse. Technology operates synchronously with signals from sensors and *eye-in-hand* VSS. Disassembling or repairing starts after assembling and the final assembled product fails the quality test. Due to the diversity of communication and control equipment such as PLCs, robots, sensors or actuators, the presented technology, although it works on a laboratory structure, has applications in the real world and meets the specific requirements of Industry 4.0.

Keywords: mechatronics line; visual servoing system; wheeled mobile robot; industrial robotic manipulator; Industry 4.0

Citation: Filipescu, A.; Ionescu, D.; Filipescu, A.; Mincă, E.; Simion, G. Multifunctional Technology of Flexible Manufacturing on a Mechatronics Line with IRM and CAS, Ready for Industry 4.0. *Processes* **2021**, *9*, 864. https://doi.org/10.3390/pr9050864

Academic Editor: Sergey Y. Yurish

Received: 16 April 2021
Accepted: 12 May 2021
Published: 14 May 2021

Publisher's Note: MDPI stays neutral with regard to jurisdictional claims in published maps and institutional affiliations.

1. Introduction

The main contribution of the paper is the overall proposed approach: a multifunctional technology and a flexible manufacturing that works on laboratory system and integrates several subsystems, namely an assembly/disassembly mechatronics line (A/DML), an A/D flexible cell (FC) with an integrated 6-DOF IRM and CAS consisting of an autonomous robotic system (ARS), which is a WMR equipped with a 7-DOF RM and an *eye-in-hand* VSS located on the end effector. All these subsystems are equipped with PLCs, wired and wireless communication devices, infrared, inductive, and optical sensors, and electric and pneumatic actuators. The technology allows the assembly of two different products and complete disassembly or repair of the product that fails quality tests. Components resulting from disassembly or repair are recovered by CAS and deposited for reuse. The main elements of originality and contributions are concentrated in designing the architecture of the entire system to allow flexible manufacturing, multifunctionality, communication, synchronization of signals from sensors, distributed control, and image processing for precise positioning. All these actions and operations are found in technologies from the real industrial world, technologies that are in connection with the digitization, communication, control, and automation of processes, concepts with Industry 4.0 specificity.

A flexible production line represents all workstations and cells, measuring and data acquisition equipment, WMRs, RMs, transport, storage, monitoring and control systems, is capable of performing tasks for component assembly or processing operations, including a reconfigurable manner that confers reversibility, repeatability and, finally, flexibility [1–3]. The concept of FML was designed and developed to manufacture different products, in small or medium batches.

A flexible manufacturing line (FML), able to perform A/D assisted by robots, consists of the following subsystems [4–8]:

- IRMs required for handling operations (require precision, trajectory control systems, sensors, and transducers) [9];
- WMRs for transport (require trajectory control systems, guidance systems, position, and navigation sensor systems) [10–14];
- A/D equipment, IRMs, stations and manufacturing cells [9–14];
- Component and/or subassembly storage warehouses necessary to ensure a continuous flow of A/D [2,10–13];
- Transport system (conveyor belts) necessary for the transport of components or subassemblies from one flexible cell to another [2];
- Reconfigurable workstations and cells with necessary equipment for A/D operations [15,16];
- Sensors and transducers placed in a distributed network on FML, WMRs, and RMs [13,17–20];
- Monitoring and control equipment in distributed and centralized structure [20,21];
- Compatibility equipment between FML, robotic, and computing systems [22,23];
- Data acquisition and communication equipment.

The general structure of an FML allows for highlighting the general functions of the system [2,15–17].

- Function of automatic processing of parts or subassemblies;
- Automatic storage, transport, and handling function [18–22];
- Function of automatic control of all system components and of automatic supervision, control, and diagnosis. This function is realized with the help of one or more PLCs in various configurations, centralized or distributed, or process computers working in real time or local control equipment (PLC for handling and transport systems, microcomputers for automatic warehouse control, etc.). Computer programs provide the entire system with the information needed to control the processing process and to control production (ordering parts and tool warehouses, ordering the transmission system, etc.). The information to perform these sub functions is obtained from the system using transducers, sensors, measuring devices, etc. and is transmitted in reverse to the process computer, AP, PLC, or local microcomputer [3,5,12];
- Automatic processing function is performed within the technological subsystem of FML, consisting of workstations (cells), means for handling parts and tools. The achievement of this function supposes the automatic supply with parts and tools of the machine tool, the actual processing in numerical control and the capability of the optimization of the control process on the machine tool. Assembly/disassembly devices may also be included here, some of which have special functions [20–23];
- Automatic storage, transport and handling function refers to the automatic flow of tools, parts, components, and subassemblies required by FML and this include several partial functions: automatic storage of parts, tools, devices, and auxiliary materials; identification and delivery in the system of the part or subassemblies automatically; automatic transport of parts, tools, devices and auxiliary materials between warehouses and workstations. The main condition in the operation of the storage and transport subsystem is that the transfer of materials is always carried out at the right place and time: handling parts, subassemblies, tools, and devices in warehouses and between workstations;

- Command, monitoring, control, and diagnostic function in an FML is performed by the information subsystem through the information flow which is transmitted in two directions: first, forward direction consisting of command information, second, reverse direction, consisting of monitoring, control, and diagnostic information.

Visual servoing is a fusion of the results obtained from several research areas such as real-time image analysis and processing, robotics, control theory and systems, and real-time application design. One of the fundamental components of a robot is the visual sensor, which allows the investigation of the working environment without contact with its elements. Visual servoing systems behavior is mainly influenced by the type of visual features used to generate control law. The control architectures corresponding to the servoing systems are divided into three categories:

- Position Based Visual Servoing (PBVS) [24–28];
- Image Based Visual Servoing (IBVS) [24,29];
- Hybrid Visual Servoing (HVS) [13,17,24,25,30].

In this paper, a hybrid architecture is used to control the *eye-in-hand* VSS mounted on the robotic manipulator of the CAS.

The rest of the paper is organized as follows. Hardware structure of the A/D/RML assisted by CAS is laid out in Section 2; flexibility and multifunctionality of the A/D/R/ML, together with task scheduling are presented in Section 3; Industry 4.0 based A/D/RML and CAS communication, control, and synchronization are presented in Section 4; Section 5 presents real-time control of multifunctional flexible manufacturing technology, only for repair function; and some remarks on technology control and supervision can be found in Section 6, Discussion. In the final section, Conclusions, the goals pursued by the approach and research in the paper are stated.

2. A/D/RML Assisted by CAS

2.1. Hardware Architecture

The basic design concept consists of three main components/subsystems which are synchronized to work together and act as a flexible manufacturing line that performs several operations such as the assembly of two different products (workpieces) with disassembly, repair, and recover functionality.

The structure of the A/D/RML is shown in Figure 1. The major components are:

- FC with 6-DOF ABB IRB120 IRM station used for assembly, disassembly, and repair of the workpieces with buffer, handling, processing, and transport capability;
- A/DML 6-WS Hera&Horstmann ML based on laboratory mechatronic system, used for assembly and transport of the workpieces with checking and storage facility [10]. A/DML has some capabilities of disassembly but are not used in this paper;
- CAS PeopleBot WMR equipped with a 7-DOF Cyton 1500 RM used for recovery and transport/return operation of the dismantled workpart [13].

The A/D/RML, as described above, is characterized by a modular structure. The hardware structure consists of two PLCs controlled subsystems/modules with specific tasks for all the manufacturing stages.

- FC Siemens S7-1200 PLC controls assembly/disassembly unit which handles the supply of workparts for the workpiece product type 1 and disassembly or repair for the workpiece type 2;
- 6-WS Hera&Horstmann ML PLC (Siemens S7-300 series) has a predefined role as a logistics unit that assemblies individual workparts, transports between modules, and stores the assembled workpieces into the final storage place.

The PLC-based hardware and software structure, as seen in Figure 2, has hybrid architecture features of distributed and centralized/decentralized architecture.

- Distributed structure, by means of separate PLCs for each of the two subsystems, to automate their respective areas with visualization or operation facilities.

- Centralized/decentralized architecture, where the FC PLC (Siemens S7 1200) besides the local control role, acts as master PLC for centrally control both subsystems of the entire A/D/RML, process and operation facilities, thereby coordinating control tasks as well as synchronizing the operations of the CAS which include a running hardware multimedia interface (HMI) KTP 700 as the main visualization and operator control.

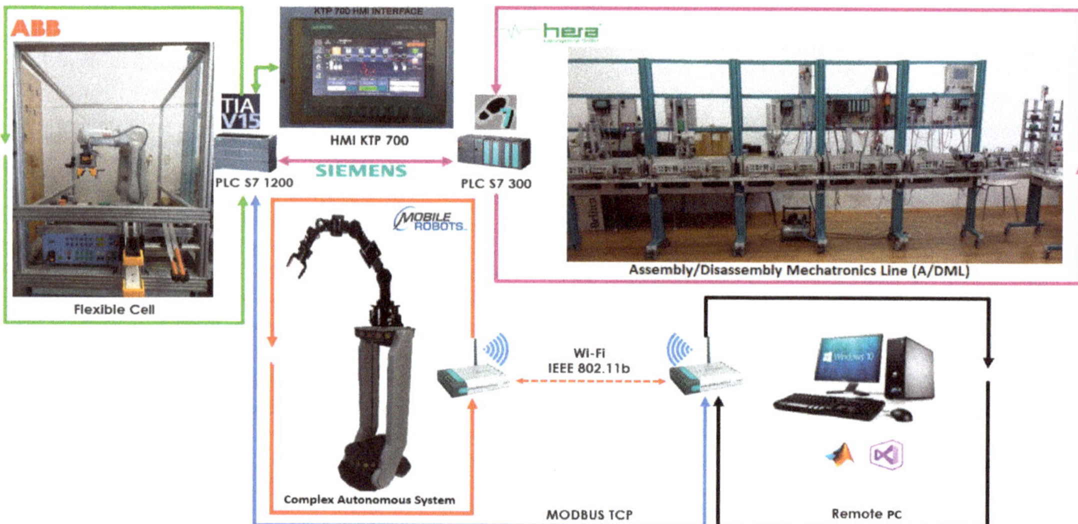

Figure 1. Control structure of A/DML Hera&Horstmann, FC with ABB IRM and CAS with PeopleBot WMR and Cyton 1500 RM.

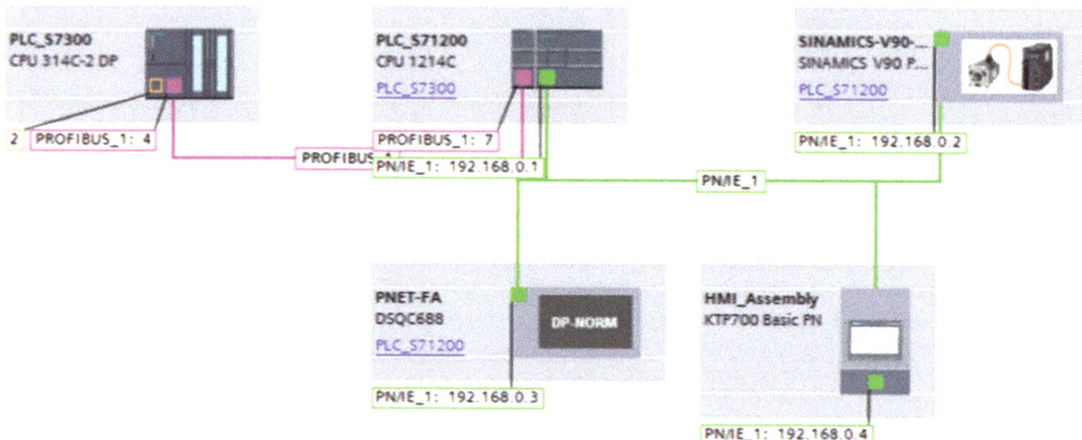

Figure 2. Flexible Cell Siemens S7-1200 PLC hardware structure.

Each PLC, including Hera&Horstmann ML Siemens S7-300 PLC, hosts several control programs whose selection is made remotely, via an HMI or via the master PLC (Siemens S7-1200). The A/D tasks are controlled strictly through this master PLC, which acts as a Central System that handles visualization and operation of the complete A/D/RML. The Hera&Horstmann ML Siemens S7-300 PLC is connected to I/O points via Profibus (magenta line in Figure 1). Profibus topology is used to communicate and control the transporting conveyor belt drives, workpiece positioning, and synchronization methods as well as to handshake and signal interface exchange with the FC via the Profibus adapter. An additional HMI (Siemens TP 177) is connected for process visualization purposes only.

FC communication is based on the industrial Ethernet network Profinet technology (green line Figure 1), to communicate with the main HMI (Siemens KTP 700), ABB IRB120 IRM controller and Intelligent Siemens Servomotor Drives having their own positioning control functionality. Compatibility between FC and Hera&Horstmann ML, by means of communication is executed, as mentioned before, via a Profibus adapter to bridge (interconnect) the two different communication technologies: Profinet (protocol based on the Industrial Ethernet) and Profibus (protocol based on serial communication).

When disassembly or repairing action is performed in the FC (IRM disassembles or repairs the workpiece by replacing the bad components), the standby Cyton RM, part of the CAS system, will grab the recovered workparts to transfer them into the designated storage locations. Several synchronization signals are transferred between the master PLC and CAS, by means of Modbus TCP protocol, a standard communications protocol widely used in industrial automation. Some of these signals (e.g., acknowledge signals) are sent when FC accomplishes the repair/dismantle action, and the replaced component (workpart) is released and ready to be picked up by the CAS. Synchronization acknowledges signals will be returned, when CAS is busy during handling, picking up, transporting, or releasing a part. After that, CAS becomes available again.

Several algorithms have been developed using Siemens programming packages such as Totally Integrated Automation (TIA) Portal, Step7 Manager, as well as WinCC Flexible for the HMIs, for the inner loop of the developed strategy (PLC level). In both PLCs, modular programming is used, function blocks or functions are created as an entity, performing a particular functionality (Assembly, Disassembly, Transport, Storage) or controlling a particular type of device in the system (ABB Robot, conveyors motors, storage, electrical and pneumatic actuators). During each scan, the PLC reads all local and remote inputs, executes every function block and function in a predefined order (using IRQ), and updates all outputs at the end of each scan. An additional part of a PLC program is the communication between master PLC (FC S7-1200 PLC) and CAS via Modbus TCP link. For that, a Modbus TCP Server is configured and programmed in the Main Routine of the Siemens master PLC at the beginning of the scan prior to the program execution to establish and maintain a stable connection and quick data exchange-synchronization signals with the CAS.

As previously described, a separate Profibus communication link is used to interface data between PLCs (FC and A/DML). This data must be sent and received between master PLC and Siemens S7-300 PLC via Profibus communication adapter, as shown in Figure 2.

2.2. Flexible Cell with ABB IRM

FC is a laboratory integrated ABB IRB120 Robot Training Station, shown in Figure 3, which consists of the following major components:

- 6-DOF ABB IRB120 IRM with electric gripper;
- PLC Siemens S7-1200 series-CPU 1214C;
- HMI Siemens KTP700, Colour Basic PN;
- Switch Siemens SCALANCE XB005;
- Conveyor Belt with Sinamics V90 Servo Drive;
- Compact storage&unloading units corresponding to each of the five-part workpiece to be assembled.

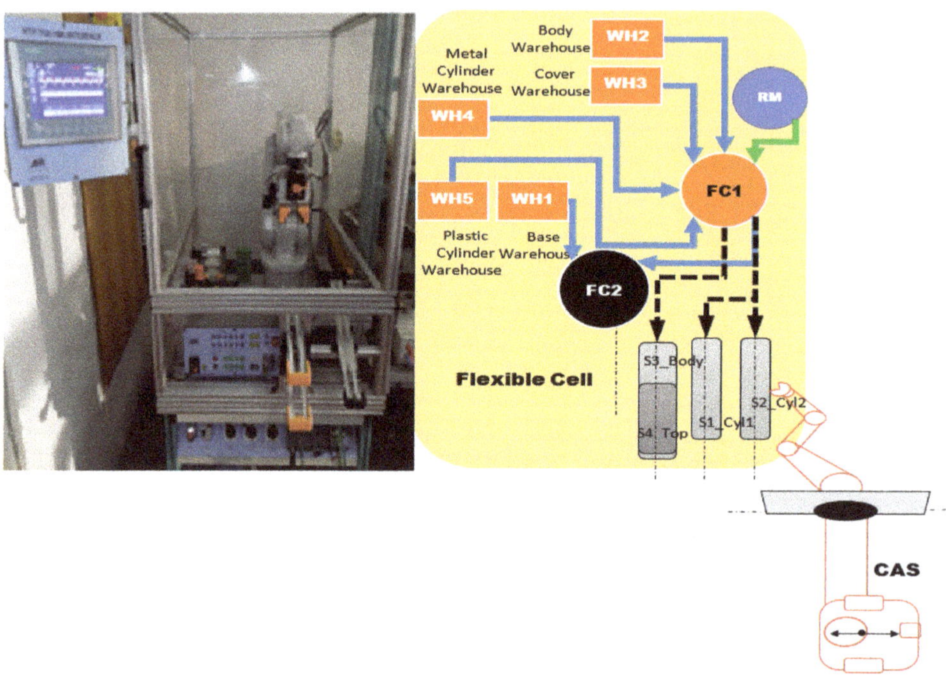

Figure 3. Flexible Cell Station with 6-DOF ABB IRB120.

Profinet communication link is used to interconnect and control all the above-mentioned devices of the FC. For the FC hardware structure, the following Profinet profiles are applicable:

- Profinet-IO, Distributed I/O (Remote I/O), in which the user data from the field devices are periodically sent to the control system process model. This can be considered an evolved Profibus protocol on the TCP layer. Profinet-IO is used to link HMI, PLC CPU, and ABB IRM Controller;
- PROFI drive, implemented for drives application scenarios and covers from simple frequency converters to intelligent servo drivers. This Profinet profile is used in Flexible Cell station to control the Conveyor Belt with Sinamics V90 Servo Drive.

ABB Robot Controller has the hardware capability to communicate with third party devices via Profinet protocol, as mention before. For that, a dedicated board AnybusCC Profinet slave (DSQC 688) is inserted into an expansion board on top of the main computer unit in the ABB Robot Controller. This Profinet Anybus device, DSQC 688, requires the Robot Controller DSQC1000 (main computer). With the Profinet Anybus Device option, the ABB IRM controller can act as a slave on the Profinet network.

2.3. A/DML Hera&Horstmann ML

The A/DML (Figure 4) includes six individual workstations with different tasks; each of them ensures the fulfilment of the operations for different stages: carrying and transporting, pneumatic workstations, conveyor belt, sorting unit, test station, and warehouse [6,10,11]. The five-part workpiece enables workflow operations such as assemblies, testing, sorting, storage, and disassembling. The components to be assembled are: workpart carrier (Base), Body, Top, Metal cylinder, and Plastic cylinder.

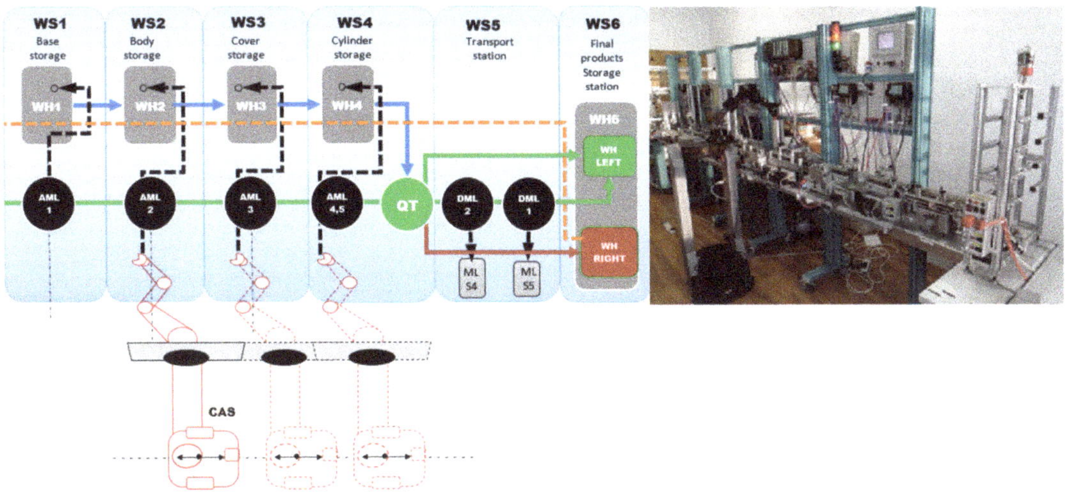

Figure 4. A/DML Hera&Horstmann ML with symmetrical final product storage.

2.4. Hardware Structure of the CAS

The CAS, shown in Figure 5, is composed of the following elements: a 7-DOF Cyton 1500 RM equipped with an *eye-in-hand* type of VSS using a high-definition camera, both being connected to a computer via USB and synchronously communicating with the A/D/RML over Wi-Fi. The RM is placed on the ARS PeopleBot, which is a WMR with two driving wheels and one free wheel (2 DW/1 FW). The CAS is used to transport the recoverable pieces picked-up by the Cyton 1500 RM to the appropriate storage depots if the assembled piece has failed the quality test and has been disassembled or repaired.

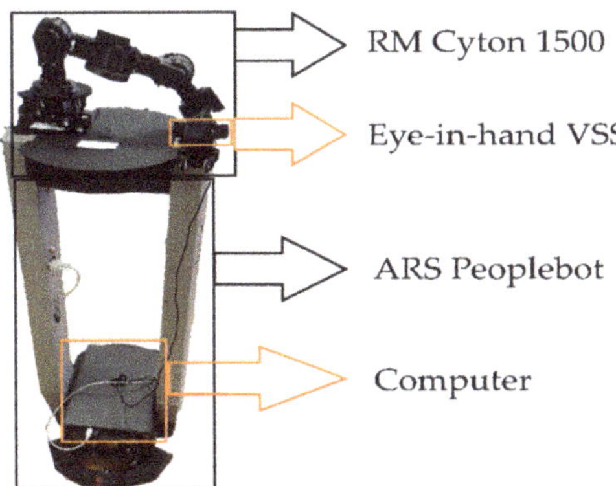

Figure 5. The CAS composed of ARS, RM, *eye-in-hand* VSS, and a computer.

The control of the CAS is carried out wirelessly using a router that is placed inside the WMR, through dedicated functions from ARIA (Advanced Robotic Interface for Applications) running on the same computer the Cyton RM is connected to.

2.5. Eye-In-Hand VSS

The *eye-in-hand VSS* is a system where the video sensor is placed on the last link of the RM, also known as the end-effector [13,17,24]. For this type of VSS, 2D image information is used to control the motion of the robot in the workspace. The object tracking and the robot positioning are achieved using the comparison between the current visual features, extracted from the images captured by the camera, and the desired visual features. The obtained difference is used to minimize the error between the present position of the piece and the anticipated location. Moreover, *eye-in-hand* type VSS indicates that the movement of the RM also induces motion of the mounted camera. One of the most utilized components in object detection and classification are called image moments. These image moments are commonly used in the robotics field because of their efficiency and simplicity in implementation. The image moments contain information about the region of interest, the coordinates of the gravity center of the piece, and the positioning of the image.

3. Flexibility and Multifunctionality of the A/D/R/ML

3.1. Flexibility

A/D/RML is a flexible manufacturing line because it assembles two different products, referred to as workpiece 1 (WP1) and workpiece 2 (WP2). WP1 is the workpiece with the Top part having triangular edges (Figures 6 and 7a) and is assembled in the FC with the ABB IRM. WP2 is the workpiece with the Top part having round edges (Figure 7b,c) and is assembled on the Hera&Horstmann ML.

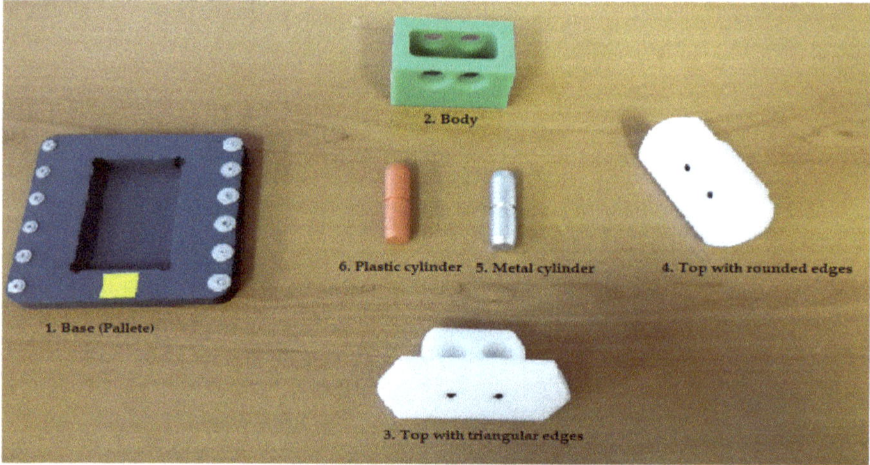

Figure 6. Workpiece parts.

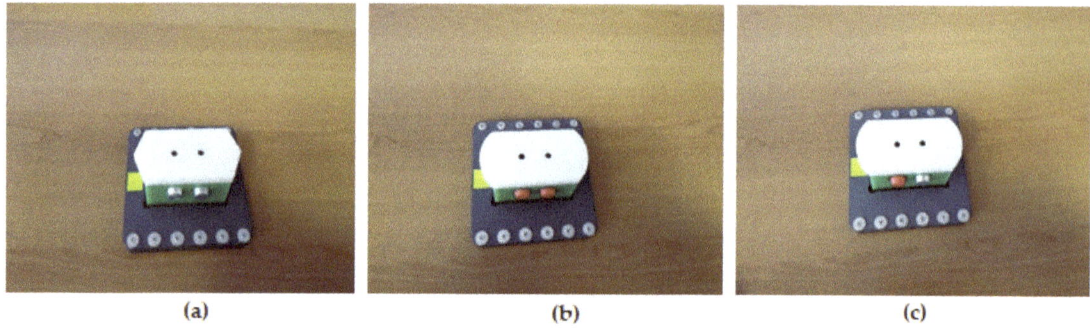

Figure 7. Assembled workpieces: (**a**) workpiece with metal cylinders; (**b**) workpiece with plastic cylinders; (**c**) workpiece with different material cylinders.

3.2. Multifunctionality

Assembly. The assembly of WP1 is made by the ABB IRM, taking from CF warehouses the components in order (Figure 7a: Base, Body, Top, and cylinders, Metal or Plastic. First, the Base is positioned on the conveyor belt (on FC2), then the rest of the product is assembled in a separate location of the FC (on FC1), then it is moved by the ABB IRM on the Base (on FC2). Finally, WP1 moves along the Hera&Horstmann ML and is stored on the left side of the WS6 station. The graphical user interface (GUI), on the HMI pen, allows selection for assembly between plastic cylinders and metal cylinders. Due to this fact, the WP1 product is of good quality and, for this reason, is stored in the rack on the left side of the WS6 station. The WP2 product is randomly assembled with the two cylinders and is subjected to the quality test on the WS4 station. To evaluate the quality for the WP2 product, the convention is that a WP2 product assembled with both metal cylinders is considered of good quality and is stored on the left side of the WS6 station. The WP2 product that contains both plastic cylinders (Figure 7b) is considered scrap product and it is stored in the rack on the right of the WS6 station. This WP2 will be disassembled for component recovery. The WP2 product having different material cylinders (Figure 7c) is also deposited in the rack on the right and it will be repaired by replacing the plastic cylinder with a metal one (Figure 8).

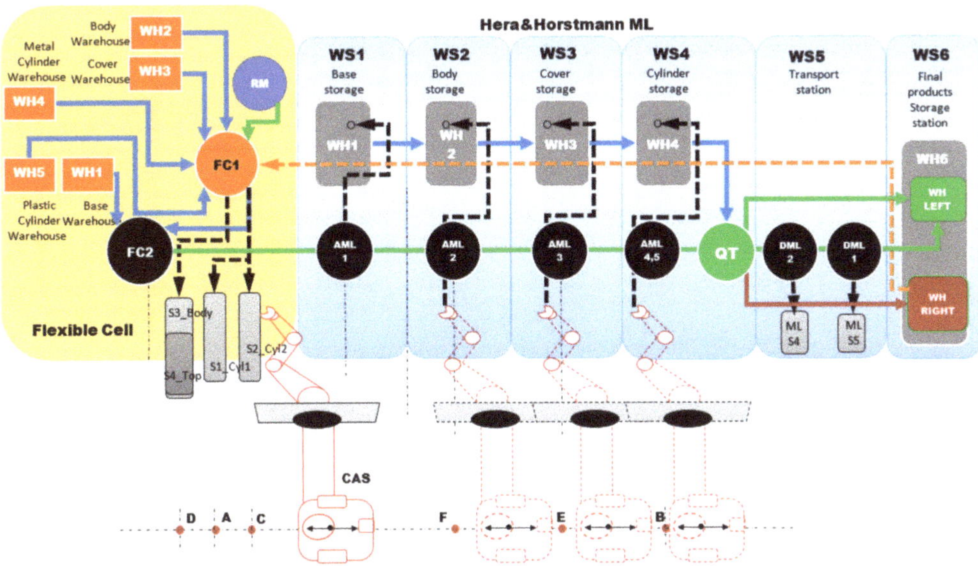

Figure 8. A/D/RML assisted by CAS.

Disassembly. WP2 being considered scrap (it has two plastic cylinders, Figure 7b) is taken over by the elevator of the WS6 and positioned on the transport station WS5. It is transported along the Hera&Horstmann ML to the FC (FC2). The ABB IRM disassembles it in the established order: Cylinder 1 (left), Cylinder Two (right) Top, and Body (on FC1), letting them slide on the corresponding trough. The Base is transported back to WH1 located on ML, where the piston pushes it into the storage warehouse. CAS takes over each component in order, Cylinder 1, Cylinder 2, Body, and Top, transporting it to the appropriate storage warehouse on the Hera&Horstmann ML. The precision positioning of the CAS is performed with the *eye-in-hand* VSS (Figure 8).

Repair. WP2, having cylinders of different materials (Figure 7c), is taken over by the elevator of WP6 and positioned on WS5. It is transported along the Hera&Horstmann ML to the FC (FC2). The ABB IRM disassembles the plastic cylinder (on FC1), letting it slide on the exhaust chute and replaces it with a metal cylinder taken from the corresponding warehouse of the FC. CAS takes over the cylinder, in any position, 1 or 2, transporting it to the appropriate storage warehouse on the Hera&Horstmann ML. WP2, now having both metal cylinders, is a good quality product and it is transported from FC, along the Hera&Horstmann ML, to the WS6 station and stored on the left side (Figure 8).

3.3. Assumptions

Specialized technology on this type of A/D/RML represents the base of a multifunctional flexible industrial production line that gives a high range of standard products. Furthermore, by disassembly or repair, the products or parts can be recovered and brought to the quality standards. The technology on A/D/RML assisted by CAS and *eye-in-hand* VSS, developed below, depends on aspects such as operation modes, operation lengths, and types of finished products (Figure 8) [1,2,4,8]. Therefore, for FC, A/DML, CAS, and VSS some assumptions must be established for controlling whole system.

Assumption 1. *The A/D/RML is a single-model line, by the nature of the product, paced line (transfers between the workstations are synchronous), by the operation mode, and deterministic line, by the nature of operation times (times known certainly).*

Assumption 2. *The number of the A/D/RML workstations involved in A/D/R is previously known and will remain unchanged (FC with ABB IRM and 6 workstations A/DML, Hera&Horstmann ML.*

Assumption 3. *Two types of workpieces are assembled, WP1 in FC with ABB IRM, WP2 in Hera&Horstmann ML.*

Assumption 4. *All conditions and parameters of the technology are initially known, including task durations.*

Assumption 5. *The workstations of the A/D/RML have a linear distribution, FC and WS1 to WS6.*

Assumption 6. *The assembly operations of WP1 are executed in FC. The assembly operations of WP2 are executed on Hera&Horstmann ML.*

Assumption 7. *The left side (in green WH left) of the WS6 station is the warehouse where good products are stored, while the right side (in red WH right) is the warehouse where products that do not pass the quality test are stored, need to be disassembled or repaired.*

Assumption 8. *The disassembly and repair operations of WP2 are executed on FC.*

Assumption 9. *Disassembly and repair start after the WP2 is assembled and it fails the quality test.*

Assumption 10. *WP2 which fails quality test is stored in the right side of WS6 and its disassembly or repair starts immediately after.*

Assumption 11. *By convention, it is assumed that the WP2 fails the quality test if it contains either plastic cylinders or different materials.*

Assumption 12. *One CAS assists the A/D/RML, having mounted a RM, used for picking up, transport and depot the workparts.*

Assumption 13. *One eye-in-hand VSS camera is mounted on the RM.*

Assumption 14. *CAS displacement is without obstacles and with the same constant speed.*

3.4. Tasks Scheduling

Presented below are the block diagrams with the scheduling of tasks for each functionality, assembly in Figure 9, disassembly in Figure 10, and repair in Figure 11, respectively.

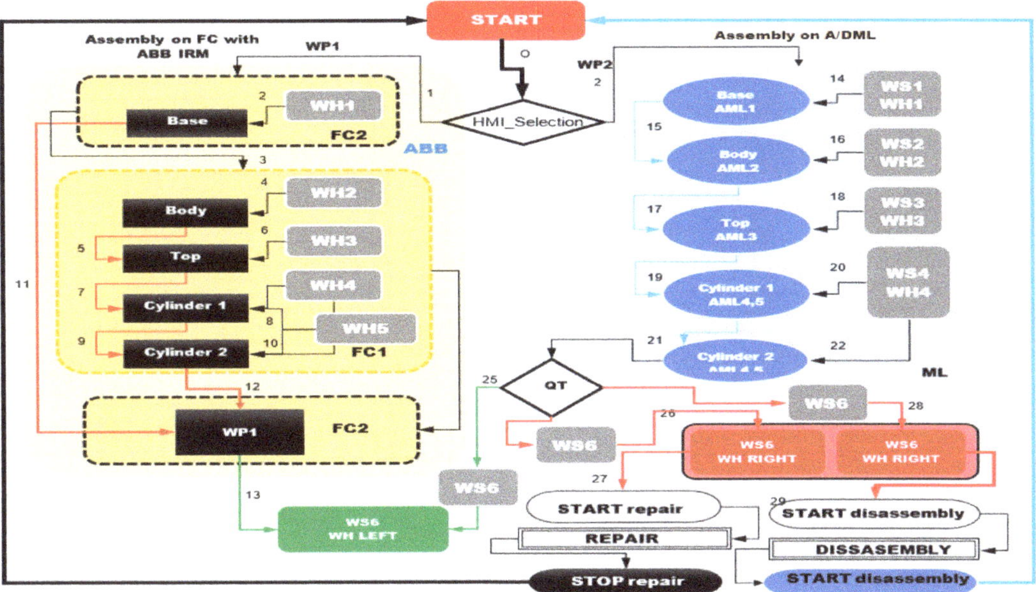

Figure 9. Task scheduling for assembly functionality.

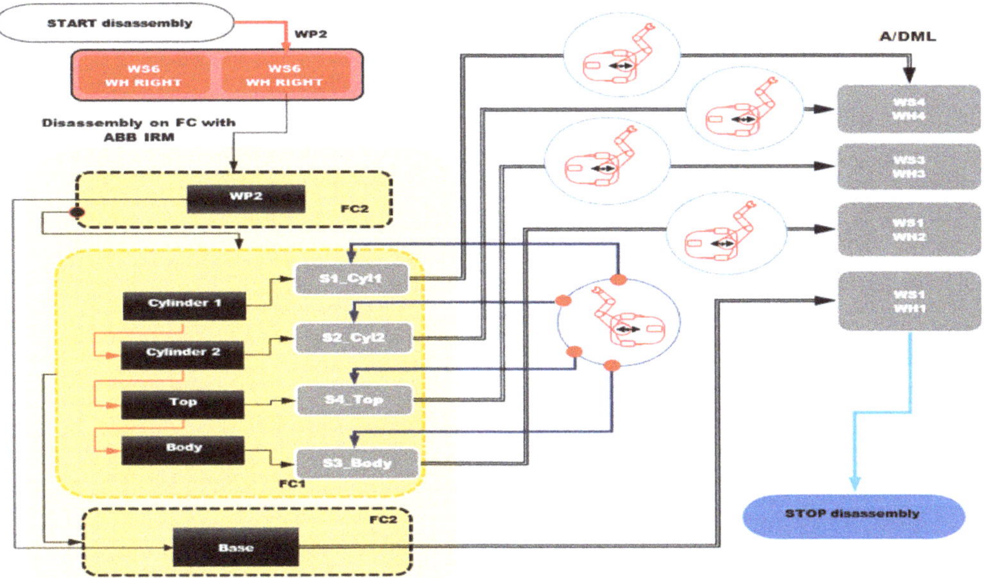

Figure 10. Task scheduling for disassembly functionality.

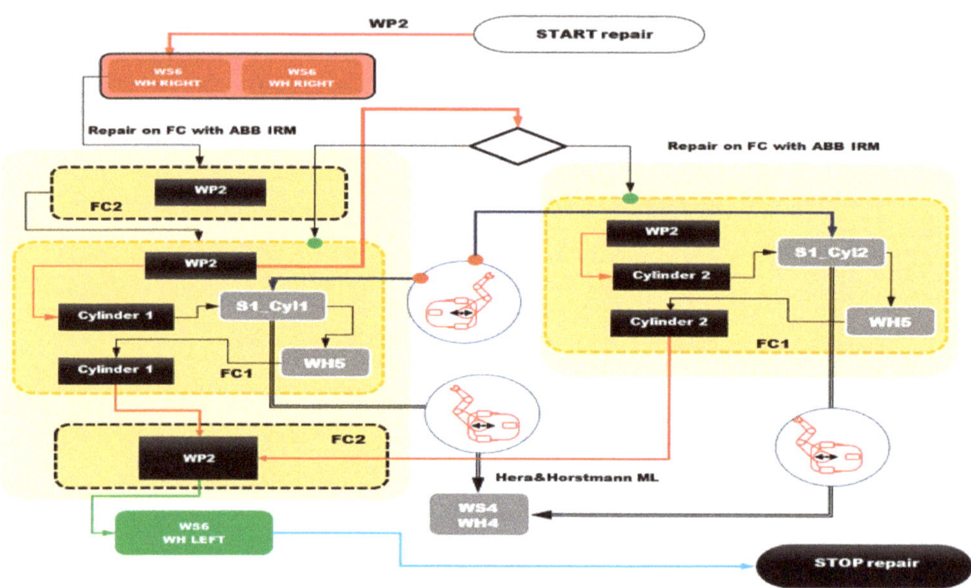

Figure 11. Task scheduling for repair functionality.

4. Industry 4.0 Based A/D/RML and CAS Communication, Control, and Synchronization

4.1. Communication and Control of A/D/RML

The A/DML (Hera&Horstmann ML) is controlled using a Siemens Simatic S7-300 Programmable Logic Controller (PLC), with five distributed modules connected by Profibus DP.

Profibus DP (Decentralized Periphery) is a RS-485 serial-based communication protocol, which ensures cyclic data exchange between PLCs (master) and devices (slaves). It polls slave distributed devices: master sending outputs and receiving inputs from all its devices, and then repeating the cycle.

Every remote I/O Workstation node and field IO device is grabbing info, writing or reading I/Os, device parameters, acting as a Profibus slave and sending response messages to the master PLC using bus cycle at regular intervals but also a cyclically on master device initiative (PLC controller).

Profibus complies with IEC 61158 and IEC 61784 standards and is oriented to the OSI (Open System Interconnection) reference model per international standard ISO 7498. Profibus is a deterministic protocol due to cyclic (periodic) polling mechanism between master and slaves. It uses transmission speeds from 9.6 kbps up to 12 Mbps.

The hardware structure of Hera&Horstmann ML (Figure 12) is based on a distributed architecture, integrating the process peripherals such as signals and function modules in Remote I/O stations on the Profibus link and consists of a Siemens Simatic S7-300 series PLC, processor type CP 314C-2 DP and Siemens CP 343-2 communication module for Profibus link. It uses Profibus DP interface, with defined speed of 12 Mbit/s and connects all six workstations Remote IOs (Siemens ET200S communication modules), which enhance flexibility and performance of flexible assembly/disassembly system within this decentralized architecture.

The main PLC rack consists of several analogue and digital IO cards, dedicated cards for counting and frequency measurement with 60 kHz, pulse width modulation with 2.5 kHz switching frequency as well as positioning cards with analogue output or digital outputs for the conveyor encoders used for precise motion control.

Each of the six Siemens ET200S Remote IO modules provide hardware-near signal, handling all the digital and analogue signals from transducers and to the actuators, as well as measuring and position detection for handling workpiece transportation and process through all line sections.

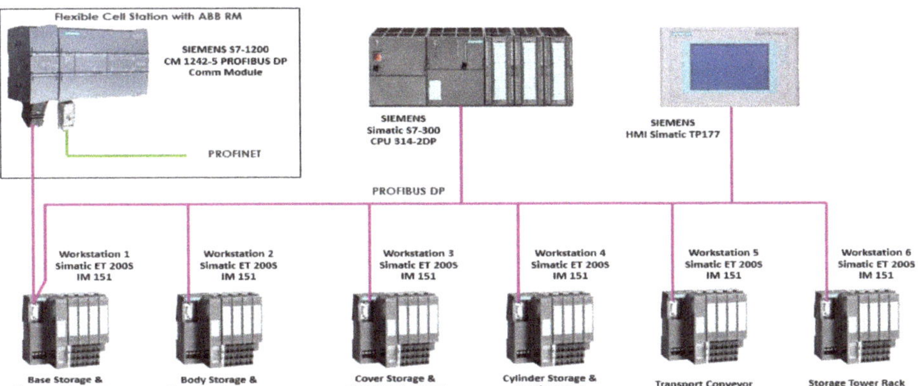

Figure 12. PLC hardware structure of the Mechatronics line Hera&Horstmann.

A Siemens HMI TP 177 operator touch panel is connected to the Profibus DP bus for presentation purposes only, allowing the operator to view the status of the mechatronics line and the current execution step of the assembly or disassembly process.

A communication adapter (Figure 12) Siemens CM 1242-5 attached to S7-1200 PLC is used for connecting the newer generation Siemens master PLC from FC, via Profibus link, to the A/DML.

This module is used to connect and integrate SIMATIC S7-1200 into an automation solution as a Profibus DP slave. The CM 1242-5 works as a DPV1 slave in accordance with IEC 61158, handles data traffic completely autonomously, and thus relieves the CPU of communication tasks.

Additionally, this communication module operates at two levels, i.e., physical layer and data link layer, converting and regenerating the signal it receives or sends and supports cyclic communication for the transfer of process data between Profibus DP slaves and DP master (Mechatronics Line S7-300 PLC). Cyclic communication is handled by the operating system of the CPU.

4.2. Moment-Based Image Method for VSS Modeling and Control

The structure of the *eye-in-hand* VSS contains the subsequent components: an autonomous system composed of a WMR equipped with a 7-DOF Cyton 1500 RM, a controller, and a visual sensor.

The most important part of this type of architecture, the image-based controller, needs deductive information about the environment of the system to minimize the error between the actual configuration of the visual features, f, and a desired configuration, f^*. To model the open loop servoing system, the components of the fixed part must be analyzed individually; these components are the RM and the visual sensor.

The purpose of the *eye-in-hand* VSS is to minimize the error between the real and the desired features extracted by the video sensor [13,17,25].

The control structure of the VSS is shown in Figure 13. The signal associated to the input control of the CAS is v_c^*, and represents the reference speed of the camera with the following structure: $v_c^* = (v^*, \omega^*)^T$, where $v^* = \left[v_x^*, v_y^*, v_z^*\right]^T$ and $\omega^* = \left[\omega_x^*, \omega_y^*, \omega_z^*\right]^T$ are defined as the linear and angular speed. The signal v_c^* is expressed in the Cartesian space and requires a transformation to be applied to the RM.

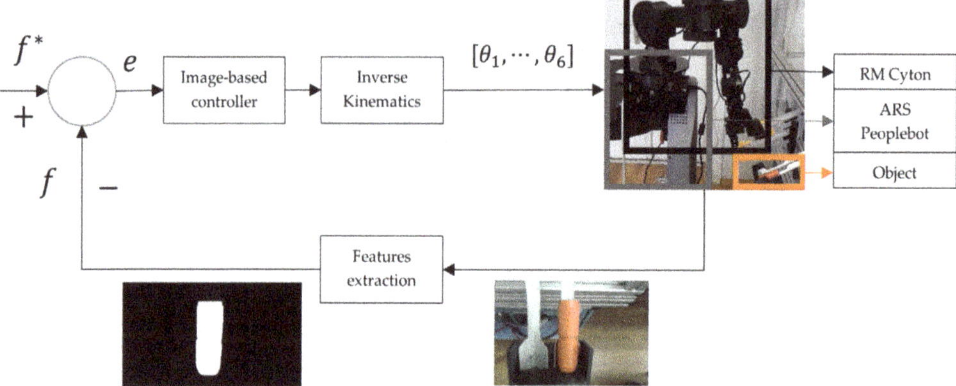

Figure 13. Closed-loop control of the RM Cyton based on eye-in-hand type VSS.

The posture is defined by the integration of the reference speed of the camera, v_c^*, and is noted with $s = [s_1, s_2, s_3, s_4, s_5, s_6]^T$ defining the robot Jacobian as follows:

$$J_r = \left[\frac{\partial s_i}{\partial q_j}\right], \; i, j = 1, \cdots, 6, \tag{1}$$

where $q_{i,j} = 1, \cdots, 6$ signifies the states of the RM's joints. Consequently, the signal transformation of v_c^* from Cartesian space to robotic joint space is J_r^{-1} and the interaction matrix. The interaction matrix must fulfil a series of properties with the purpose of obtaining the ideal performance for a VSS, such as being non-singular and diagonal. The moments m_{ij} are considered a set of visual features with the analytic form for the time variation, $\dot{m}_{ij}$, consequent to the moments of order $(i + j)$, differing depending on the speed of the camera v_c^* according to the equation:

$$\dot{m}_{ij} = L_{m_{ij}} v_c, \tag{2}$$

where $L_{m_{ij}} = \left[m_{v_x} m_{v_y} m_{v_z} m_{\omega_x} m_{\omega_y} m_{\omega_z}\right]$ is the interaction matrix.

Based on the theory presented in [13,17,24], the interaction matrix corresponding to a set of image moments $f = [x_n, y_n, a_n, \tau, \xi, \alpha]^T$ for n points is computed as follows:

$$L_f = \begin{bmatrix} -1 & 0 & 0 & a_n e_{11} & -a_n(1 + e_{12}) & y_n \\ 0 & -1 & 0 & a_n(1 + e_{21}) & -a_n e_{11} & -x_n \\ 0 & 0 & -1 & -e_{31} & e_{32} & 0 \\ 0 & 0 & 0 & \tau_{\omega_x} & \tau_{\omega_y} & 0 \\ 0 & 0 & 0 & \xi_{\omega_x} & \xi_{\omega_y} & 0 \\ 0 & 0 & 0 & \alpha_{\omega_x} & \alpha_{\omega_y} & -1 \end{bmatrix}. \tag{3}$$

4.3. Control Input

The most common procedure to generate a control signal to the robots is the proportional control.

The *eye-in-hand* VSS can be interpreted as a minimization problem which calculates the path of the visual sensor using the minimum of the cost function attached to the error vector. The notations used are as follows: f^* is the desired features vector, f is the current features vector, and $r(t)$ is the relative position between the camera and the object at a specific time, t.

The features variation reported to the relative movement between the workspace and the video sensors is noted with $f(r(t))$ and its variation is described in the equation below:

$$\dot{f} = \frac{\partial f}{\partial r}\frac{dr}{dt} + \frac{\partial f}{\partial t} = L_f v_c + \frac{\partial f}{\partial t}. \tag{4}$$

For a static object, the time variation of the features reported to the motion is equal to zero, $\frac{\partial f}{\partial t} = 0$, and this implies that Equation (4) becomes:

$$\dot{f} = L_f v_c, \tag{5}$$

where v_c is the vector that depicts the relative speed between the object and the video sensor and L_f is the interaction matrix from Relationship (3). To define the control law, it is mandatory to define an error function between the target features f^* and the current features, f:

$$e = f - f^*. \tag{6}$$

Since most of implementations of VSS disregard the dynamics of the robot, equaling the dynamics to one, then $v_c = v_c^*$ and Equation (5) becomes:

$$\dot{f} = L_f v_c^*. \tag{7}$$

From (6) and (7) the time variation of the error is expressed as:

$$\dot{e} = L_f v_c^*, \tag{8}$$

Because the robot control input is defined by v_c^* and an exponentially negative minimization of error is expected, $\dot{e} = \lambda e$, $\dot{e} = L_f v_c^* = -\lambda e$, from the previous Equation (8) results in the control law below:

$$v_c^* = -\lambda L_f^+ e, \tag{9}$$

where L_f^+ is the pseudoinverse of the interaction matrix and is computed as the following:

$$L_f^+ = \left(L_f^T L_f\right)^{-1}. \tag{10}$$

Because in real-time *eye-in-hand* VSS, the Z distance between the points of interest and the reference system attached to the camera is not accurately known, L_f^+ will be estimated, referred as $\hat{L}_f^+$.

The estimation of the matrix is based on the pseudo-inverse of the desired features interaction matrix $\hat{L}_f^+ = L_f^{*+}$ with $\hat{L}_f^+ = \frac{1}{2}\left(L_f + L_f^*\right)^+$. Because the matrix remains constant during the control algorithm execution, the control law results as follows:

$$v_c^* = -\frac{1}{2}\lambda\left(L_f + L_f^*\right)^+ e \tag{11}$$

4.4. Communication and Synchronisation

The remote PC computes control input and sends it to WMR. The remote PC also sends the data to the assembly line PLC (Figure 14) [10,12,19].

To control the CAS and the movement between the parking, grabbing, and placing positions, dedicated functions from ARIA programming package are used and the trajectory-tracking sliding-mode control (TTSMC) method is implemented.

As mentioned before, centralized architecture is used, where the FC PLC (Siemens S7 1200) acts as master PLC and synchronizes the operation with the CAS to perform the recovering process (recovering cylinders). Communication between master PLC (FC S7-1200 PLC) and CAS is executed via Modbus TCP Link.

In this application, remote PC is the Modbus master (on the Wireless TCP Network) and Siemens S7-1200 is the slave (Modbus Server considered in Siemens approach) (Figures 15 and 16).

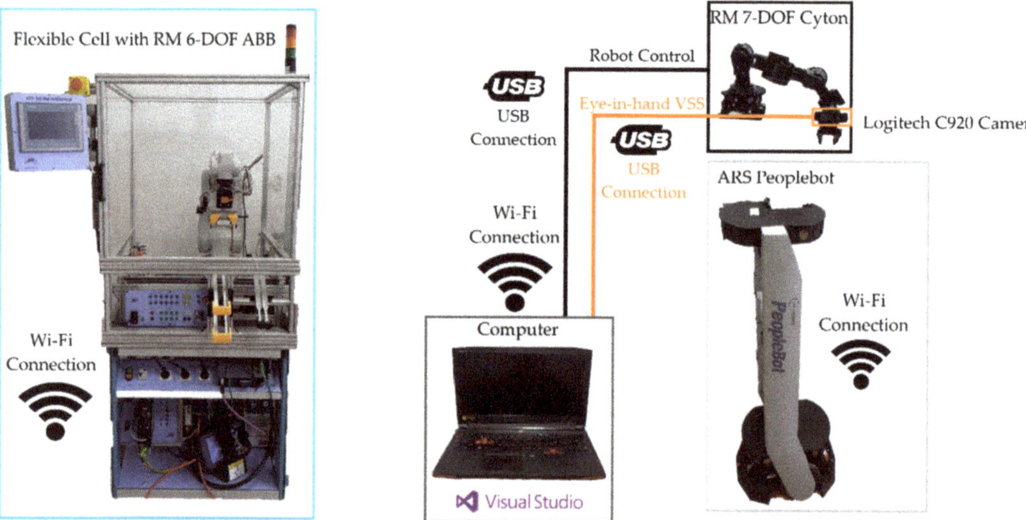

Figure 14. Communication block set of the computer between the FC, ARS PeopleBot WMR equipped with the Cyton RM and *eye-in-hand* VSS.

PC/Mobile Robot ---> S7 1200					
Signal Name	**PC/Mobile Robot**	**S7 1200**	**Type**	**ModBus Code**	**Description**
MB_MRobot_Ready	Coil 9	Q 11.0	BOOL	05-Write bit ; Device Id 1	MB In<-> Mobile Robot-Ready for Command
MB_MRobot_Cylinder1_Busy	Coil 10	Q 11.1	BOOL	05-Write bit ; Device Id 1	MB In<-> Mobile Robot Job Take 1st Cylinder-Busy
MB_MRobot_Cylinder2_Busy	Coil 11	Q 11.2	BOOL	05-Write bit ; Device Id 1	MB In<-> Mobile Robot Job Take 2nd Cylinder-Busy

S7 1200 ---> PC/Mobile Robot					
Signal Name	**S7 1200**	**PC/Mobile Robot**	**Type**	**ModBus Code**	**Description**
MB_MRobot_Cylinder1_TakeCmd	Q 10.0	Coil 1	BOOL	01-Read bit ; Device Id 1	MB Out <-> Start Mobile Robot (Take 1st Cylinder)
MB_MRobot_Cylinder2_TakeCmd	Q 10.1	Coil 2	BOOL	01-Read bit ; Device Id 1	MB Out<-> Start Mobile Robot (Take 2nd Cylinder)

Figure 15. Modbus message interface (Modbus Map).

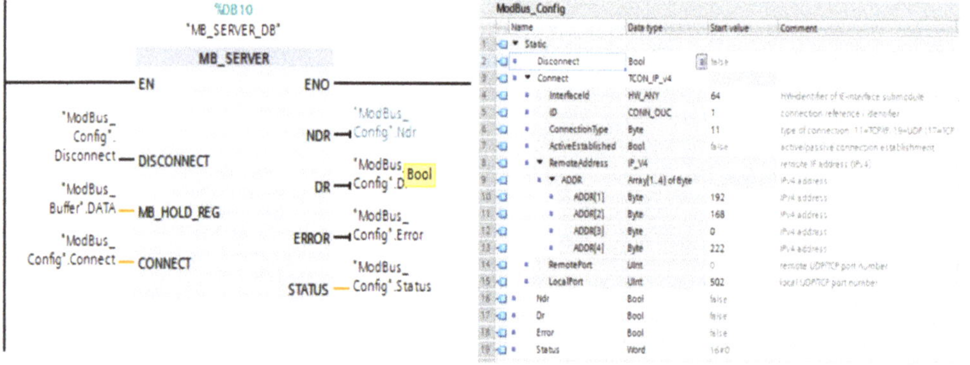

Figure 16. Siemens Modbus Server configuration in main routine.

Depending on the task carried out by the A/D/RML, repair process (one cylinder released), or disassembly process (two cylinders released), two separate command signals are used for the interface master (from PLC S7 1200 to CAS):

- start Job CAS: Recover Process Cylinder 1;
- start Job CAS: Recover Process Cylinder 2.

In the same way CAS, must acknowledge that the received command/action from A/D/RML line is handled; therefore three synchronization signals are used (from CAS to master PLC S7 1200):

- CAS Ready for Command;
- CAS Job started: Recover Process Cylinder 1;
- CAS Job started: Recover Process Cylinder 2.

Although Modbus TCP is a deterministic protocol, the handling process commands between CAS and A/D/RML line are considered a critical control application, therefore a handshake exchange signal is applied to this communication interface—a synchronization between subsystems. In the first step, when a job command-task is sent to CAS for processing and recover a workpart, CAS must acknowledge when an action is performed. Second, when CAS finished processing a task or is in the Standby State, the *Ready for Command* signal is sent back to the master PLC.

5. Real-Time Control of the Repair Function

The states and duration of the transitions on A/D/RML to the real-time management related to the repair function, from the takeover of WP from WH right to its repaired storage in WH left, are shown in Figure 17, and Video S1.

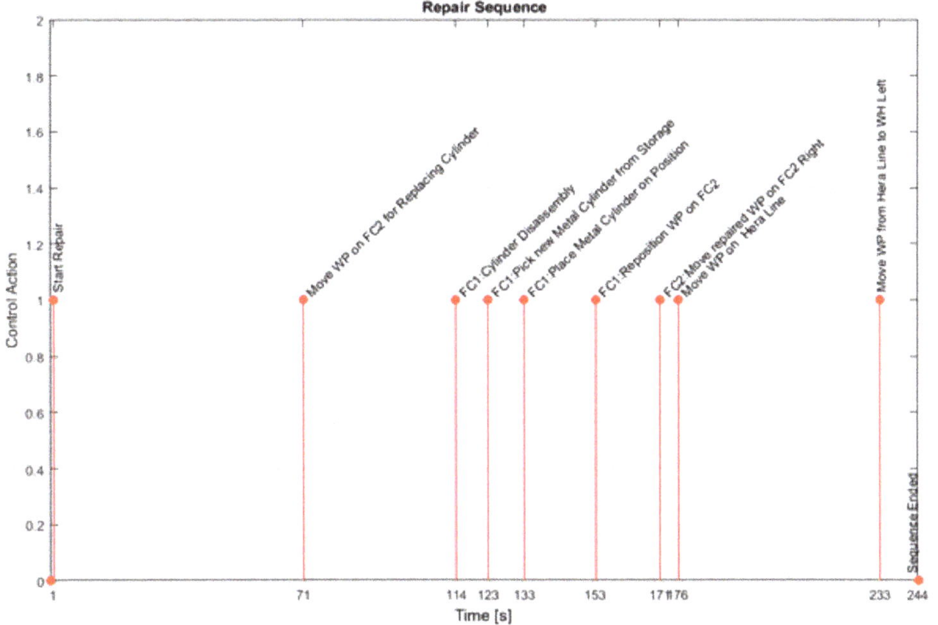

Figure 17. A/D/RML state transitions of repair function.

5.1. CAS Control Loops

The mobile part of the A/D/RML, referred to as CAS, picks the pieces from the FC in the case of a repair or disassembly function and transports them to their proper storage warehouses (Figures 18–20). The control of the mobile part is based on three control loops.

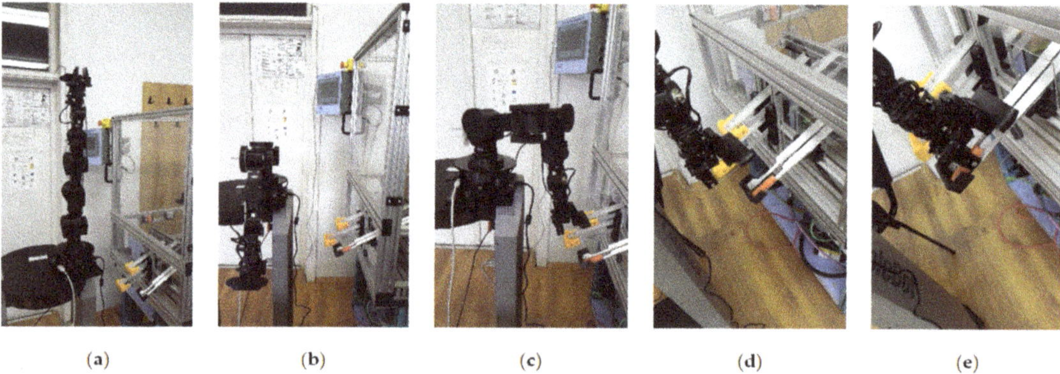

(a) (b) (c) (d) (e)

Figure 18. Real-time control of the 7-DOF Cyton 1500 RM located at the FC with the following sequences: (**a**) home position, (**b**) intermediary position, (**c,d**) scanning position, (**e**) picking object.

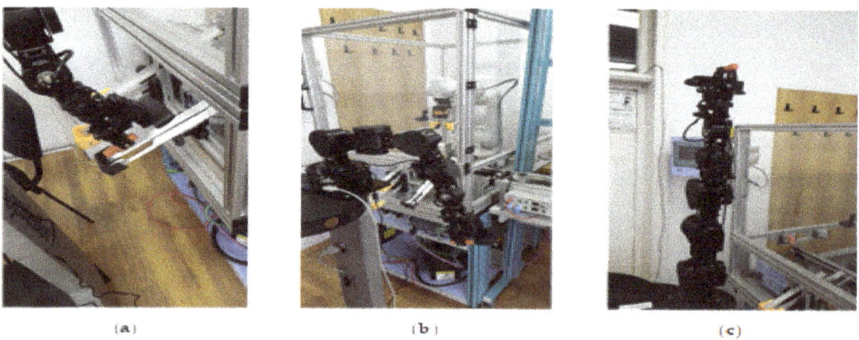

(a) (b) (c)

Figure 19. Real-time control of the RM Cyton 7-DOF located at the FC with the following sequences: (**a**) lifting object, (**b**) moving to the intermediate position, (**c**) parking position with the piece in the gripper.

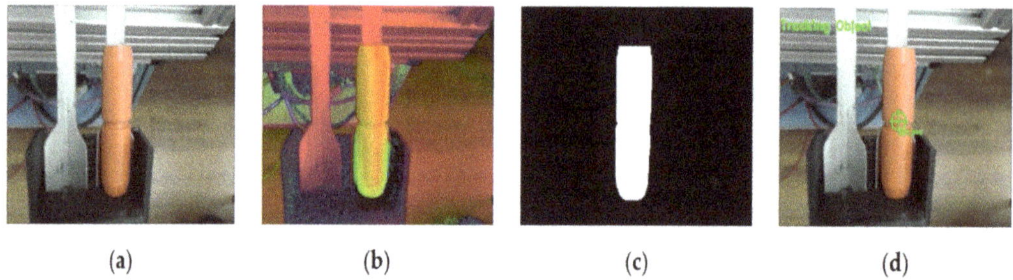

(a) (b) (c) (d)

Figure 20. Object Detection with the following steps: (**a**) raw RGB image taken from the camera, (**b**) conversion from BGR to HSV (**c**) image segmentation after HSV limits are set, with morphological operations of erosion and dilation, (**d**) object is detected and the centroid is being tracked.

1. Control loop for the synchronization between the Modbus PLC of the FC and the Cyton RM;
2. Control loop of the Cyton RM with the *eye-in-hand* VSS for accurate positioning to pick up the objects from the FC and place them in the warehouses;
3. Control loop of the PeopleBot WMR based on trajectory tracking sliding mode control (TTSMC) [19,31].

All three control loops communicate through one computer which contains the GUI and controls of the ARS, *eye-in-hand* VSS, and Cyton 1500 RM, and manages the synchronization with the FC.

Specific programming packages and libraries have been used with Microsoft Visual Studio to control the entire system. As it can be seen in the Figure 14, the communication between the Cyton RM, *eye-in-hand* VSS, and the computer is executed with USB connections, while the communication with FC is carried out wirelessly using a TCP/IP protocol [31].

The coordination between the control loops has been realized using the open-source library specialized in image processing, OpenCV, the control input defined in Equations (9) and (11), functions from Aria Mobile Robots, and synchronization with the FC's Modbus PLC, all combined in Microsoft Visual Studio with the C++ programming language.

Figure 18 illustrates a series of images captured of RM Cyton located at the FC which show: (a) the home position, when all the joints have 0 radians value, (b) the intermediate position between the home and the scanning position, (c) the scanning position, where the *eye-in-hand* VSS is used for accurate localization of the object, (d) the error between the actual features extracted from the VSS and the desired features has been minimized, and (e) after the recoverable piece has been picked up by the RM. Figure 19 illustrates the images captured: (a) after the piece has been lifted by the gripper, (b) when the RM moves to the intermediary position with the recoverable object grasped, and (c) the trajectory to the parking position and starting the sequence of TTSMC for ARS PeopleBot. The reason for introducing the intermediate position is because the space between the FC and the RM Cyton is tight, so it is better and safer for it to move first to the right then descend, rather than just move down right to the scanning position.

5.2. Object Detection, Image Processing and CAS Control for Repair Function

The main stages involved in object detection and tracking are shown in Figure 20. These stages are happening between steps (c) and (d) from Figure 18.

1. The raw RGB (Red, Green, Blue) images are taken from the camera with a resolution of width 640 and of height 480; the camera has a fixed focus and white balance, so that it does not interfere with the colors;
2. Conversion from BGR (Blue, Green, Red) is carried out in OpenCV to HSV (Hue, Saturation or Brightness);
3. After the conversion, HSV limits are imposed, so that only the objects of a specific color between those limits can be detected. Morphological operations are additionally carried out at this step so that below elements are eroded while above elements are fixed so they can be seen more easily;
4. If a group of pixels has an aria in between the minimum and maximum values set in the program, then it will be considered an object and contour detection will start so that object detection becomes more precise. Finally, the centroid will be tracked using image moments method, shown in the image with the $\oplus$ symbol. The RM Cyton will move based on the features tracked and once the error between the desired and real piece has been minimized enough, it will move above the piece and pick it up, then will turn to the intermediary position and finally the parking position.

If a group of pixels is below the minimum aria, then it will not be counted towards objects detected and if a group of pixels is larger than the maximum aria, the user is notified that the object detected is possibly eroded and will be considered noise.

Figure 21 shows some frames with the main steps performed by the RM Cyton at the warehouse where the object will be placed: (a) represents the RM with the object grasped in the gripper, (b) is the scanning position and positioning of the end-effector based on VSS, (c) shows that after the end-effector has been positioned, it moves above the warehouse, (d) is when the object is placed in the warehouse, (e) is when the RM returns to the home position. The steps illustrated in Figure 22 are happening between Figure 21b,c. Once the VSS sequence starts, (a) first it must take raw RGB images from the visual sensor, (b) convert them from BGR to HSV color space, so that in (c) segmentation of the desired color is executed more smoothly; finally, (d) after the operations are complete, the centroid will be detected and tracked with ⊕. The centroid will be used by the RM Cyton to position above the warehouse to place the object and then move back to the home position.

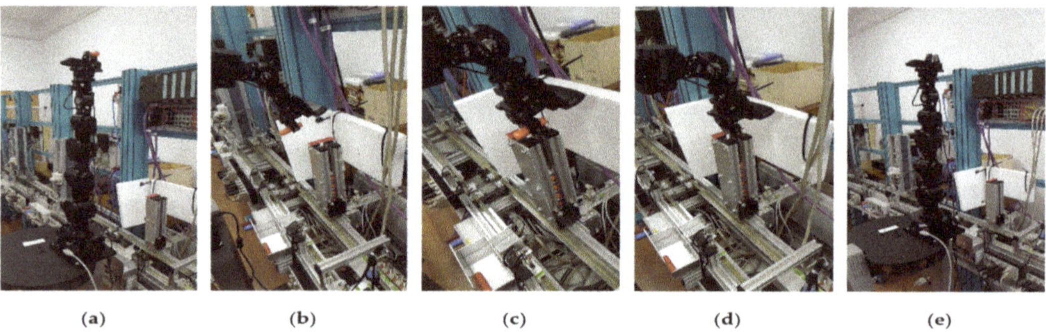

(a)   (b)   (c)   (d)   (e)

Figure 21. Real-time control of the RM Cyton located at the cylinders warehouse which has the resulting actions: (**a**) parking position with the object in the gripper, (**b**) scanning position, (**c**) moving above the warehouse, (**d**) placing the piece in the depot, (**e**) returning to home position.

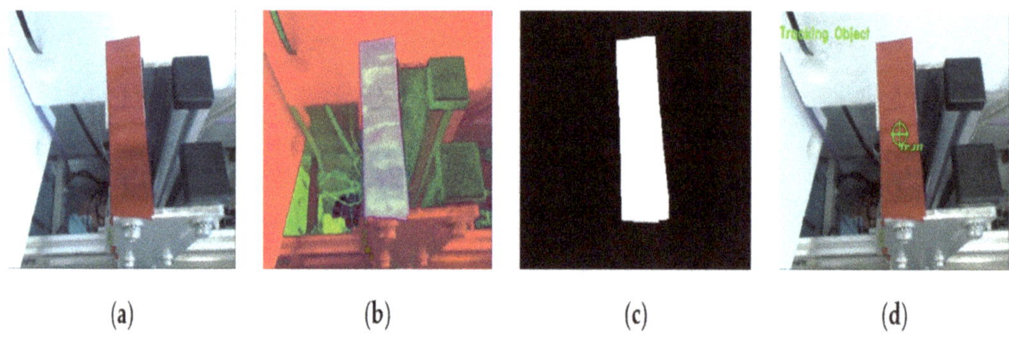

(a)   (b)   (c)   (d)

Figure 22. Warehouse reference color detection: (**a**) raw RGB image, (**b**) BGR to HSV conversion, (**c**) HSV limits are imposed and the result is shown as a white blob with a black background, (**d**) the reference color is detected, and the centroid is tracked.

Figure 23 illustrates the desired and real trajectories of the ARS PeopleBot obtained with the TTSMC in closed loop control to move from the FC to the warehouse and back to the FC in the desired time. In (a) the complete route is presented, in (b) the X axis is separated, in (c) the Y axis is separated, and in (d) the angular trajectory is so that the differences between the real and desired trajectories can be perceived easier. There are two observable deviations, one after a 90° rotation is carried out to move forward to the warehouse, as shown in Figure 23c,d between 40 and 56 seconds on the X axis, and the second one again after a 90° rotation as to move back to the FC, shown in Figure 23c,d between 78 and 90 seconds on the X axis. The tracking errors are shown in Figure 24.

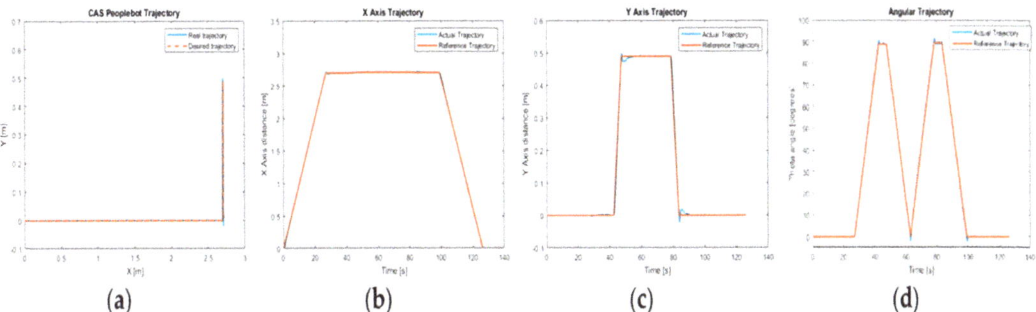

Figure 23. Desired and real trajectories of ARS PeopleBot based on Trajectory Tracking Sliding Mode Control: (**a**) complete trajectory, (**b**) X axis, (**c**) Y axis, and (**d**) angular trajectories.

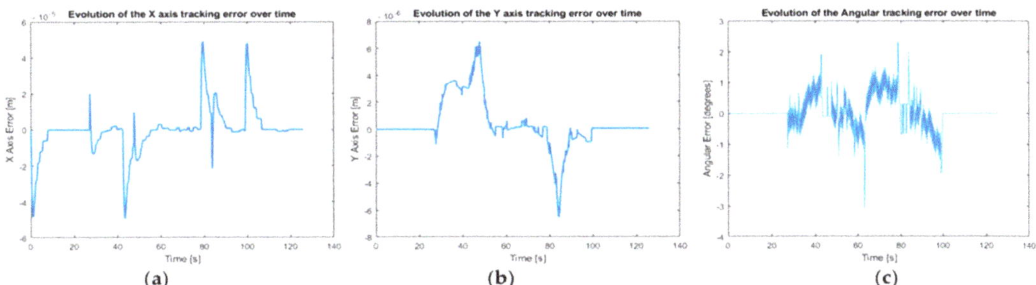

Figure 24. (**a**) X axis and (**b**) Y axis tracking errors in absolute coordinates, (**c**) angular tracking error expressed in radians per second for ARS PeopleBot.

Figure 25a depicts the movement of the RM Cyton using inverse kinematics control (IKC) from the initial position to intermediary position for safety reasons, then to the scanning position with VSS so that the error between the actual and desired visual features is minimized, and finally, above the object, making a short movement to pick the object. Figure 25b–d represents the complete trajectory separated in x, y, and z, respectively, in absolute coordinates. The VSS sequence starts at second 28, this being the reason why the errors depicted in Figure 26 appear and it takes about 5 s for the error to be minimized (a) from 11×10^{-2} on the X axis, (b) from -13×10^{-3} on the Y axis, and (c) 10 s to be minimized from $\pm 4 \times 10^{-2}$.

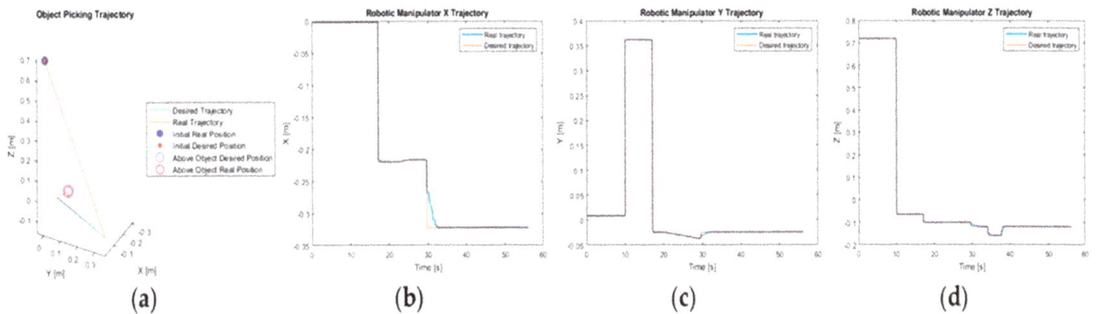

Figure 25. Desired and real object picking trajectories of the RM Cyton based on inverse kinematics control and *eye-in-hand* *VSS* for accurate positioning at the FC: (**a**) the complete trajectory, from the home position to above object position, (**b**) X axis, (**c**) Y axis, (**d**) Z axis trajectories.

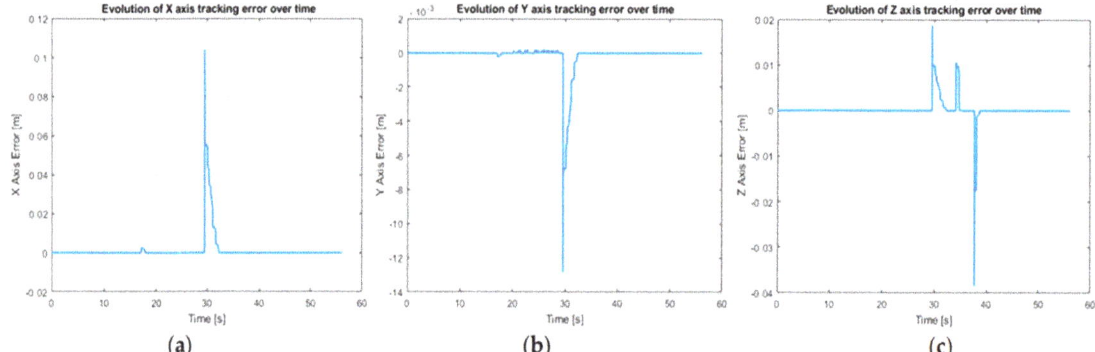

Figure 26. Tracking errors for RM Cyton picking object trajectory on: (**a**) X axis, (**b**), Y axis, (**c**) Z axis.

Figure 27a presents the trajectory of the Cyton RM based on IKC from the parking position to scanning position with the *eye-in-hand* VSS, so that the end effector is exactly above the warehouse when placing the object. In the following Figure 27b–d, the trajectories are separated on the X, Y, and Z axes so that it is easier to see the exact trajectories on the individual axis.

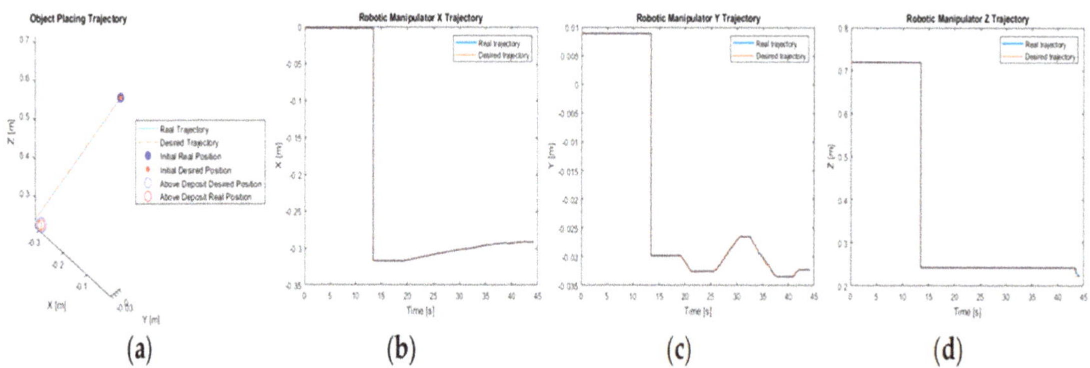

Figure 27. Desired and real object placing trajectories of the RM Cyton based on inverse kinematics control and *eye-in-hand* VSS for positioning at the warehouse: (**a**) signifies the complete trajectory, while (**b–d**) are the trajectories separated on the X, Y, and Z axes, respectively.

Since the path of placing the object is performed after the ARS PeopleBot has moved from the FC to the warehouse, external factors (such as small deviations of the WMR trajectories) can influence the time necessary for the VSS to position the end-effector of the RM exactly above the deposit, since the hole it must be put in is exactly as wide as the object itself.

Compared to the errors in the Figure 26a–c, those presented in Figure 28 are smaller, but it takes a longer time to be minimized on the X and Y axis—about 25 s and with a precision of $\pm 2 \times 10^{-4}$, as one can see in Figure 28a,b, compared to 5 s and a precision of 11×10^{-2} and -13×10^{-3} in Figure 26a,b, respectively. It is easy to see that it takes much less time to be positioned on the Z axis—2 s with a precision of 18×10^{-3}, as shown in Figure 28c, compared to 10 s and an accuracy of $\pm 4 \times 10^{-2}$ as one can see in Figure 26c.

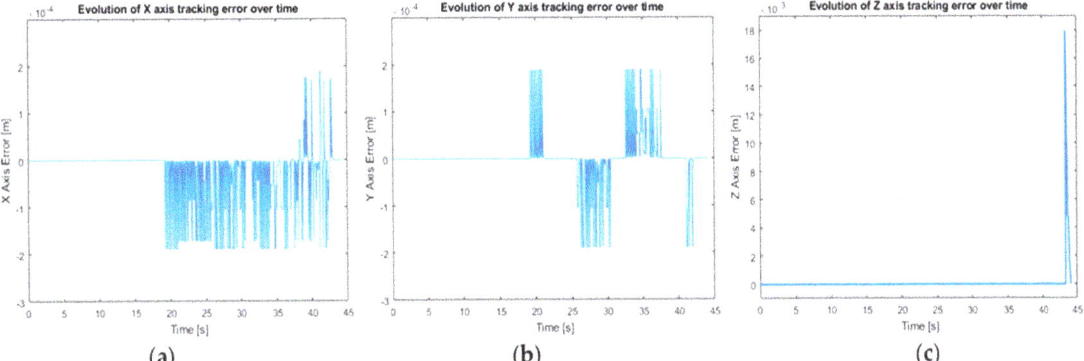

Figure 28. Tracking errors for RM Cyton placing object trajectory on: (**a**) X axis, (**b**) Y axis, and (**c**) Z axis.

6. Discussion

Unlike [10], where we approached the problem of assembly/disassembly on the Hera&Horstmann mechatronics line served by an autonomous system consisting only of the mobile platform, on which is mounted a 5-DOF robotic manipulator, this paper proposes an extension both in hardware as well as software that allows the elaboration of a flexible and multifunctional technology, fully compatible with Industry 4.0, able to manufacture different products and to recover components or to repair products that do not correspond to the desired quality. All of these functionalities are made with high precision due to the integration of an industrial robotic manipulator, a complex autonomous system that is equipped with a mobile visual servoing system, and a communication structure between the flexible cell and the mechatronics line that allows synchronizations of operations and distributed control. The control structure is hierarchical with a supervisor that monitors the process, execution, and synchronization of tasks according to the strategy. The implementation of robust control architectures to uncertainties considered for all systems are the complex autonomous system, flexible cell, and mechatronics line. The uncertainties considered are faulty sensors/actuators, route/storage space blockage, and payload variation. Finally, the research is focused on developing a cyber-physical system, i.e., a multifunctional and flexible manufacturing system that integrates computational, networking, and physical components within a single functional environment.

7. Conclusions

The presented research is in progress, the final objective being the fully automated control, without the intervention of a human operator, of the multifunctional technology of flexible manufacturing for a given production volume, with recovery, reuse of sub-assemblies, and repair of inadequate quality WPs. The research is aimed at dual purposes, one educational and another as close as possible to the real world. The educational goal aims to familiarize the system designer with everything that defines Industry 4.0 and the cyber-physical system. The educational goal is achieved by addressing the following topics: SCADA systems, communication and synchronization of tasks based on signals from sensors and actuators, PLC programming, precise positioning by visual servoing systems, control of mechatronics lines, mobile robots, and robotic manipulators. All of these bring the technology closer to the concepts and attributes of Industry 4.0. Regarding correspondence with the real industrial world, most manufacturing lines are assisted by robotic systems that have a fixed position. Through this study, we extended the degree of automation and efficiency of these production lines using mobile robotic systems equipped with manipulators and visual servoing systems. Thus, the manufacturing lines become multifunctional, able to recover and reuse components and subassemblies, if the final product does not meet quality requirements. Although this technology is implemented on a laboratory system, it can still be applied in sorting, dosing, sealing, and packaging operations in the industries of pharmaceutics, food, and consumer goods. We mention the following as being exclusively the contributions of the authors: hardware design, configuration, ABB IRM and PLC programming, and graphical interface of the FC; hardware configuration and PLC programming of the Hera&Horstmann ML; coupling and compatibility between CF, Hera&Horstmann ML and CAS; CAS hardware configuration and control; formulating the set of assumptions so that the entire system corresponds to the requirements of the multifunctional technology of flexible manufacturing; and real-time control of the entire system.

Supplementary Materials: The following are available online at www.cidsacteh.ugal.ro/video/Video_PPB_Cyton.mp4, Video S1: Multifunctional Flexible Manufacturing Technology on a Mechatronics Line with Integrated Industrial Robotic Manipulator Assisted by a Complex Autonomous System, repair function (cylinder replacement).

Author Contributions: Conceptualization, A.F. (Adriana Filipescu), D.I., A.F. (Adrian Filipescu), E.M. and G.S.; methodology, A.F. (Adriana Filipescu), A.F. (Adrian Filipescu), E.M. and D.I.; software, D.I., G.S.; validation, A.F. (Adrian Filipescu) and E.M.; formal analysis, D.I. and G.S.; writing—original draft preparation, A.F. (Adriana Filipescu) and A.F. (Adrian Filipescu); writing—review and editing, A.F. (Adriana Filipescu) and A.F. (Adrian Filipescu); supervision, A.F. (Adrian Filipescu); project administration, A.F. (Adrian Filipescu); funding acquisition, A.F. (Adrian Filipescu) and D.I. All authors have read and agreed to the published version of the manuscript.

Funding: This research was funded by the Romanian Executive Unit of Funding Higher Education, Research, Development, and Innovation (UEFISCDI), project number: PN-III-P1-1.2-PCCDI-2017-0290, project title: *Intelligent and distributed control of 3 complex autonomous systems integrated into emerging technologies for medical-social personal assistance and servicing of precision flexible manufacturing lines.*

Institutional Review Board Statement: Not applicable.

Informed Consent Statement: Not applicable.

Acknowledgments: This article (APC) will be supported by Doctoral School of Fundamental Sciences and Engineering, "Dunărea de Jos" University of Galati.

Conflicts of Interest: The authors declare no conflict of interest.

References

1. Chryssolouris, G. *Manufacturing Systems—Theory and Practice*; Springer: New York, NY, USA, 2005.
2. Filipescu, A. Contributions to Electric Drive of the Flexible Manufacturing Lines and Integrated Robots. Ph.D. Thesis, University of Galati, Galati, Romania, 2017.
3. Carlos-Mancilla, M.A.; Luque-Vega, L.F.; Guerrero-Osuna, H.A.; Ornelas-Vargas, G.; Aguilar-Molina, Y.; González-Jiménez, L.E. Educational Mechatronics and Internet of Things: A Case Study on Dynamic Systems Using MEIoT Weather Station. *Sensors* **2021**, *21*, 181. [CrossRef] [PubMed]
4. Florescu, A.; Barabas, S.A. Modeling and Simulation of a Flexible Manufacturing System—A Basic Component of Industry 4.0. *Appl. Sci.* **2020**, *10*, 8300. [CrossRef]
5. Berriche, A.; Mhenni, F.; Mlika, A.; Choley, J.-Y. Towards Model Synchronization for Consistency Management of Mechatronic Systems. *Appl. Sci.* **2020**, *10*, 3577. [CrossRef]
6. Radaschin, A.; Voda, A.; Minca, E.; Filipescu, A. Task Planning Algorithm in Hybrid Assembly/Disassembly Process. In Proceedings of the 14th IFAC Symposium on Information Control Problems in Manufacturing, Bucharest, Romania, 23–25 May 2012.
7. Kallrath, J. Planning and scheduling in the process industry. In *Advance Planning and Scheduling Solution in Process Industry*; Springer: Berlin/Heidelberg, Germany, 2003; pp. 201–227.
8. Tolio, T. *Design of Flexible Production Systems—Methodologies and Tools*; Springer: Berlin/Heidelberg, Germany, 2009.
9. Peters, A.A.; Vargas, F.J.; Garrido, C.; Andrade, C.; Villenas, F. PL-TOON: A Low-Cost Experimental Platform for Teaching and Research on Decentralized Cooperative Control. *Sensors* **2021**, *21*, 2072. [CrossRef] [PubMed]
10. Minca, E.; Filipescu, A.; Voda, A. Modelling and control of an assembly/disassembly mechatronics line served by mobile robot with manipulator. *Control Eng. Pract.* **2014**, *31*, 50–62. [CrossRef]
11. Filipescu, A.; Filipescu, A., Jr. Simulated Hybrid Model of an Autonomous Robotic System Integrated into Assembly/Disassembly Mechatronics Line. *IFAC Proc. Vol.* **2014**, *47*, 9223–9228. [CrossRef]

12. Dragomir, F.; Mincă, E.; Dragomir, O.E.; Filipescu, A. Modelling and Control of Mechatronics Lines Served by Complex Autonomous Systems. *Sensors* **2019**, *19*, 3266. [CrossRef] [PubMed]
13. Filipescu, A.; Mincă, E.; Filipescu, A.; Coandă, H.-G. Manufacturing Technology on a Mechatronics Line Assisted by Autonomous Robotic Systems, Robotic Manipulators and Visual Servoing Systems. *Actuators* **2020**, *9*, 127. [CrossRef]
14. Stoll, J.T.; Schanz, K.; Pott, A. Mechatronic Control System for a Compliant and Precise Pneumatic Rotary Drive Unit. *Actuators* **2020**, *9*, 1. [CrossRef]
15. He, Y.; Stecke, K.E.; Smith, M.L. Robot and machine scheduling with state-dependent part input sequencing in flexible manufacturing systems. *Int. J. Prod. Res.* **2016**, *54*, 6736–6746. [CrossRef]
16. Guiras, Z.; Turki, S.; Rezg, N.; Dolgui, A. Optimization of Two-Level Disassembly/Remanufacturing/Assembly System with an Integrated Maintenance Strategy. *Appl. Sci.* **2018**, *8*, 666. [CrossRef]
17. Filipescu, A.; Minca, E. Mechatronics Manufacturing Line with Integrated Autonomous Robots and Visual Servoing Systems. In Proceedings of the 9th IEEE International Conference on Cybernetics and Intelligent Systems, and Robotics, Automation and Mechatronics (CIS-RAM 2019), Bangkok, Thailand, 18–20 November 2019; pp. 620–625.
18. Minca, E.; Filipescu, A.; Coanda, H.G.; Dragomir, F.; Dragomir, O.E.; Filipescu, A. Extended Approach for Modeling and Simulation of Mechatronics Lines Served by Collaborative Mobile Robots. In Proceedings of the 22nd International Conference on System Theory, Control and Computing (ICSTCC), Sinaia, Romania, 10–12 October 2018; pp. 335–341.
19. Ciubucciu, G.; Filipescu, A.; Filipescu, A., Jr.; Filipescu, S.; Dumitrascu, B. Control and Obstacle Avoidance of a WMR Based on Sliding-Mode, Ultrasounds and Laser. In Proceedings of the 12th IEEE International Conference on Control and Automation (ICCA), Kathmandu, Nepal; 1–3 June 2016; pp. 779–784.
20. Gasparetto, A.; Zanotto, V. A new method for smooth trajectory planning of robot manipulators. *Mech. Mach. Theory* **2007**, *42*, 455–471. [CrossRef]
21. Barczak, A.; Dembińska, I.; Marzantowicz, Ł. Analysis of the Risk Impact of Implementing Digital Innovations for Logistics Management. *Processes* **2019**, *7*, 815. [CrossRef]
22. Fan, Y.; Lv, X.; Lin, J.; Ma, J.; Zhang, G.; Zhang, L. Autonomous Operation Method of Multi-DOF Robotic Arm Based on Binocular Vision. *Appl. Sci.* **2019**, *9*, 5294. [CrossRef]
23. Ravankar, A.; Ravankar, A.A.; Kobayashi, Y.; Hoshino, Y.; Peng, C.-C. Path Smoothing Techniques in Robot Navigation: State-of-the-Art, Current and Future Challenges. *Sensors* **2018**, *18*, 3170. [CrossRef] [PubMed]
24. Copot, C. Control Techniques for Visual Servoing Systems. Ph.D. Thesis, Technical University of Iasi, Iasi, Romania, 2012.
25. Petrea, G.; Filipescu, A.; Solea, R.; Filipescu, A., Jr. Visual Servoing Systems Based Control of Complex Autonomous Systems Serving a P/RML. In Proceedings of the 22nd IEEE, International Conference on System Theory, Control and Computing, (ICSTCC). Sinaia, Romania, 10–12 October 2018; pp. 323–328.
26. Song, R.; Li, F.; Fu, T.; Zhao, J. A Robotic Automatic Assembly System Based on Vision. *Appl. Sci.* **2020**, *10*, 1157. [CrossRef]
27. Lan, C.-W.; Chang, C.-Y. Development of a Low Cost and Path-free Autonomous Patrol System Based on Stereo Vision System and Checking Flags. *Appl. Sci.* **2020**, *10*, 974. [CrossRef]
28. Deng, L.; Wilson, W.; Janabi-Sharifi, F. Dynamic performance of the position-based visual servoing method in the cartesian and image spaces. In Proceedings of the IEEE/RSJ International Conference on Intelligent Robots and Systems, Las Vegas, NV, USA, 27–31 October 2003; pp. 510–515.
29. Gans, N.; Hutchinson, S.; Corke, P. Performance tests for visual servo control systems, with application to partitioned approaches to visual servo control. *Int. J. Robot. Res.* **2003**, *22*, 955–981. [CrossRef]
30. Corke, P.I.; Spindler, F.; Chaumette, F. Combining Cartesian and polar coordinates in IBVS. In Proceedings of the 2009 IEEE/RSJ International Conference on Intelligent Robots and Systems, St. Louis, MO, USA, 11 December 2009; pp. 5962–5967.
31. Tatipala, S.; Wall, J.; Johansson, C.; Larsson, T. A Hybrid Data-Based and Model-Based Approach to Process Monitoring and Control in Sheet Metal Forming. *Processes* **2020**, *8*, 89. [CrossRef]

 *processes*

MDPI

Article

General Approach for Inline Electrode Wear Monitoring at Resistance Spot Welding

Christian Mathiszik *, David Köberlin, Stefan Heilmann , Jörg Zschetzsche and Uwe Füssel

Institute of Manufacturing Science and Engineering, Chair of Joining Technology and Assembly, Technische Universität Dresden, 01069 Dresden, Germany; david.koeberlin@tu-dresden.de (D.K.); stefan.heilmann@tu-dresden.de (S.H.); joerg.zschetzsche@tu-dresden.de (J.Z.); uwe.fuessel@tu-dresden.de (U.F.)
* Correspondence: christian.mathiszik@tu-dresden.de; Tel.: +49-351-463-34346

Abstract: Electrodes for resistance spot welding inevitably wear out. In order to extend their service life, the tip-dressing process restores their original geometry. So far, however, the point in time for tip-dressing is mainly based on experience and not on process data. Therefore, this study aims to evaluate the in-situ or inline wear during the welding process without using additional sensors, and to base the timing for tip-dressing on continuous process monitoring, extending electrode life even further. Under laboratory conditions, electrode wear is analyzed by topographical measurements deepening the knowledge of the known main wear modes of resistance-spot-welding electrodes, mushrooming and plateau forming, and characterizing an electrode length delta over the number of spot welds. In general, electrode wear results in deformation of the electrode contact area, which influences process parameters and thereby weld quality. The conducted tests show correlation between this deformed contact area and the electrode length delta. The study shows that this electrode length delta is visible in actual process data, and can therefore be used as a criterion to characterize the wear of electrodes. Furthermore, this study gives reason to question commonly used spot-welding quality criteria and suggests different approaches, such as basing spot-welding quality on the possibility of nondestructive testing.

Keywords: steel alloys; resistance spot welding; RSW; electrode wear; electrode tip-dressing; process monitoring; mushrooming; plateau forming; quality control

Citation: Mathiszik, C.; Köberlin, D.; Heilmann, S.; Zschetzsche, J.; Füssel, U. General Approach for Inline Electrode Wear Monitoring at Resistance Spot Welding. *Processes* **2021**, *9*, 685. https://doi.org/10.3390/pr9040685

Academic Editor: Sergey Y. Yurish

Received: 25 March 2021
Accepted: 9 April 2021
Published: 13 April 2021

Publisher's Note: MDPI stays neutral with regard to jurisdictional claims in published maps and institutional affiliations.

1. Introduction

Resistance spot welding (RSW) is a widely used welding process characterized by short processing times of less than 1 s and a very high degree of automation. In addition, no filler metal material is necessary. These are only a few of the essential characteristics of why RSW is one of the most important joining processes in the thin-sheet-metal processing industry. The field of application ranges from manual spot welding in metalworking shops to highly automated areas such as automotive body-in-white manufacturing. These include daily kitchen utensils, such as kitchen sieves, to white goods, such as washing machines, to complex and safety-relevant applications, such as motor vehicles, where resistance spot welding has been successfully used for over 100 years [1,2]. Since the automotive industry has the highest quality standards of spot welds combined with their high number, roughly between 3000 spot welds for a small passenger car [3,4] up to 9500 for a transporter [5], further explanations mainly deal with this challenging application.

To assess electrode wear, it is necessary to understand the high complexity of different electrode-wear mechanisms; therefore, it is important to explain the fundamentals and main principles of the RSW process. A standard RSW process welds two or three steel alloy sheet metals together. The heating of the material can be described by Joule's law represented by Equation (1), where heat Q is generated by welding current I_w and total resistance R_{tot} over welding time t_w.

$$Q = \int_{t_0}^{t_w} I_w^2(t) R_{tot}(t)\,dt \tag{1}$$

Figure 1 shows the schematic process flow of RSW. An external electrode force F_{el} is applied to the work pieces via two water-cooled and opposing copper electrodes. In general, the welding process can be divided into three phases. These phases are referred to as squeeze time t_s, weld time t_w, and hold time t_h. During squeeze time, the force build-up takes place up to the preset electrode force F_{el}, followed by weld time, where welding current I_w flows through the work pieces from one electrode to another, causing them to heat up as a result of resistance heating according to Equation (1). Hold time begins after the welding current is switched off. During this time, the electrode force is still applied while the molten material cools down. The duration of hold time should at least be set until the weld completely solidifies. The solidified structure is called a nugget, and its diameter d_n is one of the most important quality criteria.

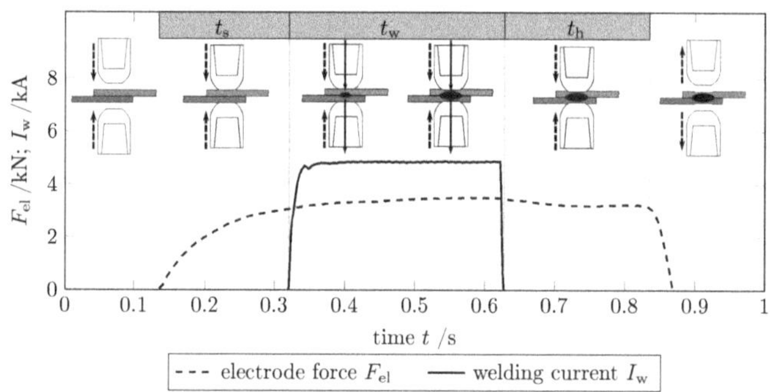

Figure 1. Course of applied electrode force F_{el} and welding current I_w over time t of a common resistance-spot-welding (RSW) process with three main stages: squeeze time t_s, weld time t_w, and hold time t_h (schematic).

Equation (1) shows that welding current I_w has a major effect on heat development. Therefore, the welding process and its result can be significantly influenced by the choice of amperage. Total resistance R_{tot} is the sum of contact resistances R_{1-3} and individual material resistances R_{4-7}, as shown in Figure 2a.

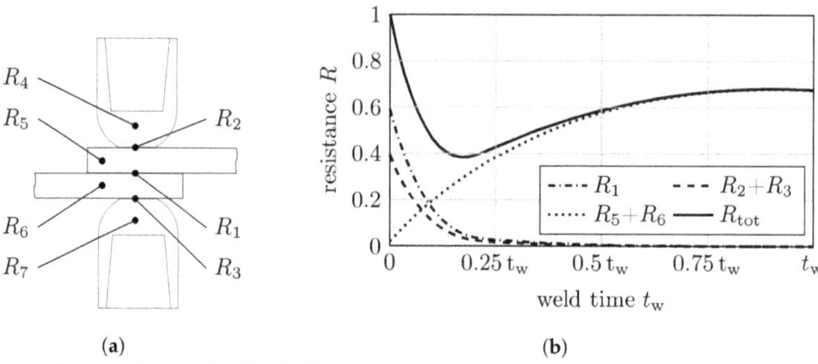

(a)　　　　　　　　　　　(b)

Figure 2. (**a**) Contact resistances R_{1-3} and individual material resistances R_{4-7} at RSW; (**b**) dynamic behavior of resistances during typical spot-welding process with resulting total resistance R_{tot} (schematic, standardized).

Figure 2b shows the dynamic behavior of individual resistances R_{1-7}, and the resulting R_{tot} resistance R_1 between sheet metals must be significantly greater than other contact resistances $R_{2,3}$, so that a spot weld is created between the sheets. However, this is only achieved at the beginning of the welding process, since surface roughness greatly reduces the actual contact area. As a result of the continuous heating of the material, the roughness and contact resistances decrease. At the same time, material resistances $R_{5,6}$ increase as they are dependent on work-piece temperatures, and thus significantly contribute to the formation of the welded joint. To reduce contact resistances, electrode force must be increased. This is especially important at the contact areas between electrodes and work pieces in order to avoid increased electrode wear [6]. Electrode force also prevents the melt from running out of the joint plane and locally limits the welding current [6].

The temperature at the electrode–sheet interface depends on contact resistances R_2 and R_3. Those depend on the coating system of the sheets and their surface condition (contaminated with dust, oil, etc.), the applied electrode force, and the wear condition of the electrodes. The acting mechanical loads caused by the applied electrode force can be as high as $300\,\mathrm{N\,mm^{-2}}$, and temperatures of $500\,^{\circ}\mathrm{C}$ and above [7,8] are reached at the interfaces. For this reason, electrode tips are also actively cooled from inside with water. In most applications, the electrode material used at RSW for the different steel grades is copper alloy CuCr1Zr [9]. At these temperatures, loss in Young's modulus E is around 20% [10], and in compressive strength of about 30% [11]. Thus, electrodes are subject to major wear due to high thermal and mechanical loads. For CuCr1Zr, many investigations describing the wear mechanisms exist. These include diffusion processes that cause both an increase in the alloy layer and a brazing of electrodes to the sheet metal, which may lead to a partial break-out of this layer from the electrodes. These factors define the resulting wear mechanism. So far, wear mechanisms for the mentioned application can be divided into two major wear modes. Mode 1 is known as mushrooming, whereas Mode 2 can be described as trimming [12] or plateau formation. Both modes are shown in Figure 3.

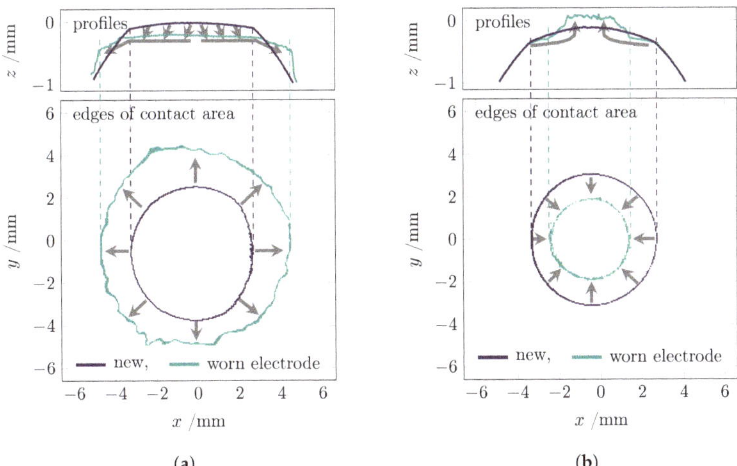

Figure 3. Typical modes of electrode wear and their mechanisms indicated by arrows at RSW of coated steel alloys: (**a**) mushrooming due to radial material flow and (**b**) plateau formation due to axial material flow.

The wear mechanism by mushrooming is well-known [11,13–20], as it has been around since RSW has been used. Briefly explained, due to thermomechanical stress on the electrode tips, radial material flow can be observed, as shown in Figure 3a. This acts together with a loss of material from the electrode tip surface to cause a decrease in electrodes length [11,19–21]. Material loss is repetitive by effects of the local melting, peeling, or breaking out of the brittle alloy layer, called pitting [15,22,23]. Analytical models describing and predicting mushrooming are presented in [21,24,25]. For trimming or plateau forming as the electrode wear mechanism, much less research can be found. One of the first publications on this was by Chang et al. [12], describing trimming. Here, electrode length decreases due to the increasing number of spot welds. The mechanism of plateau formation was deeply investigated in [11,19,20], where an increase in electrode length was shown. This wear mode occurs especially on applications with advanced high-strength steels (AHSS) such as hot-formed 22MnB5 with an aluminum–silicon coating (AlSi). Klages [19] proved that the plateau is not created by the formation of an alloy layer, but by a consecutive deformation process of the electrode during welding. The wear mechanism is shown in Figure 3b. A higher temperature development in the center of the electrode contact surface locally decreases the strength of the electrode material. The surrounding material retains its strength. By displacing the material softened by the heating in the region of the nugget, the electrode material flows towards the direction of the nugget, and the plateau is formed.

Regardless of wear mode, thermal and mechanical stresses lead to diffusion processes and deformations of the electrode contact surface. The result is increased and even accelerated electrode wear and a reduction in process stability. Process instabilities fluctuate the nugget diameter and lead to insufficient weld quality. Since the nugget diameter is one of the most important quality criteria, process capability and monitoring must ensure a high-quality spot weld at any time. To maintain a stable process, electrodes are cyclically dressed. During dressing, the diffusion layer at the contact area of the electrodes is removed, and the original physical properties of the contact area are restored. Timing and volume to be removed are based on experience. This experience can be gained through experiments to determine the electrode life. According to ISO 8166 [26] or SEP-1220-2 [27], the life is reached when

- ISO 8166 [26]: 3 out of 5: $d_\mathrm{w} < 3.5\sqrt{t}$;
- SEP-1220-2 [27]: 3 out of 7: $d_\mathrm{w} < 4\sqrt{t}$

of a test sheet, where d_w is the weld diameter after destructive testing (DT), and t is the thickness of the thinner sheet metal. Since experiments are carried out under laboratory-like conditions, the timing of the tip dressing at production is chosen long before the life-cycle limit of the electrodes is reached. This usually results in an excessive amount of the material being removed. To address the right timing for tip dressing, continuous process monitoring is necessary. This can be performed during or after the welding process. Monitoring or quality assessment after welding always brings a delay and additional process steps. For quality assessment by DT and nondestructive testing (NDT), using manual ultrasonic testing, was established [28]. Both variants are labor-intensive and expensive. Hence, the in-situ or inline process monitoring of electrode wear is preferred. On the basis of derived results from extensive studies on electrode wear, this paper presents a methodology to assess the wear mechanism by different measuring concepts. Investigations include data analysis of the RSW process and three-dimensional topographical measurements of the electrodes. To present the high industrial potential of the elaborated results, possible solutions using only already industrial integrated measurements are shown.

2. Materials and Methods

2.1. Material Combinations

The used material combinations (MC) are shown in Table 1. MC01 and MC02 were selected in such a way that both electrode-wear mechanisms were represented as well as possible. MC01 consisted of two 2 mm thick galvanized HX340LAD steels. It is a microalloyed steel with high yield strength that is mainly used for cold-formed parts of a car body [29]. MC02 consisted of two hot-stamped steels, 22MnB5 + AlSi. Due to the combination of deformation and hardening abilities, those steels are used for load-bearing bodies and safety-relevant parts in the automotive industry [30].

Table 1. Material combinations (MC) for experimental investigations.

Description	MC01	MC02
Upper sheet, Anode (t_u)	HX340LAD + Z100 (2.0 mm)	22MnB5 + AS150 (1.0 mm)
Lower sheet, Cathode (t_l)	HX340LAD + Z100 (2.0 mm)	22MnB5 + AS150 (1.0 mm)
Coating	zinc, $100\,\mathrm{g\,m^{-2}}$	AlSi $150\,\mathrm{g\,m^{-2}}$

2.2. Experiment Setup

The experiment setup was a configuration of three individual modules interacting with each other. The first module was the welding gun installed on a six axis industrial robot (Figure 4), the second module was the tip dresser, and the third was a chromatic confocal microscope. The robot could move the welding gun to the tip dresser and directly into the measuring range of the microscope. The latter made it possible to measure the 3-dimensional (3D, x–y plane, height z) topographies of both electrodes while being installed on the welding gun. For this purpose, the microscope was equipped with a special rotating measuring head. This unique experiment setup allowed for the continuous wear assessment of the electrodes. Figure 4 shows the robot in all three positions for welding, tip-dressing, and topographical measurements.

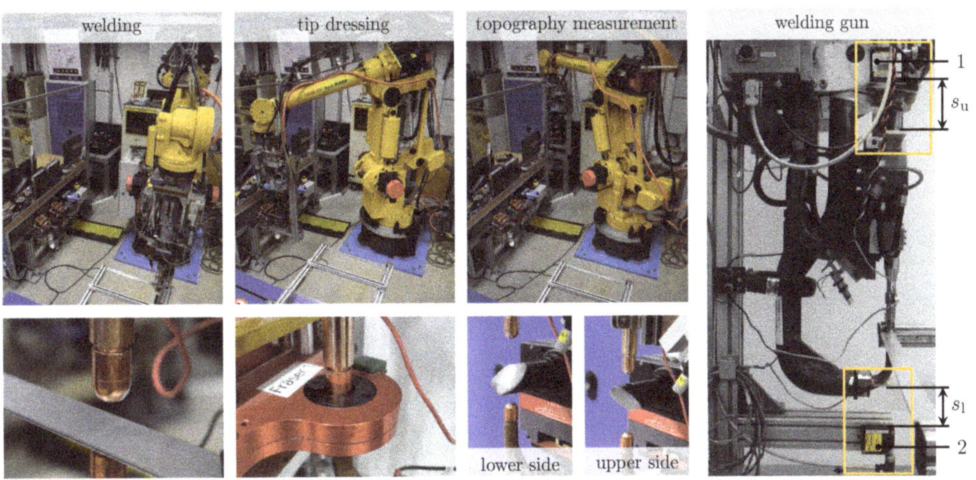

Figure 4. Experiment area for RSW for steel alloys at Technische Universität Dresden with welding gun positioned at all three modules for investigations in welding , tip-dressing, and topographical-measurement positions of upper and lower electrodes, and detailed view of welding gun with highlighted laser-triangulation sensors (1, 2) for measuring the displacement of upper electrode s_u and lower electrode s_l.

2.3. Test Procedure

The test procedure and sequence are shown in Figure 5. Before each test set (TS), electrodes were initially dressed, followed by 3D topographical measurements. Afterwards, test sheets and wear sheets were alternately welded. Test sheets were used to assess wear by determining weld diameters d_w from the second to the penultimate spot weld by means of a destructive chisel test according to ISO 10447 [31]. After each test sheet, topographical measurements of both electrodes and both surfaces of the last spot weld of a test sheet were carried out. Wear sheets served to generate wear on the electrodes, and no weld diameters were determined. This procedure was repeated until number of spot welds P_i equaled number of predefined spot welds P_end. P_end is based on results of previous lifetime investigations carried out in [11]. Geometric dimensions and the number of spot welds P_i of the sheets can be seen in Table 2. Process data were recorded for both types of sheet metal.

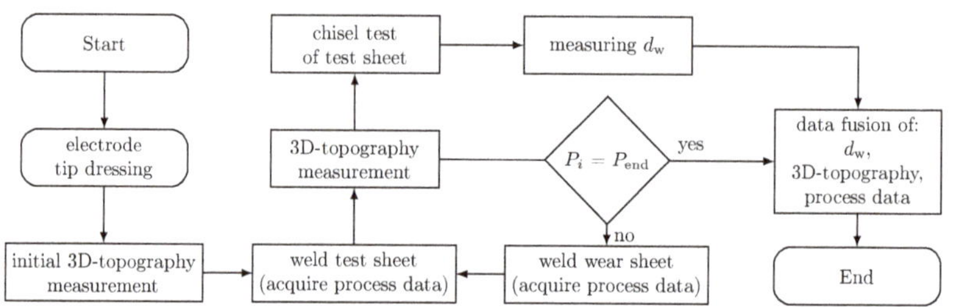

Figure 5. Test sequence of step tests with number of welded spot welds P_i, number of spot weld to be welded P_{end} for appropriate MC, and weld diameter d_w.

2.3.1. Welding Preferences

The welding equipment consisted of a servoelectric *C*-type welding gun with stiffness of $k = 2.345\,\text{kN}\,\text{mm}^{-1}$. The welding current was provided and controlled by a medium-frequency inverter of $f = 1000\,\text{Hz}$ using constant current control. The different number of spot welds of the TS between MC01 and MC02 resulted from the different geometrical dimensions and properties of the sheet metals. The target spot-weld diameter at the start of each test was chosen to be $d_w = 5.2\sqrt{t}$, where t is the thickness of the thinner sheet metal of the corresponding faying surface, determined by weldability lobes according to SEP 1220-2:2011 [27]. The chosen welding parameters are listed in Table 2. The process data for each individual weld were measured with a frequency of 100 kHz. The measured process data and starting conditions were kept constant for each MC:

- F_{el}, t_s, t_w, t_h;
- for MC02 prepulse time t_{wpp} and pause time t_p;
- cooling-water flow rate $\dot{V}$ and temperature range $T_{min} - T_{max}$;
- welding currents I_w (main pulse) and for MC02 prepulse welding current I_{wpp}.

The following process data were acquired in a time-resolved manner:

- welding current I_w;
- welding voltage U_w;
- electrode force F_{el};
- upper-electrode displacement s_u;
- lower-electrode displacement s_l;
- cooling-water flow rate $\dot{V}$ (general monitoring);
- cooling-water temperature T_c (general monitoring).

Table 2. Test setup for welding experiments.

Description	MC01	MC02
Spot welds per TS (P_{end})	TS01, TS02: 1200	TS01: 1092, TS02: 822
Electrode material	CuCr1Zr	
Electrode geometry	DIN EN ISO 5821 F1-16-20-40-6	
Geometry of test sheets in mm (P_i)	30 × 500 (8)	30 × 400 (12)
Geometry of wear sheets in mm (P_i)	500 × 360 (192)	400 × 180 (78)
Spot distance, edge distance in mm	30, 15	30, 15
F_{el} in kN	3.5	3.0
t_s, t_{wpp}, t_p, t_w, t_h in ms	400, 0, 0, 600, 400	400, 100, 30, 300, 400
I_{wpp}, I_w in kA	0, 9.2	4.5, 6.8
$\dot{V}$ in l min^{-1}, $T_{min} - T_{max}$ in °C	6.0, 20–25	4.0, 20–25

2.3.2. Tip Dressing

For the initial electrode geometry preparation, a tip dresser with a rated speed of $390\,\text{min}^{-1}$ was used. The tip dresser realized an initial electrode geometry according to DIN EN ISO 5821 F1-16-20-40-6 for all experiments, as shown in Figure 6a.

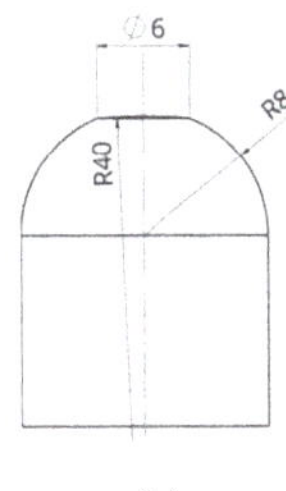

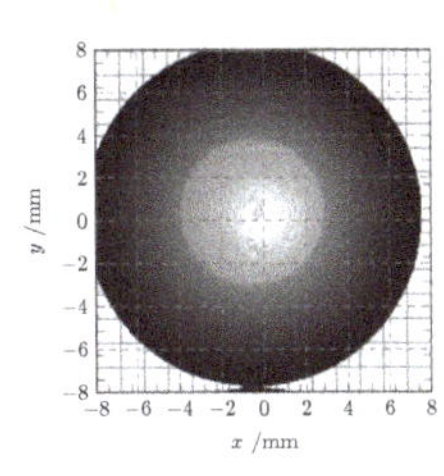

description	value
distance x	16 mm
distance y	16 mm
resolution x	0.05 mm
resolution y	0.05 mm
measuring frequency	1000 Hz

(a)   (b)

Figure 6. (a) Geometry parameters for used electrodes according to DIN EN ISO 5821 F1-16-20-40-6; (b) chromatic confocal microscope measurement after initial electrode tip dressing with parameters for topographical measurements.

2.4. Assessing Electrode Wear

The main geometric parameters to describe and assess electrode wear are shown in Figure 7. For simplicity, the spherical radius of the contact area was neglected. Parameters were also selected by Zhang et al. [32], who measured axial wear or length change Δh by the servo gun movement. In this paper, movement s_u of the upper and s_l of the lower electrode was measured by laser-triangulation sensors attached to the welding gun for each electrode. In addition, Δh was determined by 3D topographical measurements. To distinguish between the two measurement concepts for Δh, the following nomenclature was chosen:

- $\Delta h_{i.\text{topo}}$: determined using 3D topographical measurements.
- $\Delta h_{i.\text{process}}$: measured by laser triangulation during process.

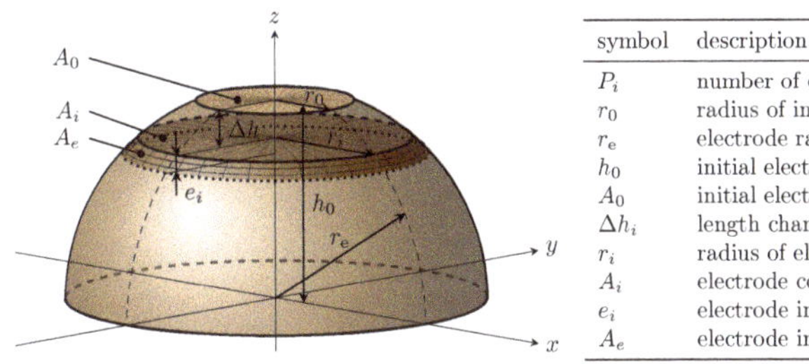

symbol	description
P_i	number of consecutive spot welds
r_0	radius of initial electrode contact area
r_e	electrode radius
h_0	initial electrode height
A_0	initial electrode contact area
Δh_i	length change after P_i
r_i	radius of electrode contact area after P_i
A_i	electrode contact area after P_i due to wear
e_i	electrode indentation after P_i
A_e	electrode indentation contact area after P_i

Figure 7. Geometric parameters of electrode tip.

For further calculations, reference was made to the following equations, which resulted from the geometrical parameters of Figure 7:

$$h_0 = \sqrt{r_e^2 - r_0^2} \tag{2}$$

$$r_i = \sqrt{(r_e - h_0 + \Delta h_i)(2r_e - 1)} \tag{3}$$

$$A_0 = \pi r_0^2 \tag{4}$$

$$A_i = \pi r_i^2 \tag{5}$$

$$A_e = 2\pi r_e e_i \tag{6}$$

2.4.1. Determining $\Delta h_{i.\text{topo}}$ by 3D Topographical Measurements

Surface-topography measurements of the electrodes were carried out with the chromatic confocal microscope mentioned in Section 2.2. These high-resolution measurements included both electrodes while they were installed on the welding gun and delivered continuous documentation of electrode wear. The first measurements on both electrodes were carried out right after tip dressing and continuously after each test sheet (8, 200, 400, ..., 1200) for MC01. Due to the expected high electrode wear of MC02, measurements were carried out after 12, 38, 64, 90, and 102 welds in the beginning. Then, every further measurement was conducted after each test sheet, as for MC01. The last spot weld of every test sheet was also measured on both sides. The combination of these measurements with the recorded electrode displacement during welding was used to estimate the contact areas between electrode and sheet metal at the respective welding end, and to characterize the wear in more detail. An example measurement after tip dressing and measurement parameters is summarized in Figure 6b. To assess $\Delta h_{i.\text{topo}}$ from the topographical measurement, each data set (DS) had to be aligned to a reference DS as shown in Figure 8. This is necessary because the positioning accuracy of the robot lays in the range of $\Delta h_{i.\text{topo}}$. The corresponding DS of P_0 was always used as a reference. The DSs were first aligned by moving them in the x–y plane and searching for the minimal difference between the two DSs (Figure 8e). To align in the z direction, profile sections were used as demonstrated in Figure 9. The DS was shifted in the z direction until uninfluenced regions had been aligned one above the other, which applied to the flanks of the electrodes (marked as dashed lines in Figure 9). $\Delta h_{i.\text{topo}}$ was then the difference of the highest point of DS of P_i minus DS of P_0.

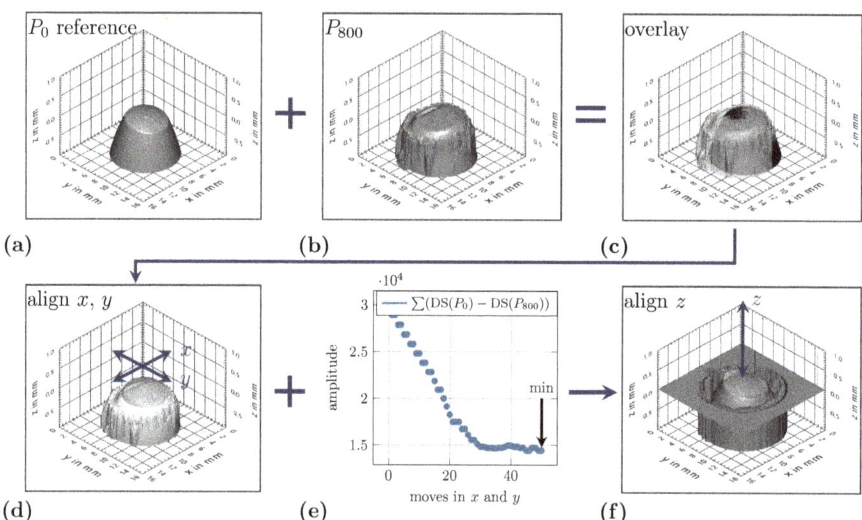

Figure 8. Procedure to align data sets (DSs) of 3D topographical measurements for comparison and determining $\Delta h_{i.\text{topo}}$ on example of MC01–TS01 $P_i = 0$ and $P_i = 800$ as 3D projection. (**a**) Plot of reference DS of P_0; (**b**) plot of DS to align; (**c**) overlay of both DSs; (**d**,**e**) aligned DS in x–y plane by determining minimal difference between two DSs; (**f**) aligned in z direction.

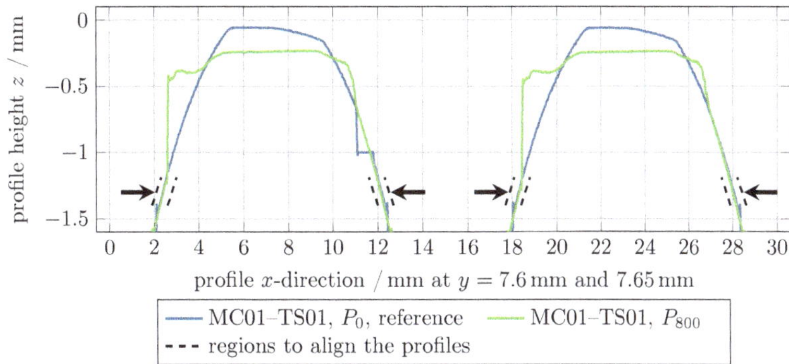

Figure 9. Exemplary comparison of selected profiles derived from topographical measurements of MC01–TS01 $P_i = 0$ and $P_i = 800$ with marked regions to align profiles.

2.4.2. Determining $\Delta h_{i.\mathrm{process}}$ during Welding Process

The total difference in electrode displacement Δs_i was calculated by Equation (7), and for each electrode by Equation (8). However, the assumption of these equations could only be made if the wear of the upper and lower electrodes was nearly symmetrical. Symmetric wear applies for symmetric MC and the same electrode geometries on both sides, which could be proven by metallographic examinations in [11], and applied for the chosen MC of this paper. Change in length $\Delta h_{i.\mathrm{process}}$ over the number of spot welds could then be calculated by Equation (9) as in [32]. $\Delta h_{i.\mathrm{process}}$ was negative for decreasing and positive for increasing electrode length. The model in Figure 7 can mainly be used for mushrooming wear mode. The model and its geometric parameters cannot be easily adapted to plateau formation, since the deformation mechanisms are much more complex. Here, topographical measurements help to explain the wear mechanism in more detail.

$$\Delta s_i = s_\mathrm{u}(P_i) - s_\mathrm{l}(P_i) \tag{7}$$

$$\Delta l_i = \frac{1}{2}\Delta s_i \tag{8}$$

$$\Delta h_{i.\mathrm{process}} = \Delta l(P_0) - \Delta l(P_i) \tag{9}$$

To evaluate the condition of the electrodes, displacement measurements were determined at certain timing positions in the process as shown in Figure 10: weld begin (WB), weld end (WE), and process end (PE). These positions show the following characteristics:

WB: beginning of welding, no nugget, no heat in components, negligible electrode indentation e;

WE: end of welding, molten nugget, components at max heat, max thermal expansion;

PE: end of process, solidified nugget, max electrode indentation e_{max}.

Figure 10 also demonstrates the course of measured process data F_{el}, I_w, s_u, s_l, and the resulting Δs_i. The parameter course of MC01–TS01 showed no evidence of problems during welding. In comparison to that, MC02–TS01 was a good example for the appearance of an expulsion. At around $t = 0.8\,\mathrm{s}$ an increase in I_w and a huge drop in F_{el} and s_l could be observed. These are typical phenomena for expulsions during welding.

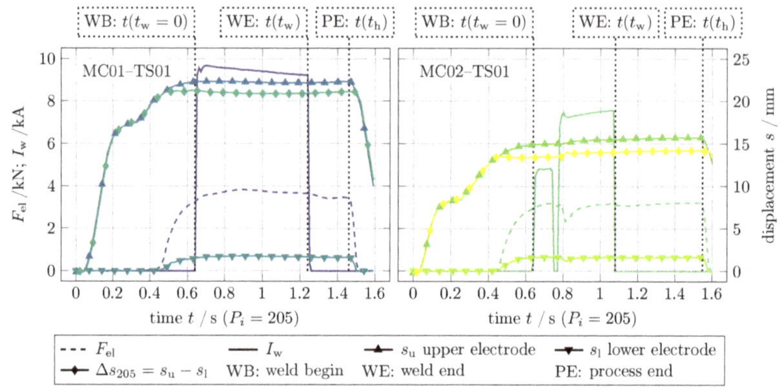

Figure 10. Exemplary process parameters of spot-weld $P_i = 205$ of MC01–TS01 and MC02–TS01 with marked timing positions weld begin (WB), weld end (WE), and process end (PE) for further evaluations.

3. Results

3.1. Welding Results

Measured weld diameters d_w after DT over the number of spot welds decreased slightly for both MC (Figure 11). While d_w of MC01 showed little variation with a small range, d_w strongly fluctuated at the beginning and from around 400 spot welds of MC02. Looking at the occurrence of weld expulsions, MC01 and MC02 showed huge differences. Whereas nearly no expulsions occurred in MC01, they appear in almost every weld of MC02 at the sheet interface.

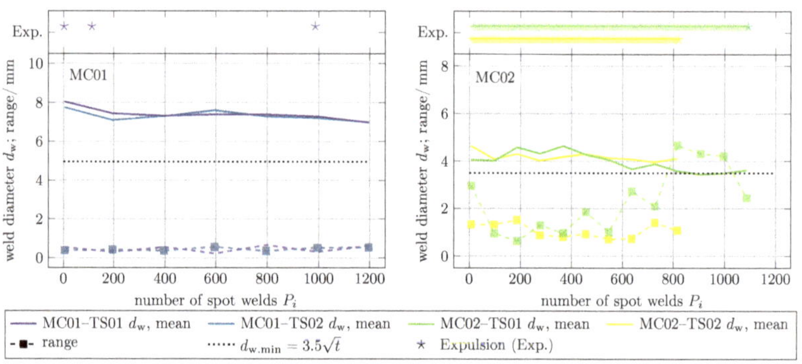

Figure 11. Development of weld diameter d_w of MC01 and MC02 determined by destructive chisel test according to ISO 10447.

3.2. Results of 3D Topographical Measurements

Figures 12 and 13 show profile sections and selected 3D topographical measurements of the electrodes of the respective TS01 at different stages. In both figures, the images of $P_i = 0$ show the electrodes right after tip dressing. The wear caused by the progressive change in the contact surface can clearly be seen with the increasing number of spot welds. In MC01, an enlargement of the electrode contact area can be seen in the radial direction. According to [32,33], the radial widening of electrodes leads to axial wear due to the law of constant volume. Crater formation can also be observed on the surfaces, especially at the upper electrode at $P_i = 800$ and $P_i = 1200$. Local melting or breaking out of the brittle alloy layer, which is known as pitting [34], forms these craters. Coating materials with a low melting temperature such as zinc ($T_m = 420\,°C$) can penetrate these craters and come into contact with the electrode material again. Mechanical stresses and the weld current flow are concentrated in the edge area of the crater, which can lead to accelerated material removal and crater growth. Besides this, the top view of the electrodes also shows a slight plateau. This is not comparable to the wear mode of plateau formation. In fact, this can be called trimming as the length of the electrode decreases [12]. In MC02, the distinctive plateau formation occurred after just a few welds. With a higher numbers of welds, the plateau increasingly moves out of the center of the electrode in the direction of the opening of the welding gun, which might be because of the elastic bending of the welding gun. Comparing the two figures, the 3D topographical measurements revealed a clear difference between the two wear modes. Mushrooming can be observed at MC01 (Figure 12) and plateau forming at MC02 (Figure 13). Both figures show symmetrical wear of the upper and lower electrodes, as assumed before. The flanks of the profile sections were not aligned one above the other as described in Section 2.4.1 in both figures. This is due to the calculation of the profile sections. Each section is the mean course of 36 individual profile sections rotating around the center of electrode with angular spacing of $\phi = 5°$.

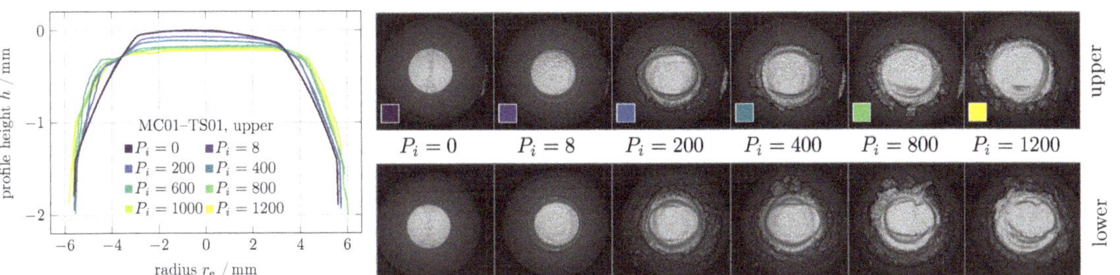

Figure 12. Wear mode of mushrooming visualized by topographical measurements of upper (anode) and lower (cathode) electrodes of MC01–TS01 by profile sections and top view on 3D topographical measurements.

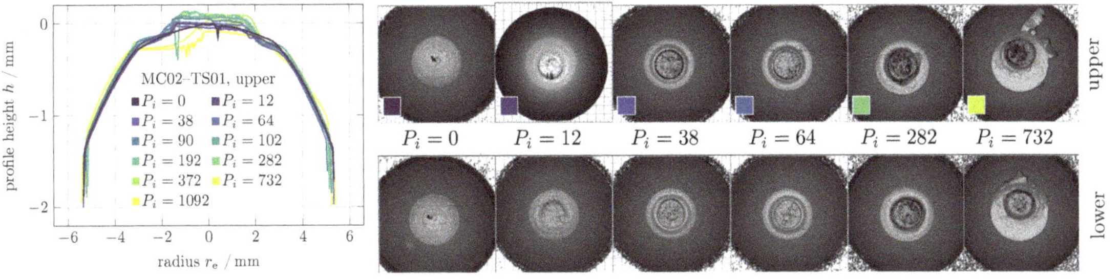

Figure 13. Wear mode of plateau forming visualized by topographical measurements of the upper (anode) and lower (cathode) electrodes of MC02–TS01 by profile sections and top view on 3D topographical measurements.

The radial flow and loss of material from the electrode surface caused the electrodes to decrease in length. Figure 14 shows a cross-section of the anode of MC01–TS01 after 1200 spot welds. The etched structure shows material flow in the radial direction near the electrode surface. Furthermore, material deposits can be seen at the edge of the electrode. Loss of material is caused by the formation and destruction of alloy layers [15].

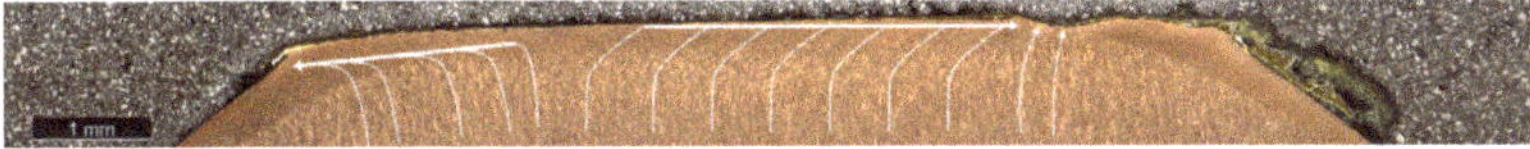

Figure 14. Mushrooming due to radial material flow indicated by white arrows of upper electrode of MC01–TS01 after 1200 spot welds in cross-section.

Axial flow that forms a plateau is shown in Figure 15, where a shift of the plateau out of the center occurred. Measuring plateau diameter d_p and height h_p proved that the plateau formation was happening after just a few spot welds, as shown in Figure 16 and already indicated in Figure 13. Plateau diameter d_p was determined in the top view of the measurements by the mean value of longest diameter d_{p1} and d_{p2} perpendicular to it (Equation (10)). The plateaus show no numerical eccentricity. This allows the area of the plateau A_p to be calculated by the conventional formula for calculating the area of a circle without any further restrictions (Equation (11)). h_p is determined via profile sections of the electrodes. It should be noted that d_p is immediately constant, whereas h_p is gradually increasing. There are no significant differences between anode and cathode at MC02–TS01 and slight differences a MC02–TS02.

$$d_p = \frac{1}{2}d_{p1}d_{p2} \tag{10}$$

$$A_p = \frac{\pi}{4}d_p^2 \tag{11}$$

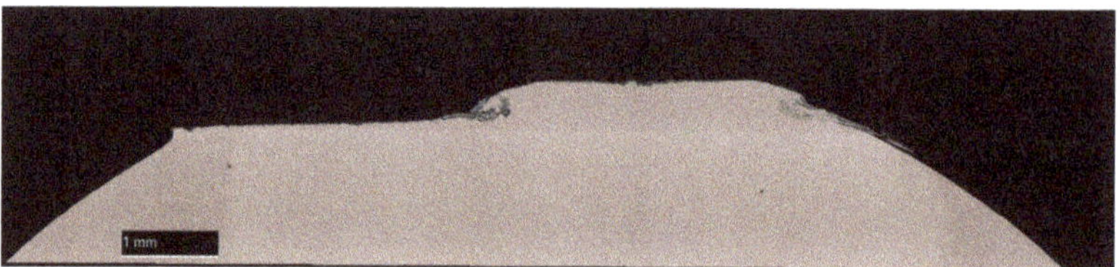

Figure 15. Plateau forming due to axial material flow of lower electrode of MC02–TS02 after 822 spot welds in cross-section [11].

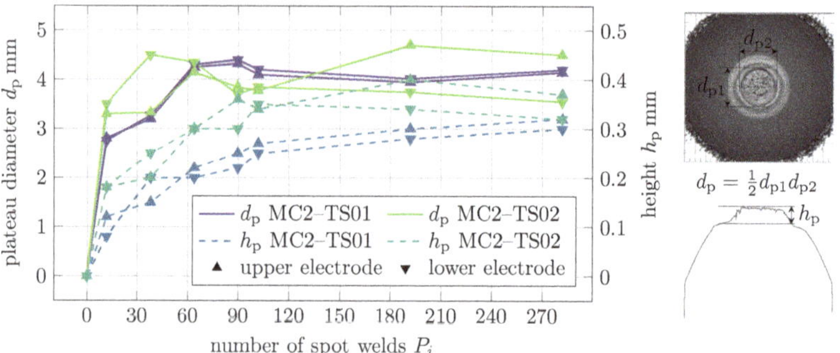

Figure 16. Plateau-forming development over number of spot welds due to axial material flow.

3.3. Assessing $\Delta h_{i.topo}$ by 3D Topographical Measurements

The results of determining electrode changes in length $\Delta h_{i.topo}$ over the number of spot welds can be seen in Figure 17. As results show, $\Delta h_{i.topo}$ decreased right from the beginning for MC01 with the wear mode of mushrooming, whereas for MC02, the length of the electrodes increased, caused by plateau forming; after about 300 spot welds, it began to slightly decrease. Constant d_p after just a few welds led to persistently high and constant current density J through plateau area A_p. In addition, the alloy layer forming on the electrodes caused poorer heat dissipation via the electrodes, resulting in more process heat remaining in the sheets, thereby leading to nuggets becoming thicker [19]. Analyzing MC02's d_w in regard to d_p and $\Delta h_{i.topo}$, two different phenomena were identified. First, d_w and d_p approached each other at a roughly 4 mm diameter and were smaller than the initial diameter of the electrode contact surface. The softening of the region close to the electrode surface of the work pieces enabled axial material flow under the pressure of F_{el}, which accelerated plateau formation but did not allow for d_p to grow. The formation of a plateau led to deep electrode indentations e. With the second phenomenon, an explanation for the variation in d_w(MC02) can be drawn by comparing the development of d_w(MC02) in Figure 11 with the course of $\Delta h_{i.topo}$(MC02). As long as $\Delta h_{i.topo}$ increased, d_w(MC02) was stable. The progressive wear resulting in a decrease in $\Delta h_{i.topo}$ starting at around 300 spot welds led to unstable d_w(MC02). Interestingly, the decrease in $\Delta h_{i.topo}$ appeared with the same gradients in both MC.

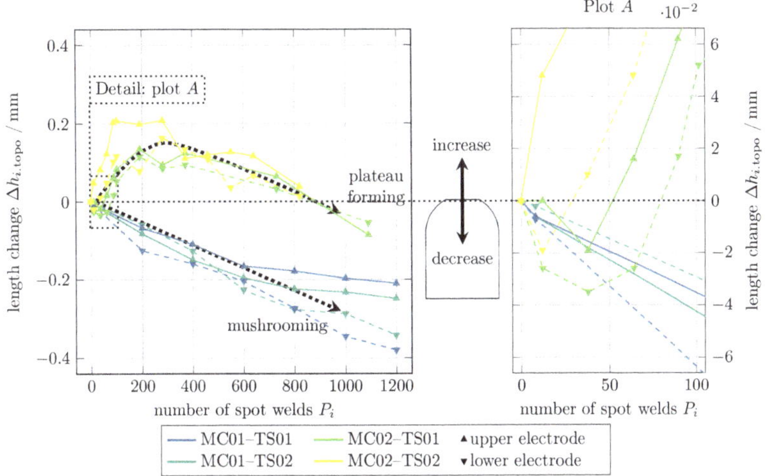

Figure 17. Different behavior in length change $\Delta h_{i.topo}$ of upper and lower electrodes from MC01 and MC02 as a result of two different wear modes, mushrooming and plateau forming.

3.4. Assessing $\Delta h_{i.process}$ during Welding Process

The change in length $\Delta h_{i.process}$ of the electrodes during the welding process could be determined by evaluating electrode position at WB (Figure 18a). Data fluctuations could occur due to adhering weld spatter or other impurities on the electrode working surfaces. However, tendencies could clearly be identified. The electrode contact area at WB and PE could be calculated from these data by Equations (13) and (14). However, the area at WE could not be exactly determined. At PE, the contact area of the electrode was supplemented by the outer surface that resulted from electrode indentation e_i, which was calculated by Equation (12) and is shown in Figure 18b. Both diagrams of Figure 18 show a clear distinction between the two wear modes, whereby $\Delta h_{i.process}$ and e_i showed the same behavior within their modes. Figure 19 shows the changes in electrode contact areas between WB and PE. Again, there were huge differences between MC01 and MC02 due to the different wear modes. For MC01, the $\Delta h_{i.process}$ was negative and decreased leading to steady enlargement of the contact areas. With the increasing number of spot welds, the contact areas of WB and PE approached each other resulting in a smaller ΔA (Equation (15)). This can be explained by the reduced surface pressure, which was confirmed by the decrease in electrode indentation depths e_i (Figure 18b). As described in Section 2.4.2, the model of Figure 7 could only be used for mushrooming; thus, A_{WB} and A_{PE} could not be calculated using $\Delta h_{i.process}$ and e_i for plateau forming. Here, contact areas were derived by using 3D topographical measurements in combination with e_i measured from the 3D topographical measurement of the test sheets. In the beginning, the gradient of $\Delta h_{i.process}$ was positive and turned negative after about $P_i = 300$. Due to plateau forming, e_i rose and corresponded with h_p (Figure 16). The same applied to A_{WB}, which resulted from d_p and A_p, respectively. The difference between contact areas at WB and PE was much larger than that in MC01. While A_{WB} was related to plateau formation, A_{PE} showed a slight increase almost with the same gradient as that of MC01. $\Delta A(MC02)$ increased in the same manner.

$$e_i = \Delta l_i(WB) - \Delta l_i(PE) \text{ , with } \Delta l_i \text{ of Equation (8)} \tag{12}$$

$$A_{\mathrm{WB}} = \begin{cases} A_i = \pi r_i^2 \text{ , for mushrooming (equation 5)} \\ A_p = \frac{\pi}{4} d_p^2 \text{ , for plateau forming} \end{cases} \tag{13}$$

$$A_{\mathrm{PE}} = \begin{cases} A_i + A_e = \pi r_i^2 + 2\pi r_e e_i \text{ , for mushrooming (Equations (5) and (6))} \\ \text{3D topographical measurements for plateau forming} \end{cases} \tag{14}$$

$$\Delta A = A_{\mathrm{PE}} - A_{\mathrm{WB}} \tag{15}$$

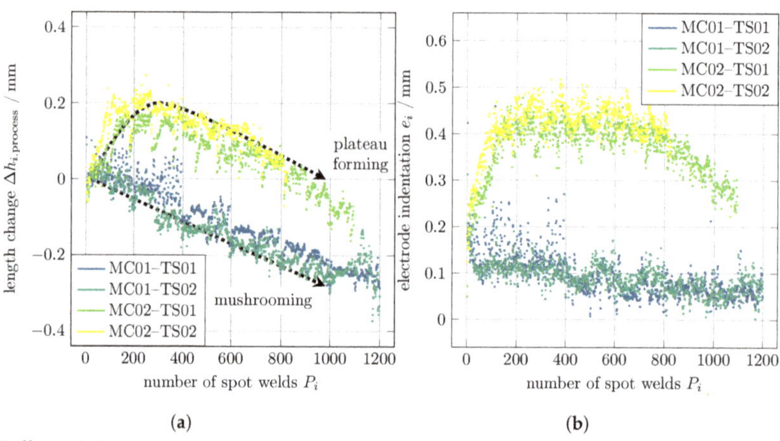

Figure 18. (a) Different behaviors in length change $\Delta h_{i.\mathrm{process}}$ of upper or lower electrodes form MC01 and MC02 as a result of two different wear modes, mushrooming and plateau forming. (b) Calculated electrode indentation e_i from displacement measurements.

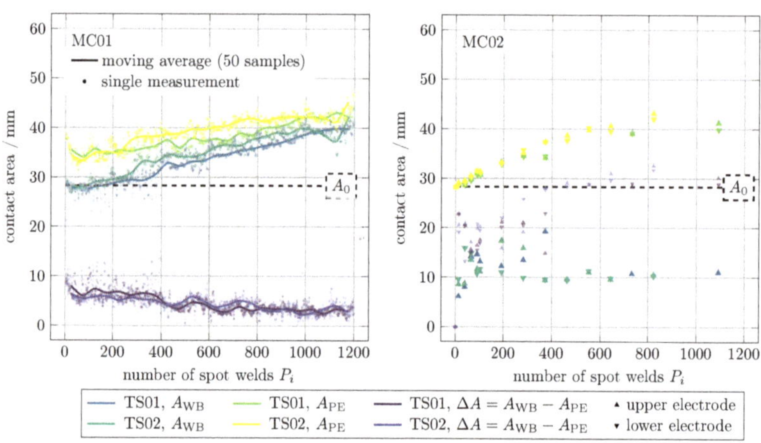

Figure 19. Comparison of contact areas A_{WB} and A_{PE} of electrodes from MC01 (left) and MC02 (right) as a result of two different wear modes, mushrooming and plateau forming.

4. Discussion

The stability of the weld diameters over the number of welds corresponded with the wear modes. Contact area A_{WB} increased steadily for MC01 and was stable for MC02 after just a few spot welds. A stable A_{WB} of MC02 may indicate good process reliability. This does not apply for plateau forming as the wear mode, since A_{WB} was much smaller with $\approx 44\%$ of A_0, resulting in much higher current densities J, triggering expulsions and deep electrode indentations. The frequent expulsions of MC02 accelerated plateau formation since material from the weld disappeared, and electrodes were pressed deeper into the sheet. Those deep electrode indentations exceeded the limit value of $e_{\max} = 0.2t$, where t is the thickness of the respective sheet metal according to ISO 14373 [35] after just a few spot welds, leading to insufficient quality. Furthermore, the lack of a corresponding calculation model for predicting the plateau-formation process can lead to unforeseen problems in the welding process. Mushrooming, in contrast, leads to a lower J, reducing the risk of expulsions. In fact, in the early stages of mushrooming, the weld process is stabilized, and electrodes are conditioned. Nevertheless, with an increase in welds, electrodes are worn out, caused by different interlocking effects as listed in Table 3. The only way to avoid expulsions is to reduce J. However, this also has a negative effect on nugget diameter. All other effects result in risks of process instabilities and expulsions. In reference to a stable nugget or weld diameter over the number of spot welds, other factors should be considered to define a worn-out electrode. Those factors might be the surface condition after spot welding, ensuring the ability for ultrasonic NDT with adequate electrode indentations.

Table 3. Interlocking effects of mushrooming at resistance spot welding (RSW).

Effect Description	Result	Risk of Expulsion
Increased number of welds	Alloy layer thickness ↑	↑
Increase in alloy layer	Material resistance R_4, R_7 ↑	↑
Increase in contact area A_{WB}	Current density J ↓	↓
	Pressure p_{WB} ↓	↑
Decrease in pressure p_{WB}	Contact resistance R_{1-3} ↑	↑
Increased material resistance R_4, R_7	Nugget diameter ↑	↑
Increased contact resistance R_1	Nugget diameter ↑	↑
Increased contact resistance R_2, R_3	Nugget diameter ↓	↑
Decreased current density J	Nugget diameter ↓	↓

The results for MC01 in Figures 17 and 18a are in accordance with the experimental tests in Rogeon et al. [36], where $\Delta h_{i.\mathrm{process}} = 0.1$ mm after 300 spot welds using zinc-coated steels similar to the steel used for MC01. In Lu et al. [21], the diameters of the electrode contact areas were measured with a result of a 32% larger diameter at the end of the electrode life. For MC01, an increase of 18% for the diameter and 42% for the contact area could be determined after only 1200 spot welds. Using the simplified model for estimating electrode face diameter for spherical shaped electrodes of Lu et al. [21] to predict the diameter development with

$$d_i = \left(K F_{\mathrm{el}} I_{\mathrm{w}} P_i + d_0^7 \right)^{1/7} \tag{16}$$

the value K could be determined with

$$K = 17\,\mathrm{mm}^7 \mathrm{kA}^{-1} \mathrm{kN}^{-1} = 0.076\,\mathrm{mm}^7 \mathrm{kA}^{-1} \mathrm{lbf}^{-1}$$

using the following weld parameters

$$F_{\mathrm{el}} = 3500\,\mathrm{N} = 786.83\,\mathrm{lbf}$$
$$I_{\mathrm{w}} = 9.2\,\mathrm{kA}$$
$$d_0 = 2r_0 = 6\,\mathrm{mm}.$$

Even though Lu et al. used many more spot welds to determine their model, this prediction is in good agreement with the experiment data of MC01, as Figure 20 shows, since K was also in the same range as in [21].

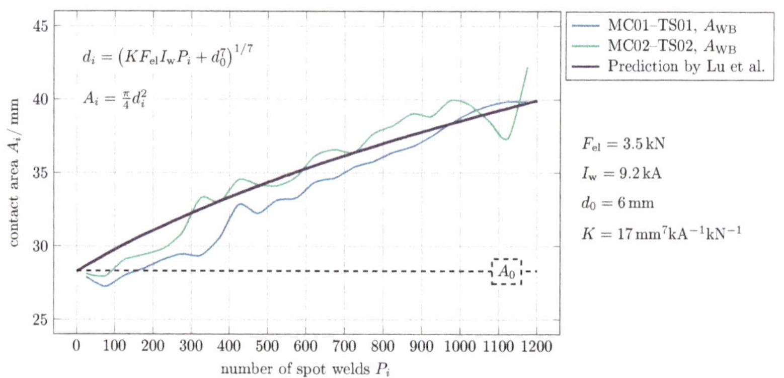

Figure 20. Comparison of A_{WB} of MC01 with prediction model of Lu et al.

The occurrence of wear mode can be attributed to the interaction of at least two factors, material strength and the dimension of the softened volume within the sheets. The softened material always has lower strength than that of the electrodes. The interaction of these two factors influences how electrodes are pressed into the material. If the sheets soften with a lateral expansion that is greater than the electrode contact surface, and the strength of the sheets is less than that of the electrodes, the electrodes can penetrate the material with the entire contact surface, and mushrooming occurs. In the case that the lateral expansion of the softened material is smaller than the electrode contact area, and the nonsoftened sheet material has higher strength than that of the electrodes, a plateau is formed. This assumption is in accordance with Klages [19]. However, no detailed investigations into borderline cases have yet been researched or carried out.

In fact, electrode wear cannot be avoided. Therefore, the question must be asked of which of the two wear modes is preferable, if it is possible to choose the process parameters in relation to one wear mode. The above discussion of the results clearly shows that the risk of process instabilities in plateau formation is higher compared to mushrooming. Looking at the tip-dressing process, the sharp edge of the plateau leads to high and sudden loads due to a punctual initial contact on the dressing tool; at mushrooming, the tool is gradually loaded and over a larger area (Figure 21). Therefore, from the point of view of tip dressing, a mushroomed electrode loads the tool more gently.

To recognize the acting wear mode, results of the experimental study show that, for evaluating Δh_i, the two wear modes of mushrooming and plateau forming can be distinguished by this. $\Delta h_{i.\mathrm{process}}$ by measuring the electrode displacement over the number of spot welds P_i was equal to $\Delta h_{i.\mathrm{topo}}$ of the high-resolution 3D topographical measurements with higher precision. This allows for wear to be assessed by evaluating the change in length $\Delta h_{i.\mathrm{process}}$. This can be performed with displacement sensors, as in this study. However, since the common industrial environment where RSW is mainly used is very rough and tough, additional sensors are not always suitable. They must have a high level of electromagnetic compatibility and should be mechanically protected. For this reason, a solution with existing system technologies is needed. A possible option is to evaluate time $\Delta t_{\text{process}}$ between the start of the welding gun movement $t_{\text{process}} = 0$ and WB. Most industrial RSW systems are triggered from outside starting the weld process, as shown in Figure 1. Since the welding process is known to the system in most applications, the thickness of the sheets to be welded and distance Δs_{open} of the open position of the electrodes and the sheets are also known. With the assumption of a constant speed at which the welding gun closes, this information can be used to determine Δh_i over the number of spot welds by evaluating the shift of $\Delta t_{\text{process}}$. For plateau formation, $\Delta t_{\text{process}}$ should therefore become smaller since electrode lengths increase resulting in a shorter Δs_{open}. Consequently, $\Delta t_{\text{process}}$ increases for mushrooming. These effects are shown in Figure 22a in detail and Figure 22b over the number of spot welds. Evaluating every single weld is not expedient. It is better to use statistical tools such as moving averages to see the trend of Δh_i to monitor electrode wear and its mode. This approach can be integrated into existing systems.

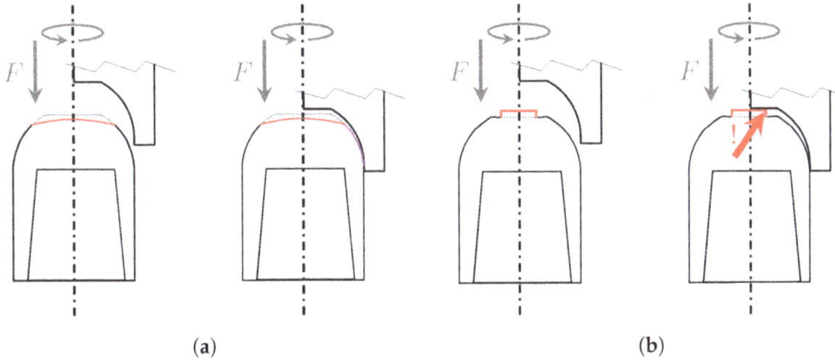

Figure 21. Tip dressing of worn electrodes: (**a**) mushrooming; (**b**) plateau forming.

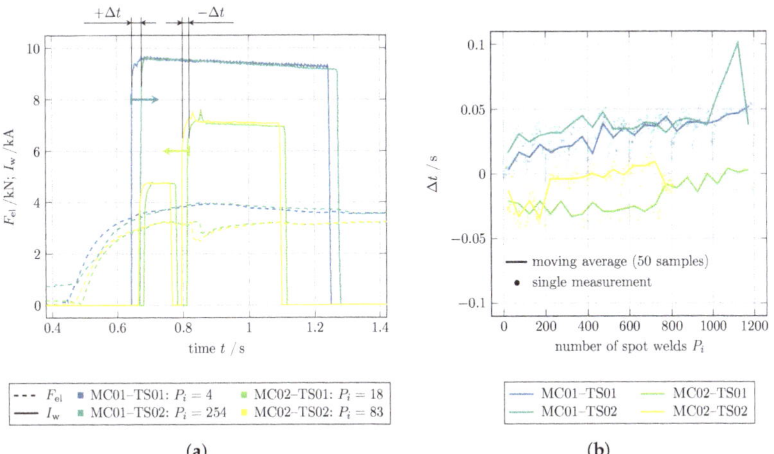

Figure 22. (**a**) Δt as shift of WB due to electrode length change in detail and (**b**) over number of spot welds.

5. Conclusions

Electrode wear is an undesirable progressive mechanical and metallurgical change to electrode tips with negative effects on the process. This paper presented the two wear modes of mushrooming and plateau formation, and their effect on the welding process. The effects of the two wear modes act together, resulting in a turning point where electrodes are worn out. It is critical to identify this turning. Thus, the industry avoids reaching this point by tip dressing much earlier. Monitoring electrode length change Δh_i over the number of spot welds can help to address this problem. This is not a problem in a laboratory. A welding system can be equipped with all kinds of sensors and technologies to monitor the RSW process. In its common application areas, such as automotive body-in-white manufacturing, the use of these additional sensors is not possible with reasonable effort. The rough and tough environment, low cycle times, and other influences of other processes result in huge challenges for most sensors to deliver trustworthy data. Signal noise is not easy to avoid compared with the needed effort. Therefore, a solution for inline electrode wear monitoring was presented on the basis of scientific investigations and evaluations. Furthermore, the possibility of completely avoiding additional sensors was shown by evaluating the time between initial electrode movement and weld begin.

The investigations and results of this paper can be the basis for numerical simulations to reduce the complexity of mechanical environments. Real data and their statistical analysis from experiments can help to support the mechanical environment of such simulation models.

For a higher degree of result generalization, further research should be extended to other steel alloys, in particular to examine borderline and transition cases of the two wear modes. When measuring electrode movements, attention should be paid to an improved signal-to-noise ratio.

Author Contributions: Conceptualization, C.M. and D.K.; methodology, C.M.; software, C.M.; validation, C.M. and D.K.; formal analysis, C.M.; investigation, C.M.; resources, C.M. and D.K.; data curation, C.M.; writing—original-draft preparation, C.M.; writing—review and editing, C.M., D.K., S.H. and J.Z.; visualization, C.M.; supervision, U.F.; project administration, J.Z.; funding acquisition, S.H. and J.Z. All authors have read and agreed to the published version of the manuscript.

Funding: This research was funded by the Deutsche Forschungsgemeinschaft (DFG, German Research Foundation) within the project "Basic investigations for the in-situ simulation of resistance spot welding processes" under grant numbers FU 307/15-1 resp. IH 124/17-1, project no.: 389519796. Open Access Funding by the Publication Fund of the TU Dresden.

Data Availability Statement: Data can be requested from the corresponding author.

Conflicts of Interest: The authors declare no conflict of interest.

Abbreviations

The following abbreviations are used in this manuscript:

RSW	Resistance spot welding
HAZ	Heat affected zone
AHSS	Advanced high strength steels
AlSi	Aluminium-silicon coating
MC	Material combinations
DT	Destructive testing
NDT	Nondestructive testing
TS	test set
DS	Data set
WB	Weld begin
WE	Weld end
PE	Process end

Nomenclature

A_0	Initial electrode contact area (mm^2)		P_{end}	Number of predefined spot welds
A_e	Electrode indentation contact area after P_i (mm^2)		p_{WB}	Contact pressure at weld begin (N mm^{-2})
A_i	Electrode contact area after P_i due to wear (mm^2)		Q	Heat (J)
A_{PE}	Contact area at process end (mm^2)		R_{tot}	Total resistance (Ω)
A_{WB}	Contact area at weld begin (mm^2)		R_{1-7}	Individual resistance (Ω)
A_{WE}	Contact area at process begin (mm^2)		r_0	Radius of initial electrode contact area (mm)
ΔA	Area change (mm^2)		r_e	Electrode radius (mm)
A_p	Plateau area (mm^2)		r_i	Radius of electrode contact area after P_i (mm)
d_0	Diameter of initial electrode contact area (mm)		Δs_i	Difference in electrode displacement (mm)
d_i	Electrode contact diameter after P_i (mm)		s_l	Displacement lower electrode (mm)
$d_{\text{p,1,2}}$	Plateau diameter (mm)		s_u	Displacement upper electrode (mm)
d_n	Nugget diameter (mm)		Δs_{open}	Distance of electrodes in opened position (mm)
d_w	Weld diameter (mm)		t	Time (s), sheet thickness (mm))
E	Young's modulus (N mm^{-2})		t_u	Upper sheet (mm) (mm)
e	Electrode indentation (mm)		t_l	Lower sheet (mm) (mm)
e_i	Electrode indentation after P_i (mm)		t_h	Hold time (s)
e_{max}	Limit of electrode indentation (mm)		t_p	Pause time (s)
f	Frequency (Hz)		t_s	Squeeze time (s)
F_{el}	Electrode force (kN)		t_w	Weld time (s)
h_0	Initial electrode height (mm)		t_{wpp}	Prepulse time (s)
h_p	Plateau height (mm)		T_c	Cooling water temperature (°C)
Δh	Length change (mm)		T_{min}	Minimal temperature (°C)
Δh_i	Length change after P_i (mm)		T_{max}	Maximal temperature (°C)
$\Delta h_{i.\text{process}}$	Δh by laser triangulation during process (mm)		t_{process}	Process start time (s)
$\Delta h_{i.\text{topo}}$	Δh by 3D topographical measurements (mm)		$\Delta t_{\text{process}}$	Change in process start time (s)
I_w	Welding current (kA)		U_w	Welding voltage (V)
I_{wpp}	Prepulse welding current (kA)		$\dot{V}$	Flow rate (l min^{-1})
J	Current density (kA mm^{-2})		x	Dimension in x (mm)
K	Constant by Lu et al. [21] (mm^{7}kA^{-1}kN^{-1})		y	Dimension in y (mm)
k	Stiffness factor (kN mm^{-1})		z	Height, dimension in z (mm)
Δl_i	Individual electrode displacement (mm)		ϕ	Angle (°)
P_i	Number of spot welds			

References

1. TWI Ltd. *What Is Spot Welding?* TWI Ltd.: Cambridge, UK, 2020.
2. Chen, J.; Feng, Z. IR-based spot weld NDT in automotive applications. In Proceedings of the SPIE Sensing Technology + Applications, Baltimore, WA, USA, 20–25 April 2015; doi:10.1117/12.2177124. [CrossRef]
3. Raab, H.H. *Handbuch Industrieroboter: Bauweise · Programmierung Anwendung · Wirtschaftlichkeit*, 2., Neubearbeitete und Erweiterte Auflage ed.; Vieweg + Teubner Verlag: Wiesbaden, Germany, 1986; doi:10.1007/978-3-322-83600-7. [CrossRef]
4. Altnau, D. (Anzahl Schweißpunkte pro Karosserie Dresden, Leipzig). Personal communication, 26 March 2020.
5. Deniz, C.; Cakir, M. In-line stereo-camera assisted robotic spot welding quality control system. *Ind. Robot. Int. J.* **2018**, *45*, 54–63. [CrossRef]
6. Deutscher Verband für Schweißen und Verwandte Verfahren e.V. *Merkblatt DVS 2902-4:2001-10: Widerstandspunktschweißen von Stählen bis 3 mm Einzeldicke Grundlagen, Vorbereitung und Durchführung*; DVS Media: Düsseldorf, Germany, October 2001.
7. Cho, H.S.; Cho, Y.J. A study of the thermal behavior in resistance spot welds. *Weld. J.* **1989**, *68*, 236–244.
8. Raoelison, R.N.; Fuentes, A.; Pouvreau, C.; Rogeon, P.; Carré, P.; Dechalotte, F. Modeling and numerical simulation of the resistance spot welding of zinc coated steel sheets using rounded tip electrode: Analysis of required conditions. *Appl. Math. Model.* **2014**, *38*, 2505–2521. [CrossRef]
9. Institut für Materialprüfung, Werkstoffkunde und Festigkeitslehre; SLV München. *Standmengenerhöhung beim Widerstandsschweißen durch Elektrodenfräsen: Schlussbericht IGF-Nr. 13.134 N/DVS-Nr. 4.031*; IMWF Universität Stuttgart: Stuttgart, Germany; SLV München: München, Germany, 2004.
10. Koal, J.; Baumgarten, M.; Heilmann, S.; Zschetzsche, J.; Füssel, U. Performing an Indirect Coupled Numerical Simulation for Capacitor Discharge Welding of Aluminium Components. *Processes* **2020**, *8*, 1330. [CrossRef]
11. Füssel, U.; Jüttner, S.; Mathiszik, C.; Sherepenko, O.; Köberlin, D.; Zschetzsche, J. *Lebensdauererhöhung von Widerstandspunktschweißelektroden durch Einsatz verschleißabhängiger Fräsintervalle und Dispersionsgehärteter Kupferwerkstoffe: Schlussbericht IGF-Nr. 18.456 BR/DVS-Nr. 04.062*; Technische Universität Dresden, Professur Fügetechnik und Montage and Otto-von-Guericke-Universität Magdeburg, Institut für Werkstoff- und Fügetechnik: Dresden, Germany, 2018.
12. Chang, H.S.; Cho, Y.J.; Choi, S.G.; Cho, H.S. A Proportional-Integral Controller for Resistance Spot Welding Using Nugget Expansion. *J. Dyn. Syst. Meas. Control* **1989**, *111*, 332–336. [CrossRef]

13. Marek, U. Beitrag zur Klärung der Legierungsschichtbildungs- und Verschleißvorgänge an Widerstandspunktschweißelektroden beim Schweißen Feuerverzinkter Stahlbleche: Zugl.: Aachen, Techn. Hochsch., Diss., 1995. In *Aachener Berichte Fügetechnik*, als ms. gedr ed.; Shaker: Aachen, Germany, 1996; Volume 96.

14. Holliday, R.; Parker, J.D.; Williams, N.T. Relative contribution of electrode tip growth mechanisms in spot welding zinc coated steels. *Weld. World/Le Soudage Dans Monde* **1996**, *4*, 186–193.

15. Matsuda, H.; Matsuda, Y.; Kabasawa, M. Study of electrode wear mechanism in consecutive spot welding: Spot weldability of Zn-Ni electrogalvanised steel sheet and organic-silicate composite steel sheet (1st Report). *Weld. Int.* **1997**, *11*, 860–867. [CrossRef]

16. Williams, N.T.; Parker, J.D. Review of resistance spot welding of steel sheets Part 2 Factors influencing electrode life. *Int. Mater. Rev.* **2004**, *49*, 77–108. [CrossRef]

17. Deutscher Verband für Schweißen und verwandte Verfahren e.V. *Merkblatt DVS 2920:2000-02: Widerstandspunkt-, Buckel- und Rollennahtschweißen von Stahlblechen bis 3 mm mit metallischen Überzügen*; DVS Media: Düsseldorf, Germany, February 2000.

18. Galler, M.; Vallant, R. The influence of Zinc coatings on the electrode wear during resistance spot welding of sheet steel. In *Verarbeitungs- und Gebrauchseigenschaften von Werkstoffen—Heute und Morgen*; Mayr, P., Ed.; Verl. der TU Graz: Graz, Austria, 2008; pp. 117–119.

19. Klages, E.C. Beurteilung der Beanspruchung von Elektrodenkappen Beim Widerstandspunktschweißen von höher—und höchstfestem Stahl. PhD Thesis, Logos Verlag Berlin GmbH, Hannover, Germany, 2014.

20. Großmann, C. Nutzung vorhandener Standmengenpotentiale, Verschleißverringerung durch Angepasste Elektrodenwerkstoffe und Elektrodenverschleißdiagnose beim Widerstandspunktschweißen. Ph.D. Thesis, Technische Universität Dresden, Dresden, Germany, 2019.

21. Lu, F.; Dong, P. Model for estimating electrode face diameter during resistance spot welding. *Sci. Technol. Weld. Join.* **1999**, *4*, 285–289. [CrossRef]

22. Dong, P.; Victor Li, M.; Kimchi, M. Finite element analysis of electrode wear mechanisms: Face extrusion and pitting effects. *Sci. Technol. Weld. Join.* **1998**, *3*, 59–64. [CrossRef]

23. Kondo, M.; Konishi, T.; Nomura, K.; Kokawa, H. Degradation mechanism of electrode tip during alternate resistance spot welding of zinc coated and uncoated steel sheets. *Sci. Technol. Weld. Join.* **2010**, *15*, 76–80. [CrossRef]

24. Li, W. Modeling and On-Line Estimation of Electrode Wear in Resistance Spot Welding. *J. Manuf. Sci. Eng.* **2005**, *127*, 709–717. [CrossRef]

25. Zhang, J.; Zhang, P.X.; Xu, X.J. A Model for Predicting the Wear Degree of Electrode Tip. *Appl. Mech. Mater.* **2014**, *574*, 292–297. [CrossRef]

26. DIN Deutsches Institut für Normung e.V. *DIN EN ISO 8166:2003-09: Verfahren für das Bewerten der Standmenge von Punktschweißelektroden bei Konstanter Maschinen-Einstellung*; DIN Deutsches Institut für Normung e.V.: Berlin, Germany, September 2003.

27. Stahlinstitut VDEh. *Prüf- und Dokumentationsrichtlinie für die Fügeeignung von Feinblechen aus Stahl—Teil 2: Widerstandspunktschweißen*; Stahlinstitut VDEh: Düsseldorf, Germany, August 2011.

28. Deutscher Verband für Schweißen und verwandte Verfahren e.V. *Merkblatt DVS 2916-5:2017-09: Prüfen von Widerstandspresssschweißverbindungen: Zerstörungsfreie Prüfung*; DVS Media: Düsseldorf, Germany, September 2017.

29. Voestalpine Stahl GmbH. *Micro-Alloyed Steels: High-Strength Steels with Yield Strengths up to 550 MPa: Data Sheet Micro-Alloyed Steels*; Voestalpine Stahl GmbH: Linz, Austria, November 2020.

30. Salzgitter Flachstahl GmbH. *22MnB5 Borlegierte Vergütungsstähle*; Salzgitter Flachstahl GmbH: Salzgitter, Germany, 2014.

31. Deutsches Institut für Normung. *DIN EN ISO 10447:2015-05: Widerstandsschweißen—Prüfung von Schweißverbindungen—Schäl- und Meißelprüfung von Widerstandspunkt- und Buckelschweißverbindunge*; Deutsches Institut für Normung: Berlin, Germany, May 2015. [CrossRef]

32. Zhang, X.Q.; Chen, G.L.; Zhang, Y.S. Characteristics of electrode wear in resistance spot welding dual-phase steels. *Mater. Des.* **2008**, *29*, 279–283. [CrossRef]

33. Zhang, X.Q.; Chen, G.L.; Zhang, Y.S. On-line evaluation of electrode wear by servo gun in resistance spot welding. *Int. J. Adv. Manuf. Technol.* **2008**, *36*, 681–688. [CrossRef]

34. Wang, S.C.; Wei, P.S. Modeling Dynamic Electrical Resistance during Resistance Spot Welding. *J. Heat Transf.* **2001**, *123*, 576–585. [CrossRef]

35. Deutsches Institut für Normung. *DIN EN ISO 14373:2015-06: Widerstandsschweißen—Verfahren zum Punktschweißen von Niedriglegierten Stählen mit oder ohne Metallischem Überzug*; Deutsches Institut für Normung: Berlin, Germany, 2015; doi:10.31030/2238811. [CrossRef]

36. Rogeon, P.; Carre, P.; Costa, J.; Sibilia, G.; Saindrenan, G. Characterization of electrical contact conditions in spot welding assemblies. *J. Mater. Process. Technol.* **2008**, *195*, 117–124. [CrossRef]

 *processes*

MDPI

Case Report

Processing Technologies for Crisis Response on the Example of COVID-19 Pandemic—Injection Molding and FFF Case Study

Bogna Sztorch [1], Dariusz Brząkalski [2], Marek Jałbrzykowski [3] and Robert E. Przekop [1,*]

[1] Centre for Advanced Technologies, Adam Mickiewicz University in Poznań, Uniwersytetu Poznańskiego 10, 61-614 Poznań, Poland; bogna.sztorch@amu.edu.pl

[2] Faculty of Chemistry, Adam Mickiewicz University in Poznań, Uniwersytetu Poznańskiego 8, 61-614 Poznań, Poland; dariusz.brzakalski@amu.edu.pl

[3] Faculty of Mechanical Engineering, Bialystok University of Technology, Wiejska 45C, 15-351 Białystok, Poland; m.jalbrzykowski@pb.edu.pl

* Correspondence: r.przekop@gmail.com or rprzekop@amu.edu.pl

Abstract: The paper presents a comparison of two methods of manufacturing utility objects made of plastics, applied to the emerging immediate need in the field of quick provision of personal protective equipment for medical services. The traditional processing method, which is injection molding (IM), and a modern rapid prototyping method, which is fused filament fabrication (FFF) 3D printing, were compared in terms of unit costs and production possibilities at various timeframes. The paper presents the effects of launching two production processes of protective helmets (face shields) using the example of real cases implemented ad hoc during the epidemic development. The implementation of the protective helmet production project based on polyamide-6 processing showed the real possibilities of quickly launching the rapid production of protective equipment with the aid of mold injection technology.

Keywords: COVID-19; FDM; 3D printing; injection molding; personal protection; rapid prototyping; protective face shields

Citation: Sztorch, B.; Brząkalski, D.; Jałbrzykowski, M.; Przekop, R.E. Processing Technologies for Crisis Response on the Example of COVID-19 Pandemic—Injection Molding and FFF Case Study. *Processes* **2021**, *9*, 791. https://doi.org/10.3390/pr9050791

Academic Editors: Sergey Y. Yurish and Maximilian Lackner

Received: 24 March 2021
Accepted: 26 April 2021
Published: 30 April 2021

Publisher's Note: MDPI stays neutral with regard to jurisdictional claims in published maps and institutional affiliations.

1. Introduction

The COVID-19 epidemic has caused a sharp increase in the demand for individual personal protective equipment, such as masks and protective helmets, on an unprecedented scale. Supply options have been significantly limited by the travel restrictions introduced to combat the pandemic. In addition, the production capacity in most countries for these protective measures was not adapted to the existing demand, which resulted in the need for non-standard solutions and small-batch processes requiring minimal set-up preparations. Currently, the development of new technologies driven by the 4th generation industrial revolution results in the development and application of new solutions increasingly based on additive technologies [1], both in the media area and in the academic world. Among the key technologies are additive techniques, popularly called 3D printing, which have been developing dynamically for over a decade, when the patent protections developed by the creators of the technology expired [2]. Initially, this technology did not arouse much interest—in the Gartner report from 2012 it was included in the area of the so-called 'Trough of Disillusionment' [3]. However, its development was so rapid that already in 2013 this statement changed and in 2015 Gartner published a separate report dedicated to 3D printing technology [4]. According to Gartner's estimates, by 2021 as many as 20% of key global companies would create departments specializing in the application and use of 3D printing. We are already witnessing the intensive creation of start-ups related to this particular branch of the industry, improvement of 3D printers (ever better parameters and printing possibilities), and the emergence of new printing techniques and improvement of existing techniques. Currently, additive technologies can be divided into: photochemical (digital light processing (DLP) [5], stereolithography (SLA) [6–8]), laser-mediated selective laser sintering (SLS) [9], thermal fused deposition modeling/fused filament fabrication (FDM/FFF) [10,11] or laminated object manufacturing (LOM) [12].

Nowadays, there are still serious concerns about 3D printing techniques regarding the quality of produced objects, the significantly high process failure rate, or significantly higher unit production time and cost when compared to mature production processes, such as injection molding. Nevertheless, there are products in which the functionality itself becomes more important and constitutes the essence of the product, which means the visual quality ceases to be of decisive importance. In such cases, starting production can be much faster and cheaper. Examples of such products are the components of protective face shields. Such products, especially in the event of an exceptional situation in a global pandemic, are faced with two requirements: they are needed in high volumes and in a short delivery time, and their main function is protection against deadly microorganisms.

With the abovementioned elements in mind, the authors attempted to compare two extremely differently characterized techniques—injection molding and FFF 3D printing. Calculations were made to provide a simulation for a comparison of these two methods under different conditions (time of process run (see Figure 1), single- and multiple-socket forms for IM, single appliance vs. printer farm for FFF). The purpose of this comparison is to show the possibility of quick production of finished products by an industrial method without the need to use rapid prototyping techniques. It is a special case of the injection molding process, in which the visual quality is a negligible aspect in a situation determined by the function of the product. In other words, when it becomes more important to provide protection against the pathogenic effects of the surrounding environment, the visual quality of the products ceases to matter at a critical level, especially in a global pandemic situation. The problem analysis allowed for a rational response towards the pandemic crisis and a quick launch of production of protective helmets for Polish medical services, schools, and universities within the infrastructure of Centre for Advanced Technologies (AMU).

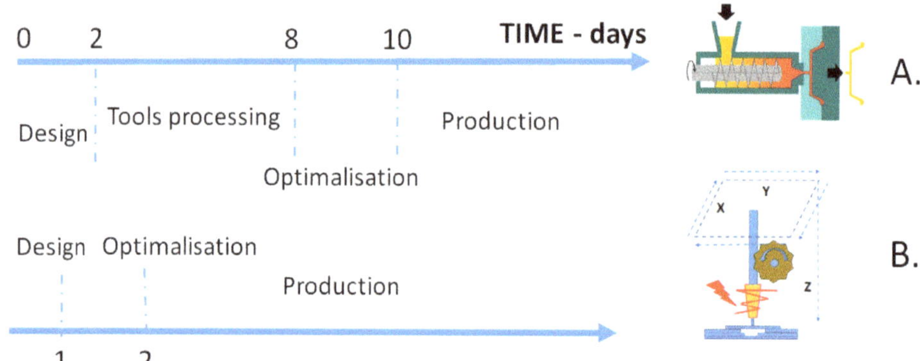

Figure 1. Comparison of the time needed to start the production process for injection molding technology (**A**) and rapid prototyping—3D printing, e.g., FFF (**B**).

2. State of the Art

2.1. Literature Review of FFF 3D Printing

Along with the rapid development of machines dedicated to any process from the vast group of 3D printing techniques, it becomes necessary to design new materials that will allow the improvement of key parameters such as: layer-to-layer adhesion, print integrity and sealing, or other performance parameters of the material. The traditional industry is still approaching this subject with distance and caution, using incremental technologies mainly in the area of design or rapid prototyping, mainly due to the precise mapping of the model geometry and the possibility of obtaining complex shapes, while the production process of high-quality and accuracy products is still based on subtractive technologies, such as CNC machining [13].

3D printing has many features that constitute an advantage and improvement over traditionally used techniques, primarily:

- No need to design and make a tool, e.g., an injection mold—it allows you to save time and avoid high costs of expensive tools—especially important in small-lot production or in the case of a model consisting of several different elements, where traditional techniques would require the use of more than one mold or a complex supply chain [14].
- Personalized products are an example—production of models for a specific application, tailored to the user's preferences, along with quick model optimization—especially in the sports industry or in industrial use in maintenance. 3D printing gives the possibility of producing spare parts for machines directly in the plant, which significantly contributes to reduction of production losses related to equipment stoppage, waiting for transport, etc. [15,16].
- The possibility of constructing complex, non-standard models with varying shapes and arrangement of planes/curves, as well as selected internal geometry (the object may have a full or openwork filling of a different arrangement), impossible or difficult to obtain with traditional techniques, as well as the possibility of changing the model or improving it at any time of production [14].
- Launching the production cycle immediately after the design stage, usually from a few hours to two days [17].

The versatility and availability of the method makes it applicable in large production plants, micro-enterprises or by an individual customer [18], and equipment is relatively cheap when compared to standard industrial processing equipment, e.g., injection molding machines [17].

Despite the many advantages of FFF technology, it has many pitfalls, either remaining to be solved or constituting dead-ends associated with the technology concept itself. The biggest shortcomings of the FFF and relating techniques include:

- Increased energy consumption per product unit mass—3D printing technologies require significant amounts of energy to apply a thin layer of material each time [19].
- Increased waste production related to factors resulting from the technology itself (unsuccessful prints, supports printing, post-processing); currently there is a significant problem of plastic waste production and the recycling thereof, which also applies to materials used for 3D printing. The popular PLA, which despite being considered 'bio-friendly', is not effectively recycled or composted as of today [20,21].
- Very long production cycle compared to the injection molding process, longer by at least two orders of magnitude [22] (Atzeni shows an even greater difference, but in comparison between HPDC and SLS techniques for aluminium processing [23]).
- Increased processing time per production unit increases the likelihood of a process fatal error [24].
- Lower mechanical strength, especially in the Z axis, resulting from, among others, poor layer-to-layer (interlayer) adhesion, the presence of air gap between the layers, and inter-layer distortion—many high-strength materials still do not meet this criterion [25]; fibre-reinforced compositions were proposed as a solution for some of these issues; however, this method is still not perfect and causes problems related to the formation of voids between the layers, poor adhesion of fibres and matrix, and greater wear on the nozzle [26].
- The number of currently available materials for FFF printing is still limited, and each individual polymer has its own specificity of work, which can be problematic to control. Most of the thermoplastics on the market are dedicated to traditional processing techniques, such as injection/extrusion/blow molding, and the modification of plastics for FFF is still a niche topic explored mainly in the scientific community.

- For materials with high processing shrinkage, there is a need to heat the table or the entire printer chamber, which significantly increases the energy consumption per production unit [19].
- Noticeably lower visual quality of products, with the exception of systems with a very small nozzle diameter and parameters optimized for a specific printing material and product, which increases both the preparation time for printing and the printing process itself. Alternatively, the product requires additional post-processing [27].

One of the weaknesses of printing with the FFF technique is, above all, the lack or insufficient amount and availability of construction materials that would allow the elimination of poor characteristics of the finished models. In addition, the choice of materials in the context of the types of polymer matrices and their grades is limited. Improvement of the materials for printing is expected mainly in the processing and utility areas, e.g., increased interlayer adhesion by better melting behaviour of the material (temperature–rheology optimization), thus also increasing the mechanical strength, improving flexibility while maintaining other parameters, increasing the integrity of the layers (reducing air pockets), and water resistance. Finally, an improvement of aesthetic values is sought, which will contribute to saving time and reducing costs related to post-processing (ultimately it should be eliminated, so the detail obtained by 3D printing would be used "as received").

Despite the many advantages and the increasing popularity of additive techniques, they are still far from being perfect. In many respects they cannot replace traditional techniques, and it is even advisable to use well-known processing methods such as injection, calendering, extrusion, etc. Therefore, instead of considering additive technologies as a substitute for mature processing techniques, it is advisable to continue looking for new areas where the unique possibilities of 3D printing can be used to extend the use of traditional production processes. 3D printing can also be used in combination with traditional techniques, where the product is manufactured using a hybrid technique (overprinting, overmolding). In this approach, for example, the greater part of the detail, of a relatively large mass fraction (difficult to produce with the printing technique due to the weight or size and time of printer operation) and simple shapes, are injection-molded, while the part with smaller dimensions and geometry that makes it difficult to implement the injection process is made with a 3D printing technique, after which the parts are combined into a finished detail [28].

In addition, it should be noted that 3D printing becomes an unprofitable technique in a situation where there is a need to make large batches of products, even if their design is highly complicated and thus expensive or time consuming in comparison with tools/molds preparation. In such situations, it is more advantageous to use other options, e.g., injection molding. Despite expensive tools and machines, it becomes more profitable in large-scale and mass production due to short production cycles—counted in seconds or minutes.

3D printing is dedicated to the fabrication of individual products or small series of products, due to the long printing cycles counted in several hours. For this reason, 3D printing is an ineffective process, unsuitable for larger volumes of production. The "break-even point" parameter is calculated, i.e., the number of items at which it is economically viable to transfer production to a process with higher production efficiency, but with higher production implementation costs. In the case of 3D printing, the estimates give values up to 400 pieces, depending on the type of product and the parameters adopted for calculations [22]. With this in mind, it is believed that the future of 3D printing is assigned to single and non-standard applications, where it becomes unprofitable to produce an expensive injection mold for the production of individual, non-standard products. In such situations, the choice of 3D printing over different tool-dependent techniques becomes justified, as the cost of waiting several hours for the production of an unusual detail is lower than the production of an expensive mold.

Obtaining an element by injection molding requires designing an appropriate form, which is associated with 3D modelling processes; however, it is only the beginning of preparations for starting the process. Computer simulations with the use of specialized software are necessary to obtain the geometry of the mold that generates low polymer flow resistance, which facilitates the injection of the detail without defects, and also reduces mold wear, e.g., by cavitation or the abrasive effect of fillers in the polymer (especially important for aluminium molds) [29,30].

2.2. Additive Techniques Compared to the Injection Molding Method

The material for injection forming is usually one of the popular thermoplastics; additionally it can be reinforced with structural glass, carbon, Kevlar, aramid or natural fibres, as well as various mineral fillers. The injection process is a method used mainly in large-scale production (professional products, mass production). In the case of injection molding, it is not possible to freely improve the design after the mold production stage. The geometry of the product is also limited—with more complex shapes, the product must be divided into parts injected as independent details—3D printing in this area offers much more freedom due to the lack of a need for a form, from which it is necessary to remove a once produced detail. One of the greatest advantages of the process is its speed—the cycle time is usually from under half a minute up to several minutes, compared to printing, where the fabrication of a product unit usually takes from less than an hour up to several hours or days.

Injection molding itself, especially in recent years, has become considered as an engineering art [31,32].

In practical terms, the injection molding process has two main goals. The first is to obtain finished and quality products in the shortest time possible, in a fast reproducible and large-scale process. The second goal is the conscious control of the phenomena occurring in the liquid polymer melt [31–33].

The heart of the process is the tool—the injection mold. As of today, the design of injection molds and launching injection production is carried out with the inseparable aid of simulation techniques. MoldFlow or Moldex3D can be mentioned as the more well-known programs for this purpose [30,34–36].

Despite the use of modern computer simulations, starting a production of a new product, even nowadays, sometimes requires many hours of technological trials and subsequent tool corrections. However, the long period of starting production is not only due to technological problems related to the design of the tool (location of the injection point, the geometry of the gate and inlet channels, etc.), but the visual quality optimization of the final product which plays a large part here [31,37–39]. Most often, companies set very high quality criteria for products, which significantly extends the launch of production [29,31,39–48].

3. Materials and Methods

3.1. Materials and Methods for 3D Printing vs. Injection Molding Experiments

Two methods were used to produce protective face shields. The first one was FFF 3D printing, and the second one was the injection molding method. For FFF, an open-source PRUSA 3D model of the protective helmet was used. For injection molding, both the new helmet model and the required injection mold have been designed with SolidWorks software (for the technical drawing and 3D model Figures, see Supplementary Materials).

For the first processing method, FFF, the objects were printed with Dreamer printers with 230 × 150 × 140 mm printing space and double extruder printheads, using standard 1.75 mm filament. The face shields were printed using PET-G (Verbatim) filament. The optimization comprised testing different layer heights, number of shells, extruder temperature, and bed temperature. Table 1 contains all parameters for optimized 3D printing process applied for this particular process.

Table 1. Process parameters for sample printing.

Layer height	0.25 mm
Top layer height	0.25 mm
Shells	2
Top and bottom layers number	3
Bottom layers number	3
Infill density	5%
Infill pattern	Triangular
Printing speed	60 mm/s
Idle speed	80 mm/s
Extruder temp.	240 °C
Bed temp.	80 °C

The second process, that is injection molding, was carried out on an ENGEL e-victory 80/170 injection molding machine with a 25 mm screw, equipped with an aluminium single socket mold. The thermoplastic material used was a recyclate (regranulate) of PA6-GF15 (under the trade name Tarnamid T-27 GF15, Grupa Azoty S.A.), reinforced with 15% w/w glass fibre, generously provided by STER INSTITUTE company (Poland). The process optimization required testing various temperature zone setting changes, heating and cooling the movable plate to different temperatures, adjusting intrusion time and the injection and holding pressure profiles. In addition, small adjustments of the mold gates and vents were required, which were done on the completed mold mounted on the injection molding machine during technological tests. The optimized parameters for injection molding were as follows: temperature profile of plastification unit, from feed to die: 250 °C, 255 °C, 255 °C, 270 °C, mold temperature: 60 °C stationary plate, 20 °C movable plate, max. injection time: 2.7 s, intrusion time 4.3 s, cooling time: 30 s, injection pressure: 1300 bar, holding pressure profile: 575 bar/0.5s, 675 bar/1s.

Mechanical tests of the materials were done in accordance to the norm EN ISO 527-2:1996. Dumbbell samples were prepared by either 3D printing or injection molding. The speed of traverse was set to 50 mm/min. A universal testing machine Instron 5969 was used. The Charpy impact test (with no notch) was performed on a Instron Ceast 9050 impact-machine according to ISO 179−1. For all the series, 10 measurements were performed.

The data are collected in the table of see Section 5.

3.2. Simulation Model

The calculation model was designed to simulate the most important production parameters, such as: unit production cost, total production volume vs. time, and unit production cost vs. total production volume. In order to achieve this, data on a variety of production parameters (e.g., tool cost, material cost, and unit production time, etc.) are collected in Table 2. Different scenarios were assigned for simulations, i.e., operating on a single FFF 3D printer, an FFF 3D printer farm, a single socket IM machine, or a multiple socket IM machine. These scenarios were chosen to discuss some production aspects, such as the total production output, production costs, or costs per production unit, when different production set-ups are applied, and how the changes of set-up affect these parameters (e.g., switching from a single FFF printer to a printer farm). Data such as the number of product units per 24 h, percentage of defective product units, and effective product unit output per 24 h were taken from our production tests made with a small 3D printing farm (10 pcs) for FFF, and an Engel injection molding machine for injection molding. Data on tools and designing were supplied by the industrial partner specialized in injection molding, STER INSTITUTE (Poland). Basics of the tool design and cost prediction may be found in the literature [49]. Equations (1)–(9) were assigned to run calculations.

Table 2. A set of parameters for the purposes of the research simulation, based on the literature data and own experimental data.

Variable	Variable Symbol	3D Printer	Single Socket Injection Molding	3D Farm 200 pcs	Multiple Socket Injection Molding
Tool cost *	c_t	0	12,000 [1]	0	15,000 [1]
Design cost *	c_d	250 [1]	450 [1]	250 [1]	1000 [1]
Material type	-	PET	PA REC	PET	PA REC
Material cost [USD/kg]	c_m	16	3	16	3
Output [g/h]	o	15 [2]	5800 [2]	3000 [2]	23,200 [2]
No. product units per 24 h	u_d	6 [2]	2142 [2]	1108 [2]	8566 [2]
% defective product units	d	7% [2]	1% [2]	7% [2]	1% [2]
Effective product unit output per 24 h	o_e	5 [2]	2120 [2]	1030 [2]	8480 [2]
Product unit mass [g]	m_u	65 [2]	65 [2]	65 [2]	65 [2]
Cost per a unit *	c_u	1.04	0.195	1.04	0.195
Machine power * [kW]	p	0.3	19	60	28
Energy consumption per 24 h [kWh]	e_d	2.16	182.4	576	268.8
Energy consumption per unit [kWh]	e_u	0.39	0.09	0.52	0.03
Cost of energy per unit *	c_e	0.0390	0.0085	0.0520	0.0031
Total cost per unit *	c_{to}	2.1428	0.3974	2.1558	0.2563
Fixed costs *	c_f	250	12,450	250	16,000
Machinery costs *	c_{ma}	1000	75,000	200,000	90,000
Machinery costs per day of production *	c_{md}	5.5	411.0	1095.9	493.2
Machinery cost per pcs/day *	c_{mu}	1.064	0.194	1.064	0.058
Work fraction	w_f	0.3	0.4	0.4	0.4

* expense [USD]; [1] Values obtained from STER INSTITUTE calculations; [2] Values obtained from experiments (10 pcs 3D printer farm or an injection molding machine).

For effective output per 24 h:

$$o_e = o * (1 - d/100\%) \tag{1}$$

where d—% defective product units
For cost per unit

$$c_u = c_m * m_u/1000 \tag{2}$$

where c_m—Material cost [USD/kg]; m_u—Product unit mass [g]
For no. of product units per 24 h

$$u_d = 24 * o/m_u \tag{3}$$

where o—Output [g/h]
For energy consumption per unit

$$e_u = e_d/u_d \tag{4}$$

where e_d—Energy consumption per 24 h [kWh]; ud—No. product units per 24 h
For energy consumption per 24 h

$$e_d = 24 * p * w_f \tag{5}$$

where p—Machine power* [kW]; wf—Work fraction
For fixed costs

$$c_f = c_t + c_d \tag{6}$$

where c_t—Tool cost [USD]; cd—Design cost [USD]
For machinery costs per day of production

$$c_{md} = c_{ma} * 2/365 \tag{7}$$

where c_{ma}—Machinery costs [USD]
For machinery cost per unit a day

$$C_{mu} = c_{md}/o_e \tag{8}$$

For total cost per unit

$$C_{to} = c_u + c_e + c_{mu} \tag{9}$$

On the basis of these simple equations, the final model Equations (10)–(12) were assigned as follows:

For unit production costs vs. production run (see Figure 2)

$$C_{ud} = (c_f/x_d + u_d * c_{to})/u_d \tag{10}$$

where x_d—days of production

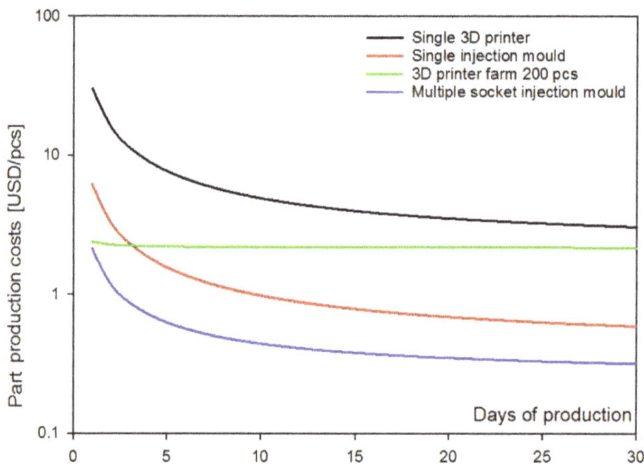

Figure 2. Comparison of unit production costs depending on the considered production period in days for various production techniques.

For the cumulative production volume vs. time (see Figure 3)

$$V_c = (x_d - n) * o_e \tag{11}$$

where $n = 1$ for 3D printing and 3D Farm printing; $n = $ for Injection molding and Multiple socket injection molding

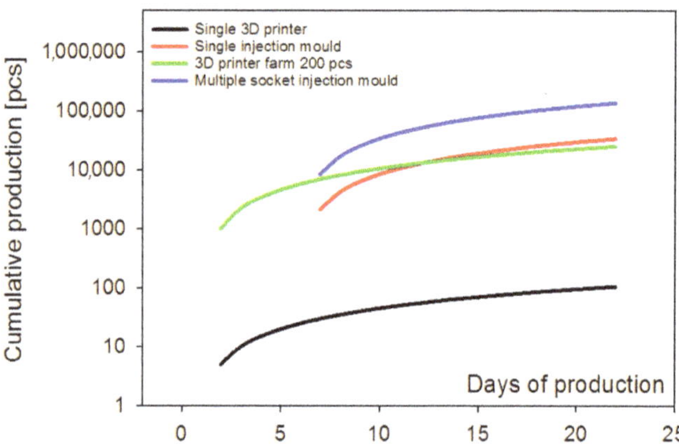

Figure 3. Comparison of the cumulative production volume of an item as a function of time for various manufacturing techniques.

For the unit production cost vs. total production volume (see Figure 4)

$$C_{uv} = (c_f + c_{to} * x_u)/x_u \tag{12}$$

where x_u—units of production

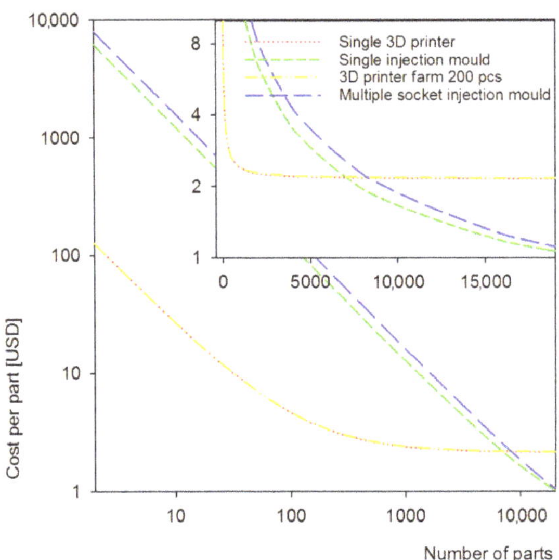

Figure 4. Cost comparison of manufacturing an element depending on the production volume for various manufacturing techniques.

4. Process Set-Up and Simulation Purpose

The main advantage of the FFF technology is the ease and speed of prototyping, which does not require the production of specialized tools in the form of molds. Most analyses state that the production process begins 1–2 days after "making a decision" (Figure 1B). However, the time of designing and optimizing the model is often omitted in the analyses and the time equal to 0 is assumed for this part of the process preparation chain, which is incorrect from the technological point of view and is purely a marketing approach. In our case, the production started within two days, which were needed to optimize the printing process towards process stability and product quality. In the case of the traditional technique, which is thermoplastic injection, the time from "making a decision" to starting production takes more than one week (under the most optimal conditions). In our case (real conditions during pandemic outbreak), the process was launched within 10 days before reaching full output (Figure 1A). This is due to the fact that from the moment of making the decision up to launching the production, it is necessary to carry out preparations such as:

(1) Designing combined with computer simulation of the injection process—this stage may take up to several days depending on the complexity of the product. Despite the fact that it requires time, it is a much cheaper operation than direct injection tests after designing the tool and subsequent reworking of a wrongly designed tool. In the case of making a tool without simulations, this stage often ends with the production of a new tool, the actual geometry of which has been determined in technological trials.

(2) Technological tests after making a tool, the critical elements of which have been verified in a computer simulation—this stage may be very short, in the order of hours or, in an unfavorable situation, up to several days. The optimization time was also included (possible tool or process correction).

(3) Production start-up—from an hour to several hours, assuming that in all previous stages the most satisfactory results were achieved. In relation to the injected product, injection tools are elements of large mass, have a precise structure and require specialist knowledge in the area of designing and operating the apparatus. In the absence of the latter, the period from the decision to the launch of production may be, in extremely unfavorable cases, up to 40–60 days for complex details to be produced.

The factors characterized above show the first feature that justifies the increasingly more and more popular use of incremental techniques—the speed of the process and little preparation needed. In addition, the investment outlays in the first phase of designing and starting the process are incomparably greater in the case of injection molding technology. For large-size objects with complex geometry, often requiring several separate processes, and thus also processing machines, the cost of tools can reach hundreds of thousands of dollars. Therefore, especially for small lot production, injection molding is much less cost effective compared to additive techniques.

As a part of activities supporting medical services during the COVID-19 epidemic in Poland, our team launched the process of manufacturing protective helmets. In the first phase of the epidemic, the production of the shields by 3D printing using FFF printers was very popular. The designs were available in an open-source format available online. Often, printer manufacturers shared their projects, such as the Czech company PRUSA. On the basis of this project, elements of protective gear were manufactured and sent to hospitals by individuals or research units. It was a measurable example of how the high flexibility and speed of FFF technology can contribute to mitigating the shortages of protective measures in the country. Our team, performing tests of manufacturing elements using the FFF technique, analyzed the functionality, durability, and performance properties of such objects made of plastics such as PET-G or PLA. With such a high demand for personal protective equipment, the printing time of one unit was too long and the results were unsatisfactory. It was therefore decided to launch an alternative project of manufacturing these by an injection molding method. The reasons for this decision were the following factors: (a) material cost analysis, (b) unsatisfactory process time of FFF, (c) unsatisfactory print quality, (d) unsatisfactory functional features and ergonomics of open-source projects.

For the purposes of the simulation, data on operational, cost and performance parameters of selected processing methods were collected. The analysis included both the work of one tool (printer, single-socket mold, etc.), as well as the tool sets, such as the printer farm or a multi-socket mold (Table 2). These scenarios were selected to compare the aspects of different production approaches side by side and stress their limitations or shortcomings. Due to the significant technological differences between the FFF and IM techniques, it was not possible to use the same materials. Plastics dedicated to injection molding and FFF differ in terms of parameters and availability—it is not possible to obtain the same material for both techniques. As a result, material type and cost are different for the simulations provided for these two techniques and favour of injection molding. Per unit of mass, plastics intended for 3D printing are also much more expensive, which significantly increases the unit price of the product in case of FFF 3D printing. It is necessary to note that the project was carried out at the peak of the first wave of the COVID-19 pandemic outbreak in Poland and Europe, when the supplies of specialized materials were suspended, and the work was carried out using on-site available materials for the processing techniques discussed. The simulation presented in the manuscript shows the minimum time in which the discussed processing techniques can be implemented, which has been proven in real operating conditions.

5. Discussion

A product design must always be compatible with the technique applied for the manufacturing process. To enable 3D printing of the protective helmet in household conditions, there was a requirement for the open-source design that could accommodate the most popular 3D printer bed sizes (it could not exceed the printing dimension limitations of common 3D printers for amateur use). For this reason, the protective helmet body of the open-source design was small and barely comfortable, and it was also not very stable on the user's head (see Supplementary Materials, Figure S1). In addition, 3D printed pieces usually required some post-processing, namely sanding, as rough surfaces were often generated during printing, which was another issue of the user comfort and it extended the time of production. Furthermore, due to the abovementioned dimensional limitations, the FFF design of the helmet body required a lot of elastic rubber band to accommodate the user head, which quickly resulted in a national shortage of this material, as it was needed to secure most of the simple design protective helmets or face shields. Our design of the protective helmet body dedicated for injection molding addressed this issue: the support band of the helmet was elongated so that it would match the head circumference better and therefore minimize the amount of required rubber band, and it was easily adjustable with simple tools, by cutting off the excess of the plastic band (see Supplementary Materials, Figure S7). In addition, from a technical point of view, when it comes to 3D printing by FFF technique and designing the object models dedicated for FFF printing, sagging is often an issue for objects with overhanging details. To avoid this, a 3D model may be provided with additional supports or redesigned so that the overhangs are eliminated by additional geometric shapes, or smoothed/reduced. This can be seen on the example of mounting pins of the protective helmet body, when comparing the FFF-printed object and the injection-molded one (see Supplementary Materials, Figures S1–S3).

In terms of unit costs, the multi-cavity injection molding technology does not have any competition (Figure 2). Calculations have shown that only for large farms of 3D printers, the unit cost can be competitive compared to a single-cavity injection mold over a period of a couple of days. Large printer farms are characterized by flat cost characteristics from the first days of starting production, which shows that only for large printer units they can be a tool for mass production. The analysis of the cumulative value of production (Figure 3) showed a relatively high competitiveness of a large printer farm in relation to a single-cavity injection molding process. The performance curves for these two manufacturing tools intersect around 10–15 days after the decision to start a given production. Comparative data on the efficiency values for a given technology allow for a precise assessment of the appropriate selection of technology for the purpose of its application. The calculations (Figure 4) show that the techniques usually used for rapid prototyping work well for short-term production. Perhaps that is why in the first phase of the pandemic, in the absence of forecasts about its possible development, much attention was paid to simple and accessible techniques such as FFF printing. This made it possible to meet many basic needs. In the later period, a gradual switch of production plants from the current production over to the missing elements needed to save human life and health was observed. This has often happened without meeting all material performance and certification standards that existed before the pandemic. Increasing the production capacity resulted in a decline in the interest in products manufactured with additive techniques due to their poor quality (Figure 5). In addition, injection molding provided products of much higher mechanical performance when compared to FFF process, as we verified the processes experimentally (Table 3). It can be seen that the large difference is based on the type of material used, but this fact also emphasizes the benefits of injection molding as a high performance process, as it can utilize materials with which FFF process would be impossible to carry out by the means of currently available 3D printers.

While in the first weeks of the pandemic, economic factors were not of primary importance, along with its development a need arose to rationalize resources for the production and optimize the use of raw materials. In order to more precisely approximate the economic profitability threshold, i.e., in the conditions of a normally functioning economy, the basic factor determining the use of a given processing technique, the change in the cost of manufacturing a product unit was compared in respect to its growing production. For the compared manufacturing techniques, the differences are significant and noticeable, especially for the first few thousand units of the product produced (Figure 4). Rapid prototyping techniques show their advantage for series up to several thousand elements, while the advantages of classic techniques are revealed in larger numbers. Taking into account the choice of manufacturing technology in emergency situations, the quality of the final detail is also of secondary importance. However, one should be aware of the advantages and disadvantages of the method used. Figure 5A–H shows the details of the products observed microscopically.

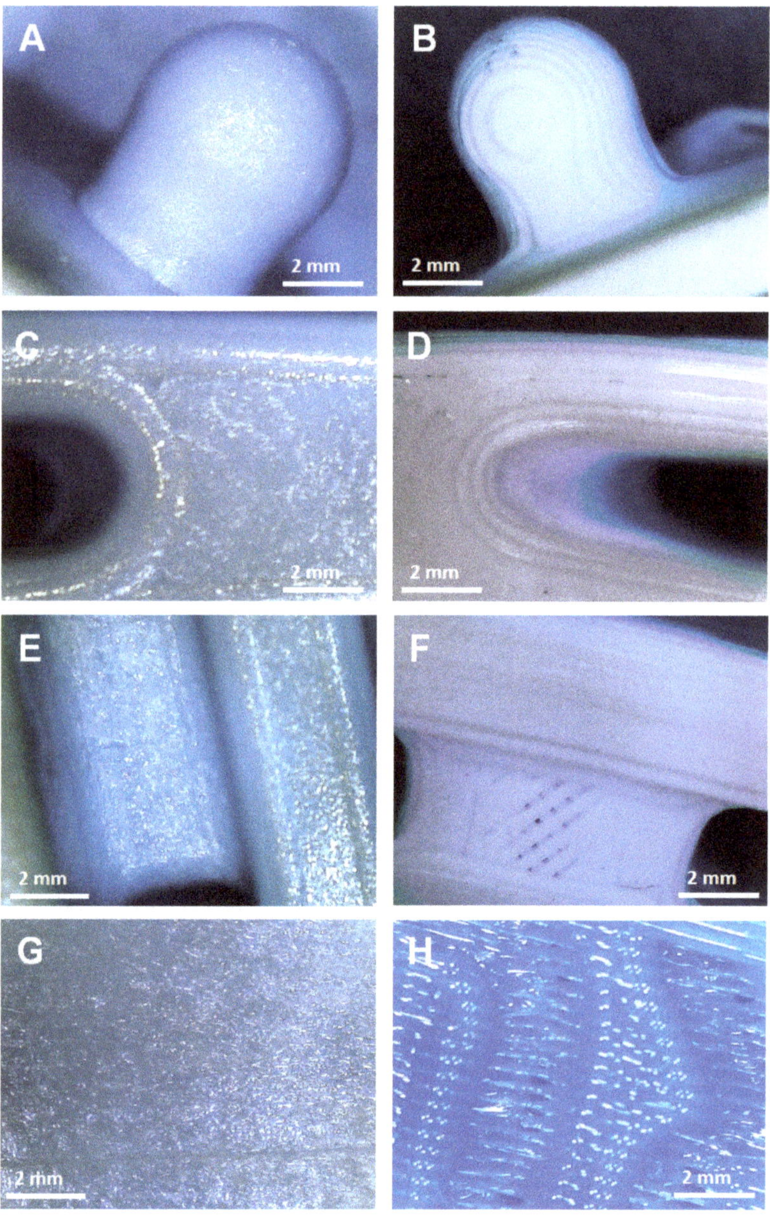

Figure 5. Comparison of manufacturing precision for traditional injection molding technology (**A,C,E,G**) and the incremental FFF technique (**B,D,F,H**).

Table 3. Data for mechanical properties comparison of the objects prepared by FFF and injection molding.

Technique	FFF Theoretical [1]	FFF Experimental [2]	Injection Molding Theoretical [3]	Injection Molding [4]
Material	PET-G	PET-G	PA6-GF15	PA6-GF15
Tensile modulus [MPa]	2020	1720 ± 84	5700	5070 ± 208
Tensile strength [MPa]	50	44 ± 6	130	112 ± 19
Charpy impact strength	8.1	4.3 ± 0.8	50	41 ± 4
Flexural modulus [MPa]	2050	1850 ± 95	5000	4260 ± 178
Flexural strength [MPa]	69	62 ± 8	180	152 ± 21
Elongation at break [%]	23	5.2 ± 1.1	1	1 ± 0.3

[1] manufacturer data for PET-G Verbatim filament; [2] data obtained experimentally for PET-G Verbatim filament; [3] manufacturer data for Tarnamid T-27 GF15; [4] data obtained experimentally for Tarnamid T-27 GF15 recycled material.

The defects of the printed objects, in particular the areas of material discontinuities (Figure 5F,H), make their usefulness in medical applications far below satisfactory. Dirt and bacteria can collect in the cavities, and the porosity makes the object a breeding ground for microorganisms. Furthermore, as these objects are elastic and meant to deform in order to accommodate the user's head, the material discontinuities may cause cracks and splitting while using the object or during helmet assembly. Such behaviour was observed during the tests of the 3D-printed protective helmet body pieces (Figure S1, Supplementary Materials). In addition, the reduced mechanical parameters of the test specimens were measured when compared to the technical reference (Table 3). The described examples of the use of elements made of plastics in crisis conditions did not require the use of plastics with approval for medical applications. The pandemic crisis has shown that it is essential to have manufacturing tools to respond when the supply chain is interrupted or disrupted. In such a situation, the availability of systems such as 3D printer farms can prevent a shortage of basic personal protective equipment. In such situations, product certification and appropriate permits become a secondary consideration, and the priority is to save human lives and ensure the collective safety of the population.

6. Conclusions

The conducted project and analyses supported by the performed simulation allowed us to collect and quantify the most important aspects which are the limitations of the FFF technique. In order for the 3D printing technology to compete more effectively with traditional methods such as thermoplastics mold injection, two basic limiting conditions must be met simultaneously or independently—increasing the speed of the process 6–10 times and reducing defect formation in the interlayer areas, which cause poor object quality, reproducibility issues, and mechanical failures. Nevertheless, large groups of FFF printers (printer farms) can successfully compete with injection machines for low-volume solutions, if a satisfactory quality of printed objects is provided. Most importantly, it should be pointed out that the FFF technique is a viable method of response to a sudden crisis situation, as was the case with the COVID-19 pandemic, as it allowed for providing the basic needs for personal protection in the first weeks of the pandemic. Individual users contributed their products to public entities, such as hospitals, which gave the time needed to launch production lines of high-quality products with the application of mature processing methods.

Supplementary Materials: The following are available online at https://www.mdpi.com/article/10.3390/pr9050791/s1, Table S1: Project chronology, Figure S1: Protective helmet body made with FFF 3D printing technique, Figure S2: Technical drawing of protective helmet body design for Injection molding production method, Figure S3: 3D model of protective helmet body design for Injection molding production method, Figure S4: technological tests and optimization of injection molding, Figure S5: Protective helmet fully assembled and tested for the comfort of using, Figure S6: The protective helmet detail as received during injection molding process (with sprue remaining),

Figure S7 : Correction of the protective helmet band for the head size of a particular user, Figure S8: The technological team dealing with the study of the production process of the protective helmets.

Author Contributions: Conceptualization, R.E.P. and B.S.; methodology, R.E.P., B.S. and D.B.; software, R.E.P.; validation, R.E.P. and M.J.; investigation, B.S., D.B. and R.E.P.; resources, B.S.; data curation, R.E.P.; writing—original draft preparation, B.S., D.B., R.E.P. and M.J.; writing—review and editing, visualization, R.E.P.; supervision, B.S.; project administration, B.S., funding acquisition, B.S. All authors have read and agreed to the published version of the manuscript.

Funding: This research was funded by National Centre for Research and Development, Poland, grant no. LIDER/01/0001/L-10/18/NCBR/2019 and from the resources Adam Mickiewicz University (Poland).

Institutional Review Board Statement: Not applicable.

Informed Consent Statement: Not applicable.

Data Availability Statement: The data provided for the calculation model were either obtained experimentally from on-site tests made with a small 3D printing farm (10 pcs) for FFF, and an Engel injection molding machine for injection molding. Data on tools and designing were supplied by the industrial partner specialized in injection molding, STER INSTITUTE (Poland).

Acknowledgments: The authors thank Bronislaw Marciniak (AMU) and Marek Nawrocki (AMU) for entrusting the task of producing protective shields. Many thanks to the STER company for technological support of the conducted research. We express our gratitude to the key industrial partner—STER INSTITUTE and CEO Maciej Szymanski for support in the field of injection mold manufacturing, injection mold design, modelling, assembly, and technical support during technological tests, as well as providing PA-6 material for injection molding. Thanks to Krzysztof J. Kurzydlowski for consultations in the field of material engineering. Many thanks to ERGIS company, Dąbrowskiego 2, Wąbrzeźno, Poland, for donating PET film for the production of protective shields.

Conflicts of Interest: The authors declare no conflict of interest.

References

1. Fawcett, S.E.; Waller, M.A. Supply chain game changers—mega, nano, and virtual trends—and forces that impede supply chain design (i.e., building a winning team). *J. Bus. Logist.* **2014**, *35*, 157–164. [CrossRef]
2. Gibson, I.; Rosen, D.W.; Stucker, B. *Additive Manufacturing Technologies*; Springer: Berlin/Heidelberg, Germany, 2010.
3. LeHong, H.; Fenn, J. *Emerging Technologies Hype Cycle*; Report; Gartner: Stamford, CT, USA, 2012.
4. Baasiliere, P.; Shanler, M. *Hype Cycle for 3D Printing*; Report; Gartner: Stamford, CT, USA, 2015.
5. Valentinčič, J.; Peroša, M.; Jerman, M.; Sabotin, I.; Lebar, A. Low Cost Printer for DLP Stereolithography. *J. Mech. Eng.* **2017**, *63*, 559–566. [CrossRef]
6. Schmidleithner, C.; Kalaskar, D.M. Stereolithography. In *3D Printing*; Cvetković, D., Ed.; IntechOpen: London, UK, 2018; pp. 1–22.
7. Kataria, A.; Rosen, D.W. Building around inserts: Methods for fabricating complex devices in stereolithography. *Rapid Prototyp. J.* **2001**, *7*, 253–262. [CrossRef]
8. Bártolo, P.J.; Gibson, I. *History of Stereolithographic Processes*; International Journal of Scientific and Research Publications; Centre for Rapid and Sustainable Product, Polytechnic Institute of Leiria: Leiria, Portugal, 2018; Volume 8, ISSN 2250-3153.
9. Fina, F.; Goyanes, A.; Gaisford, S.; Basit, A.W. Selective laser sintering (SLS) 3D printing of medicines. *Int. J. Pharm.* **2017**, *529*, 285–293. [CrossRef]
10. Upcraft, S.; Fletcher, R. The rapid prototyping technologies. *Assem. Autom.* **2003**, *23*, 318–330. [CrossRef]
11. Novakova-Marcincinova, L.; Kuric, I. Basic and Advanced Materials for Fused Deposition Modeling Rapid Prototyping Technology. *Manuf. Ind. Eng.* **2012**, *11*, 24–27.
12. Kalmanovich, G. "Curved-Layer" Laminated Object Manufacturing. In Proceedings of the 1996 International Solid Freeform Fabrication Symposium, Austin, TX, USA, 12–14 August 1996.
13. Vinodh, S.; Sundararaj, G.; Devadasan, S.R.; Kuttalingam, D.; Rajanayagam, D. Agility through rapid prototyping technology in a manufacturing environment using a 3D printer. *J. Manuf. Technol. Manag.* **2009**, *20*, 1023–1041. [CrossRef]
14. A Third Industrial Revolution. *The Economist*, 21 April 2012.
15. Khajavi, S.H.; Partanen, J.; Holmström, J. Additive manufacturing in the spare parts supply chain. *Comput. Ind.* **2014**, *65*, 50–63.
16. Murr, L.E.; Gaytan, S.M.; Ramirez, D.A.; Martinez, E.; Hernandez, J.; Amato, K.N.; Shindo, P.W.; Medina, F.R.; Wicker, R.B. Metal fabrication by additive manufacturing using laser and electron beam melting technologies. *J. Mater. Sci. Technol.* **2012**, *28*, 1–14. [CrossRef]
17. Berman, B. 3-D printing: The new industrial revolution. *Bus. Horiz.* **2012**, *55*, 155–162. [CrossRef]
18. Rayna, T.; Striukova, L. From rapid prototyping to home fabrication: How 3D printing is changing business model innovation. *Technol. Forecast. Soc. Chang.* **2016**, *102*, 214–224. [CrossRef]

19. Yoon, H.-S.; Lee, J.-Y.; Kim, H.-S.; Kim, E.-S.; Shin, Y.-J.; Chu, W.-S.; Ahn, S.-H. A Comparison of Energy Consumption in Bulk Forming, Subtractive, and Additive Processes: Review and Case Study. *Int. J. Precis. Eng. Manuf. Green Technol.* **2014**, *1*, 261–279. [CrossRef]
20. Sanchez, F.A.C.; Boudaoud, H.; Hoppe, S.; Camargo, M. Polymer recycling in an open-source additive manufacturing context: Mechanical issues. *Addit. Manuf.* **2017**, *17*, 87–105. [CrossRef]
21. Dobrosielska, M.; Przekop, R.E.; Sztorch, B.; Brząkalski, D.; Zgłobicka, I.; Łepicka, M.; Dobosz, R.; Kurzydłowski, K.J. Biogenic Composite Filaments Based on Polylactide and Diatomaceous Earth for 3D Printing. *Materials* **2020**, *13*, 4632. [CrossRef] [PubMed]
22. Franchetti, M.; Kress, C. An economic analysis comparing the cost feasibility of replacing injection molding processes with emerging additive manufacturing techniques. *Int. J. Adv. Manuf. Technol.* **2017**, *88*, 2573–2579. [CrossRef]
23. Atzeni, E.; Salmi, A. Economics of additive manufacturing for end-usable metal parts. *Int. J. Adv. Manuf. Technol.* **2012**, *62*, 1147–1155. [CrossRef]
24. Drizo, A.; Pegna, J. Environmental Impacts of Rapid Prototyping: An Overview of Research to Date. *Rapid Prototyp. J.* **2006**, *12*, 64–71. [CrossRef]
25. Sood, A.K.; Ohdar, R.K.; Mahapatra, S.S. Parametric appraisal of mechanical property of fused deposition modelling processed parts. *Mater. Des.* **2010**, *31*, 287–295. [CrossRef]
26. Heidari-Rarani, M.; Rafiee-Afarani, M.; Zahedi, A.M. Mechanical characterization of FDM 3D printing of continuous carbon fiber reinforced PLA composites. *Compos. Part B Eng.* **2019**, *175*, 107147. [CrossRef]
27. Ngo, T.; Kashani, A.; Imbalzano, G.; Nguyen, K.; Hui, D. Additive manufacturing (3D printing): A review of materials, methods, applications and challenges. *Compos. Part B Eng.* **2018**, *143*, 172–196. [CrossRef]
28. Boros, R.; Rajamani, P.K.; Kovács, J.G. Combination of 3D printing and injection molding: Overmolding and overprinting. *Express Polym. Lett.* **2019**, *13*, 889–897. [CrossRef]
29. Gordon, J., Jr. *Total Quality Process Control for Injection Molding*; John Wiley & Sons: Hoboken, NJ, USA, 2010.
30. Huamin, Z. *Computer Modeling for Injection Molding: Simulation, Optimization, and Control*; John Wiley & Sons: Hoboken, NJ, USA, 2013.
31. Kent, R. *Energy Menagement in Plastics Processing. Strategies, Targets, Techniques and Tools*; Plastics Information Direct: London, UK, 2013.
32. Jałbrzykowski, M. *Wybrane Zagadnienia Procesu Wtryskiwania Biopolimerów do Zastosowań Technicznych na Przykładzie Polilaktydu (PLA)*; Wydawnictwo Polskiego Towarzystwa Inżynierii Rolniczej, Inżynieria Rolnicza: Kraków, Poland, 2019.
33. Jałbrzykowski, M.; Obidziński, S. A comparison of the breaking strength of samples of parts made from pure and regrind polypropylene. *J. Res. Appl. Agric. Eng.* **2017**, *62*, 55–58.
34. Rytka, C.; Lungershausen, J.; Kristiansen, P.M.; Neyer, A. 3D filling simulation of micro- and nanostructures in comparison to iso- and variothermal injection moulding trials. *J. Micromech. Microeng.* **2016**, *26*, 065018. [CrossRef]
35. Martowibowo, S.Y.; Kaswadi, A. Optimization and Simulation of Plastic Injection Process using Genetic Algorithm and Moldflow. *Chin. J. Mech. Eng.* **2017**, *30*, 398–406. [CrossRef]
36. Rosato, D.V.; Rosato, D.V.; Rosato, M.G. *Injection Molding Handbook*; Springer Science & Business Media LLC: New York, NY, USA, 2000.
37. Jałbrzykowski, M.; Obidziński, S.; Minarowski, Ł. Wpływ wybranych parametrów procesu wtryskiwania na wytrzymałość na rozciąganie detali z polilaktydu starzonych w środowisku płynu Sorensena. *Przemysł Chem.* **2017**, *96*, 1869–1872. [CrossRef]
38. Kitayama, S.; Onuki, R.; Yamazaki, K. Warpage reduction with variable pressure profile in plastic injection molding via sequential approximate optimization. *Int. J. Adv. Manuf. Technol.* **2014**, *72*, 827–838. [CrossRef]
39. Sykutera, D.; Czyzewski, P.; Kosciuszko, A.; Szewczykowski, P.; Wajer, L.; Bielinski, M. Monitoring of the injection and holding phases by using a modular injection mold. *J. Polym. Eng.* **2018**, *38*, 63–71. [CrossRef]
40. Pruitt, L.A.; Chakravartula, A.M. *Mechanics of Biomaterials: Fundamental Principles for Implant Design*; Cambridge University Press: Cambridge, UK, 2011.
41. Hatta, N.M.; Azlan, M.Z.; Shayfull, Z.; Roselina, S.; Nasir, S.M. Analysis of The Shrinkage at The Thick Plate Part using Response Surface Methodology. In Proceedings of the 3rd Electronic and Green Materials International Conference (EGM), Aonang Krabi, Thailand, 30 April 2017; p. 020152.
42. Giret, A. Sustainability in manufacturing operations scheduling: A state of the art review. *J. Manuf. Syst.* **2015**, *37*, 126–140. [CrossRef]
43. Fisher, J.M. *Handbook of Molded Part Shrinkage and Warpage*; Plastic Design Library, William Andrew Inc.: Norwich, CT, USA, 2003.
44. Fernandes, C.; Pontes, A.J.; Viana, J.C.; Nobrega, J.M.; Gaspar-Cunha, A. Modeling of Plasticating Injection Molding—Experimental Assessment. *Int. Polym. Process.* **2014**, *29*, 558–569. [CrossRef]
45. Diduch, C.; Dubay, R.; Li, W.G. Temperature control of injection molding. Part 1: Modeling and Identification. *Polym. Eng. Sci.* **2007**, *44*, 2308–2317. [CrossRef]
46. Chil-Chyuan, K.; Wei-Hua, C.; Jia-Wei, Z.; Ysai, D.-A.; Cao, Y.-L.; Juang, B.-Y. A new method of manufacturing a rapid tooling with different cross-sectional cooling channels. *Int. J. Adv. Manuf. Technol.* **2017**, *92*, 3481–3487.
47. Bilewicz, M.; Viana, J.C.; Dobrzanski, L.A. Development of microstructure affected by in-mould manipulation in polymer composites and nanocomposites. *J. Achiev. Mater. Manuf. Eng.* **2008**, *31*, 71–76.

48. Eladl, A.; Mostafa, R.; Islam, A.; Loaldi, D.; Soltan, H.; Hansen, H.N.; Tosello, G. Effect of Process Parameters on Flow Length and Flash Formation in Injection Moulding of High Aspect Ratio Polymeric Micro Features. *Micromachines* **2018**, *9*, 58. [CrossRef] [PubMed]
49. Rosato, D.V.; Rosato, M.G. *Injection Molding Handbook*, 3rd ed.; Springer: Berlin/Heidelberg, Germany, 2000.